Non-coding RNAs and Epigenetic Regulation of Gene Expression

Drivers of Natural Selection

Edited by

Kevin V. Morris

The Scripps Research Institute
La Jolla, CA
USA

Caister Academic Press

Caister Academic Press
Norfolk, UK

www.caister.com

British Library Cataloguing-in-Publication Data
A catalogue record for this book is available from the British Library

ISBN: 978-1-904455-94-3

Cover art created by Paula J. Morris, http://sciencegraphics.daportfolio.com

Printed and bound in Great Britain

Non-coding RNAs and Epigenetic Regulation of Gene Expression

Drivers of Natural Selection

Edited by

Kevin V. Morris

The Scripps Research Institute
La Jolla, CA
USA

Caister Academic Press

Caister Academic Press
Norfolk, UK

www.caister.com

British Library Cataloguing-in-Publication Data
A catalogue record for this book is available from the British Library

ISBN: 978-1-904455-94-3

Description or mention of instrumentation, software, or other products in this book does not imply endorsement by the author or publisher. The author and publisher do not assume responsibility for the validity of any products or procedures mentioned or described in this book or for the consequences of their use.

Cover art created by Paula J. Morris, http://sciencegraphics.daportfolio.com

Printed and bound in Great Britain

Contents

Contributors

Mariano Alló
Laboratorio de Fisiología y Biología Molecular
Departamento de Fisiología, Biología Molecular y Celular
IFIBYNE-CONICET
Facultad de Ciencias Exactas y Naturales
Universidad de Buenos Aires
Ciudad Universitaria
Buenos Aires
Argentina

marianoallo@gmail.com

Miguel A. Andrade-Navarro
Computational Biology and Data Mining
Max-Delbrück Center for Molecular Medicine
Berlin
Germany

miguel.andrade@mdc-berlin.de

Jérôme Cavaillé
Université de Toulouse
UPS
Laboratoire de Biologie Moléculaire Eucaryote;
CNRS
LBME
Toulouse
France

cavaille@ibcg.biotoul.fr

Fabricio F. Costa
Cancer Biology and Epigenomics Program
Children's Memorial Research Center
Northwestern University's Feinberg School of Medicine
Chicago, IL
USA

fcosta@childrensmemorial.org

Hengmi Cui
Center for Epigenetics
School of Medicine
Nanjing University
Nanjing
China

hcui@nju.edu.cn

Isabelle Cui
University of Maryland School of Medicine
Baltimore, MD
USA

icui001@umaryland.edu

Kirk Deitsch
Department of Microbiology and Immunology
Weill Medical College of Cornell University
New York, NY
USA

kwd2001@med.cornell.edu

Mohammad Faghihi
Department of Psychiatry and Behavioral Sciences
Center for Therapeutic Innovations
John P. Hussman Institute for Human Genomics
Miller School of Medicine
University of Miami
Miami, FL
USA

mfaghihi@med.miami.edu

Marianne Farnebo
Department of Oncology-Pathology
Cancer Centrum Karolinska (CCK)
Karolinska Institutet
Stockholm
Sweden

marianne.farnebo@ki.se

Paul Halley
Wahlestedt Laboratory
John P. Hussman Institute for Human Genomics
Miller School of Medicine
University of Miami
Miami, FL
USA

phalley@med.miami.edu

Daniel H. Kim
Division of Biology
California Institute of Technology
Pasadena, CA
USA

dkim@caltech.edu

Hidenori Kiyosawa
Department of Animal Resource Sciences
Graduate School of Agricultural and Life Sciences
The University of Tokyo
Tokyo
Japan

akiyosaw@mail.ecc.u-tokyo.ac.jp

Chihiro Kohama
Cell-based Product Development Department
Japan Tissue Engineering Co., Ltd
Aichi
Japan

chihiro_kohama@jpte.co.jp

Alberto R. Kornblihtt
Laboratorio de Fisiología y Biología Molecular
Departamento de Fisiología, Biología Molecular y Celular
IFIBYNE-CONICET
Facultad de Ciencias Exactas y Naturales
Universidad de Buenos Aires
Ciudad Universitaria
Buenos Aires
Argentina

ark@fbmc.fcen.uba.ar

Dacia Kwiatkowski
Department of Microbiology and Immunology
Weill Medical College of Cornell University
New York, NY
USA

dak2015@med.cornell.edu

Stéphane Labialle
Université de Toulouse
UPS
Laboratoire de Biologie Moléculaire Eucaryote
Toulouse
France;
CNRS
LBME
Toulouse
France

labialle@ibcg.biotoul.fr

Jeannie T. Lee
Howard Hughes Medical Institute
Department of Molecular Biology
Massachusetts General Hospital;
Department of Genetics
Harvard Medical School
Boston, MA
USA

lee@molbio.mgh.harvard.edu

Salah Mahmoudi
Department of Oncology-Pathology
Cancer Centrum Karolinska (CCK)
Karolinska Institutet
Stockholm
Sweden

salah.mahmoudi@ki.es

Kevin V. Morris
Department of Molecular and Experimental Medicine
The Scripps Research Institute
La Jolla;
Molecular and Cellular Biology
The Beckman Research Institute at the City of Hope
Duarte, CA
USA

kmorris@scripps.edu

Enrique M. Muro
Computational Biology and Data Mining
Max-Delbrück Center for Molecular Medicine
Berlin
Germany

enrique.muro@mdc-berlin.de

Jessica M. Silva
Mayo Clinic and Foundation
Rochester, MN
USA

silva.jessica@mayo.edu

David I. Smith
Mayo Clinic and Foundation
Rochester, MN
USA

smith.david@mato.edu

Anna Vilborg
Department of Biophysics and Biochemistry
Yale University
New Haven, CT
USA

anna.vilborg@yale.edu

Patrice Vitali
Université de Toulouse
UPS
Laboratoire de Biologie Moléculaire Eucaryote;
CNRS
LBME
Toulouse
France

vitali@ibcg.biotoul.fr

Claes Wahlestedt
Department of Psychiatry and Behavioral Sciences
Center for Therapeutic Innovations
John P. Hussman Institute for Human Genomics
Miller School of Medicine
University of Miami
Miami, FL
USA

cwahlestedt@med.miami.edu

Andreas Werner
Institute for Cell and Molecular Biosciences
Medical School
Newcastle University
Newcastle upon Tyne
UK

andreas.werner@ncl.ac.uk

Xi Yang
Department of Psychiatry
Maryland Psychiatric Research Center
University of Maryland School of Medicine
Baltimore, MD
USA

jyang@mprc.umaryland.edu

Zane Zeier
Wahlestad Laboratory
John P. Hussman Institute for Human Genomics
Miller School of Medicine
University of Miami
Miami, FL
USA

zzeier@med.miami.edu

Preface

A growing body of evidence is beginning to emerge in human cells suggesting that particular species of non-coding RNAs play an active role in modulating gene transcription and epigenetic states. While such observations were in the past relegated to imprinted genes, it is now becoming apparent that several genes in differentiated cells may be under some form of RNA-based transcriptional and epigenetic regulatory control. Importantly, this form of regulation may be highly influenced by selective pressures and function in the governance of adaptability of the cell. Many studies have been carried out to date that have begun to discern the mechanism of action whereby non-coding RNAs modulate gene transcription. This volume attempts to capture a summary of the field to date. No doubt the implications that RNA can control epigenetic states will have an impact on our understanding of the basic fabric of the cell and possibly portend an era of therapeutic potential where long-term regulatory control of gene expression can be instilled on target genes.

Kevin V. Morris

Current Books of Interest

Small DNA Tumour Viruses	2012
Extremophiles: Microbiology and Biotechnology	2012
Bacillus: Cellular and Molecular Biology	2012
Microbial Biofilms: Current Research and Applications	2012
Bacterial Glycomics: Current Research, Technology and Applications	2012
Non-coding RNAs and Epigenetic Regulation of Gene Expression	2012
Brucella: Molecular Microbiology and Genomics	2012
Molecular Virology and Control of Flaviviruses	2012
Bacterial Pathogenesis: Molecular and Cellular Mechanisms	2012
Bunyaviridae: Molecular and Cellular Biology	2011
Emerging Trends in Antibacterial Discovery: Answering the Call to Arms	2011
Epigenetics: A Reference Manual	2011
Metagenomics: Current Innovations and Future Trends	2011
Nitrogen Cycling in Bacteria: Molecular Analysis	2011
The Biology of Paramyxoviruses	2011
Helicobacter pylori	2011
Microbial Bioremediation of Non-metals: Current Research	2011
Lactic Acid Bacteria and Bifidobacteria: Current Progress in Advanced Research	2011
Viruses and Interferon: Current Research	2011
Essentials of Veterinary Parasitology	2011
Hepatitis C: Antiviral Drug Discovery and Development	2011
Streptomyces: Molecular Biology and Biotechnology	2011
Alphaherpesviruses: Molecular Virology	2011
Recent Advances in Plant Virology	2011
Vaccine Design: Innovative Approaches and Novel Strategies	2011
Salmonella: From Genome to Function	2011
PCR Troubleshooting and Optimization: The Essential Guide	2011
Insect Virology	2010
Environmental Microbiology: Current Technology and Water Applications	2011
Sensory Mechanisms in Bacteria: Molecular Aspects of Signal Recognition	2010
Bifidobacteria: Genomics and Molecular Aspects	2010
Molecular Phylogeny of Microorganisms	2010
Nanotechnology in Water Treatment Applications	2010
Iron Uptake and Homeostasis in Microorganisms	2010
Caliciviruses: Molecular and Cellular Virology	2010
Epstein-Barr Virus: Latency and Transformation	2010
Anaerobic Parasitic Protozoa: Genomics and Molecular Biology	2010
Lentiviruses and Macrophages: Molecular and Cellular Interactions	2010
Microbial Population Genetics	2010
Borrelia: Molecular and Cellular Biology	2010
Influenza: Molecular Virology	2010

Part I

Non-coding RNAs: Form, Function, and Diversity

Genome-wide Analysis of Sense–Antisense Transcripts

1

Chihiro Kohama and Hidenori Kiyosawa

Abstract

Genomic DNA and full-length cDNA sequencing projects have provided fundamental data that have led us to the novel notion that antisense transcription is a universal phenomenon in many organisms, including mammals and plants. Successive expression analyses utilizing microarrays, genome-tiling arrays, and recent RNA sequencing with next-generation sequencers have supported this concept. We review how these genome-wide, natural antisense transcript analyses have proceeded, and we then introduce representative examples of the known functions and functional implications of antisense transcripts. The variety of modes in which antisense transcripts function in cells reveals that these transcripts do not constitute a uniform group, but perform various types of gene expression regulation.

Introduction

General background

Antisense transcription was once assumed to be a rare event, especially in mammals, as simultaneous transcription from both strands of DNA appeared to be unlikely (see review by Shearwin *et al.*, 2005). Antisense transcription was initially identified in bacterial extrachromosomal elements such as plasmids (Stougaard *et al.*, 1981; Tomizawa and Itoh, 1981) and transposons (Simons and Kleckner, 1983), and also in bacteriophages (Krinke and Wulff, 1987; Liao *et al.*, 1987).

In contrast to these 'compact' bacterial DNA elements, mammalian genomes were thought to be very large with the individual genes separated by intergenic spacer sequences. Each transcription unit, which defines the region of a gene to be transcribed, was presumed to be distant from other units. During the mid- to the late 1980s, however, several endogenous antisense transcription phenomena were reported in mammals – mainly in mice and humans. In the 1990s, other antisense RNAs continued to be discovered, but antisense transcription was still not regarded as universal.

This situation has changed with the advent of the 'genome era,' during which genome-sequencing projects were planned and performed in various model organisms. In the early 2000s, draft genome sequences of humans and mice were released (International Human Genome Sequencing Consortium, 2001; Mouse Genome Sequencing Consortium, 2002). Full-length cDNA projects using human and mouse resources on a large scale were also planned (RIKEN Genome Exploration Research Group Phase II Team and the FANTOM Consortium, 2001; FANTOM Consortium and RIKEN Genome Exploration Research Group Phase I and II Team, 2002; Mammalian Gene Collection Program Team, 2002; Imanishi *et al.*, 2004; Ota *et al.*, 2004; The Mammalian Gene Collection Project Team, 2004). In the course of these projects, many antisense transcripts were identified in humans and mice. It is relatively easy to identify the locations of genes (or transcription units) on chromosomes if they are ordinary protein-coding genes, especially if they form gene families. However, if the genes are non-protein-coding and/or lack ordinary exon-intron organizations, it becomes more difficult to precisely indicate the position of the transcription unit by using only the genome sequences and bioinformatics analysis. Even with protein-coding genes, the actual length of the 3′ UTR (untranslated region) is difficult to determine without full-length cDNA sequence data. Regions of overlap between sense and antisense transcripts frequently contain 3′ UTR sequences. Thus, full-length cDNA sequences make an important contribution to antisense RNA analyses.

Issues concerning the definition of 'antisense'

As genome sequence analysis by bioinformatics approaches has proceeded, the definition of antisense RNA or antisense transcripts has come under discussion: in other words, which is the sense transcript and which is the antisense? In earlier times, antisense transcription was assumed to be infrequent and exceptional; antisense RNAs were thought not to code proteins and essentially to differ from sense, protein-coding RNAs.

However, as will be introduced in later sections of this chapter, as many as half of the gene pairs producing sense and antisense transcripts consist of two protein-coding genes. The terms 'antisense' and 'antisense RNA' are often considered to imply 'non-coding' and 'non-coding RNA' (ncRNA), respectively, partly because antisense oligonucleotides (i.e. non-coding ones) have been used in gene expression inhibition experiments. This is why endogenous antisense RNA is sometime referred to as 'natural antisense RNA' or 'naturally occurring antisense RNA,' in contrast to artificially synthesized antisense RNA.

The 'sense' strand is the strand of double-stranded chromosomal DNA that has the same sequence as the transcribed RNA, with the exception of the differences between deoxyribonucleic acid and ribonucleic acid. The antisense strand is merely the opposite strand to the sense strand. Consequently, antisense RNA is a transcript produced from the strand opposite to where the sense gene is located. Thus, by definition, antisense RNAs are not necessarily ncRNAs.

As the relationship between sense and antisense transcripts is relative, it is difficult to determine which should be the sense transcript and which the antisense transcript. In this chapter, when we do not have to distinguish the sense transcript from the antisense, we usually call them 'sense–antisense transcripts' (SATs) or a SAT gene pair, i.e. a unit. When explaining or analysing particular topics, we also use the term 'antisense RNA' to describe the specific transcript produced from the antisense strand of a gene: in other words, we use the terms 'antisense RNA,' 'antisense transcript,' and 'antisense transcription' as general concepts widely accepted in this field. This nomenclature has been used for a long time, but bear in mind that the term 'antisense RNA' is not a fixed name for any particular type of RNA and is always relative.

SATs discovered before the genome era

Approximately 20–30 mammalian antisense transcription events were reported before about 2000, when large-scale genome sequence analyses started.

The first SATs were documented in 1985 by Farnham *et al.* (1985). These authors did not realize that they had discovered antisense RNA, but by RNase protection assay they observed a protected band in the antisense strand of the promoter region of the mouse *Dhfr* gene. It is now known that the *Msh* gene lies in the antisense strand of the *Dhfr* gene and that the region these authors presumed was the *Dhfr* promoter is part of a longer 5′ UTR. Thus, they detected antisense transcription between the overlapping regions of the first exons of each gene.

The first examples of SATs recognized as products of antisense transcription were *Surf-2* and *Surf-4* in mice (Williams and Fried, 1986). These authors confirmed the presence of a 133-bp overlap between the 3′ UTR sequences of these two genes.

After this discovery by Williams and Fried, other antisense transcription events continued to be reported. Representative ones are listed in Table 1.1.

Antisense RNA in the *c-Myc* gene was reported in 1986 (Nepveu and Marcu, 1986). However, to date, no protein-coding genes have been identified in the opposite strand of the *c-Myc* locus. Thus, the antisense RNA at the *c-Myc* locus appears to be

Table 1.1 Genes for which antisense RNAs have been reported

Gene name (sense strand)	Organism	Reference
Dhfr	Mouse	Farnham *et al*. (1985)
beta-globin	Mouse	Volloch (1986)
Surf-2/Surf-4	Mouse, human	Williams and Fried (1986)
c-Myc	Mouse, rat	Nepveu and Marcu (1986)
GnRH	Rat	Adelman *et al*. (1987)
c-Myb	Mouse	Bender *et al.* (1987)
Ribosomal L27	Mouse	Belhumeur *et al.* (1988)
p53	Mouse, human	Khochbin and Lawrence (1989)
THRA	Human, rat	Lazar *et al.* (1989)
ERCC1	Human	van Duin *et al.* (1989)
N-Myc	Mouse, human	Krystal *et al.* (1990)
eIF2 alpha	Human	Silverman *et al.* (1992)
SC35	Human	Fu and Maniatis (1992)
CD3 zeta/eta/theta	Mouse	Lerner *et al.* (1993)
TYMS	Human	Dolnick (1993)
BCMA	Human	Laabi *et al.* (1994)
bFGF	Human, rat	Murphy and Knee (1994)
WT1	Human	Eccles *et al.* (1994)
Hoxd3	Mouse	Bedford *et al.* (1995)
Hoxa11	Mouse, human	Hsieh-Li *et al.* (1995)
pMCH	Rat	Hervieu and Nahon (1995)
Hsp70.2	Mouse	Murashov and Wolgemuth (1996)
Pkn	Mouse	Batshake and Sundelin (1996)
MHC	Rat	Luther *et al.* (1998)
Tgf beta2	Mouse, human, rat	Coker *et al.* (1998)
SMAD5	Human	Zavadil *et al.* (1999)
HIF1 alpha	Human	Thrash-Bingham and Tartof (1999)
Ucn	Rat	Shi *et al.* (2000)

the first known instance of non-coding antisense RNA.

In the early days of antisense RNA analysis, although research on the functional aspects of antisense RNAs was limited mostly to describing the relationships between sense and antisense expression, some fundamental characteristics of antisense RNA were reported. These included:

1 nuclear localization of antisense RNAs (e.g. c-*Myc*: Bond and Wold, 1993; *p53*: Khochbin and Lawrence, 1989; *TS*: Dolnick, 1993);
2 absence of poly(A) tails (e.g. N-*Myc*: Krystal *et al.*, 1990; *HIF-1 alpha*: Thrash-Bingham and Tartof, 1999); and
3 existence of various-sized RNAs or smears on blots (e.g. N-*Myc*: Krystal *et al.*, 1990; *TS*: Dolnick, 1993).

Lastly, we must also mention a unique group of endogenous antisense RNAs: the antisense RNAs of *beta-globin, BCMA, Ucn,* and *Troponin I* (Volloch, 1986; Laabi *et al.*, 1994; Shi *et al.*, 2000; Podlowski *et al.*, 2002). These antisense RNAs are complementary to the full-length mRNA sequences of the sense-strand. Moreover, the intron sequences present in the genome sequences of the sense genes have been removed in these antisense RNA sequences. For example, in the case of *BMCA*, the consensus sequences at both ends of the intron are GT-AG. However, when the antisense RNA sequence is mapped onto the genome sequences, the putative intronic end sequences are CT-AC. The existence of these endogenous antisense RNAs has been proven by various methods, such as strand-specific RT-PCR, Northern hybridization, and RNA protection assay. The possibility that these RNAs are the products of pseudogenes has been evaluated and is considered unlikely (Laabi *et al.*, 1994).

The only possible explanation for these antisense RNAs of *beta-globin, BCMA, Ucn,* and *Troponin I* is that they are produced by RNA-dependent RNA polymerases (RdRps). Although *RdRp* genes with sequences homologous to those of viral RdRp have not been identified in mammals to date, data showing the existence of RdRp activity have been reported as part of antisense RNA analyses (Volloch, 1986; Bartsch *et al.*, 2004). Recently, genome-wide small RNA analysis has also suggested the existence of RdRp activity in mammalian cells (Kapranov *et al.*, 2010), although the *RdRp* gene itself is yet to be identified. Antisense RNA synthesis by RdRp might be a genome-wide phenomenon.

Global analysis of antisense transcripts

Since research entered the genome era, various experimental disciplines have been applied to the identification and analysis of SATs at the genome-wide level. As explained in the previous section, the data obtained in full-length cDNA projects have greatly changed and enhanced our fundamental knowledge, or concept, of 'antisense transcription' (FANTOM Consortium and RIKEN Genome Exploration Research Group Phase I and II Team, 2002; Kiyosawa *et al.*, 2003). SAGE (Serial Analysis of Gene Expression) data have also been used to identify SATs (Ge *et al.*, 2006).

Genome tiling array techniques have also identified the loci producing SATs (Kapranov *et al.*, 2002; Yamada *et al.*, 2003; Kampa *et al.*, 2004; Cheng *et al.*, 2005; Gagneur *et al.*, 2009). These methods measure direct genome transcription activity without the need to isolate cDNA clones, and they can be used to identify the genome regions producing sense and antisense transcripts. Although strand-specific microarray-based platforms are not as comprehensive as genome tiling arrays, they were designed for specifically discovering novel antisense transcripts (Ge *et al.*, 2008; Numata *et al.*, 2009).

Besides the approaches mentioned above, the following unique methods have been used to identify SATs at the genome-wide level. A method named 'GRO-seq' (global run-on sequencing), which maps the position, amount, and orientation of transcriptionally engaged RNA polymerases, estimated that as many as 58.7% of human genes have antisense transcription (Core *et al.*, 2008). He *et al.* (2008) developed a technique called asymmetric strand-specific analysis of gene expression (ASSAGE) in order to correctly distinguish sense and antisense transcripts in humans. This technique takes advantage of bisulfite treatment of RNA molecules to assign the correct orientations of transcripts on the genome sequences. With this method, they found evidence for antisense transcripts in 2900–6400 genes, depending on the cell type.

In the following sections, we describe the nature and characteristic features of SATs, as determined by using genome-wide approaches; we also show how the identification of SATs proceeded in model organisms, and we give some examples of pairs for which precise functional analyses have been performed.

Genome-wide analysis of SATs

Classification of SATs

Definition of SATs

Natural antisense transcripts are simply RNAs containing sequences complementary to other endogenous RNAs and capable of forming double-stranded RNA (dsRNA) with their sense counterparts. They can be transcribed *in cis* from opposing DNA strands in the same genomic region, or in *trans* from a separate genomic region (Fig. 1.1). *cis*-SATs have a perfect complementary

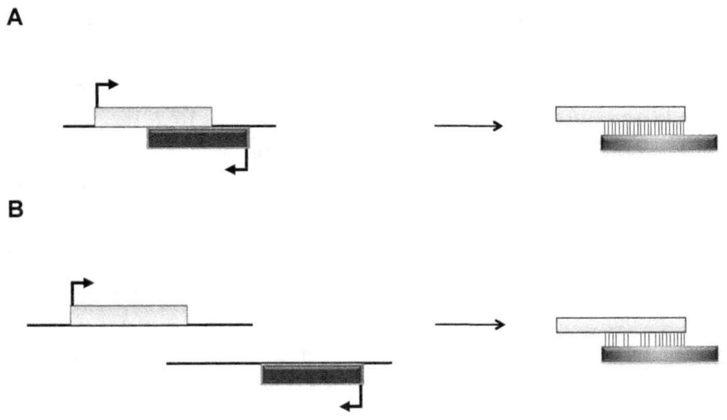

Figure 1.1 Model of *cis*-SAT and *trans*-SAT. (A) Genomic organization and transcript of *cis*-SAT gene pairs. *cis*-SATs are transcribed at the same genomic locus and have a perfect complementary overlap between sense and antisense transcripts. (B) Genomic organization and transcript of *trans*-SAT gene pairs. *trans*-SATs are transcribed from separate genomic loci and often possess imperfect complementarity. Boxes represent transcription units and arrows indicate the direction of transcription.

overlap between sense and antisense transcripts, whereas *trans*-SATs often possess imperfect complementarity.

In transcription at the same genomic locus in opposing directions (i.e. with bidirectional transcript pairs), two types of genomic arrangement are possible. One is a *cis*-SAT pair, in which the exons of the sense and antisense genes overlap, and the other one is a transcript pair, where there is no complementarity between both exons. We refer to the latter type as non-antisense bidirectional transcript pairs (NABTs). SATs are pairs capable of forming dsRNA with each other, whereas NABTs are not.

Many large-scale transcriptome analyses in the last decade have extensively characterized SATs in a variety of eukaryotic organisms. Most of the genome-wide studies of SATs have concentrated on *cis*-SATs. In contrast, not much attention has been given to *trans*-SATs, except in the case of small RNAs, such as microRNAs (miRNAs) that inhibit the translation of mRNAs and small nuclear RNAs that guide the modification of ribosomal RNAs. In this chapter we focus on *cis*-SATs.

Genomic organization of SATs

Many SATs have now been identified by genome-wide analysis, as we will discuss later. Assuming that the complementarity of SAT pairs is the major basis of the regulation and function of antisense transcripts, the genomic organization of overlaps in SAT pairs is important.

Each research group tends to have its own classification based on its own concept, but, generally, SATs can be classified according to their complementarity with each other (exon overlap, the presence or absence of introns) and the coding potential of each gene. The first genome-wide study of SATs was performed by Lehner *et al.* (2002), who, along with Shendure and Church (2002), arranged SATs in four categories: head-to-head (5′ UTR and 5′ UTR overlap, divergent); tail-to-tail (3′ UTR and 3′ UTR overlap, convergent); one transcript starting in an intron of the gene of the opposing strand; and one transcript entirely contained within the gene of the opposing strand (Fig. 1.2). The term 'UTR' is not applicable to the sequences of ncRNA, as the entire sequence of ncRNA is 'untranslated.' However, we use the terms '5′ UTR' and '3′ UTR' to illustrate overlapping patterns between SATs even if the SATs contain ncRNAs, as many references use these terms.

The studies that followed took more precise account of the exon–intron organization of SAT genes and added new categories of SATs (e.g. Zhang *et al.*, 2006). Whereas these studies have focused on SATs only, several studies have also included NABTs. We have categorized and compared SATs and NABTs in mice in order to illuminate the intrinsic characteristics specific to SATs (Kiyosawa *et al.*, 2003). Several other groups have also classified SATs and NABTs according to the patterns of overlap relative to the exon-intron organization, in an effort to define antisense regulation by dsRNA formation between complementary SATs (Fig. 1.3) (Galante *et al.*, 2007; Numata *et al.*, 2007).

To summarize, although various classifications have been proposed in genome-wide analyses of SATs, the following three subtypes are basically common: head-to-head (overlap of 5′ UTR with each other), tail-to-tail (overlap of 3′ UTR with each other), and full overlap (one transcriptional unit is completely covered by the transcriptional unit of the opposite strand) (Fig. 1.4) (Munroe, 2004; RIKEN Genome Exploration Research Group *et al.*, 2005; Lapidot and Pilpel, 2006; Werner *et al.*, 2009). Given that genomic arrangements are predominantly responsible for dsRNA-mediated transcript regulation, each category of SAT may have a different biological role. For example, tail-to-tail pairs tend to be associated with regulatory elements in the 3′ UTR, which might affect mRNA stability and alternative polyadenylation. Head-to-head pairs may support co-regulated overlapping promoters.

However, these classifications are apparently not enough to infer the relationships between genomic organization and function in SATs. All the analyses we have described above are based on the sequences of processed transcripts. However, overlaps between exon and intron sequences may form base-pairings between unspliced pre-mRNAs. In these cases, regulation by the formation of dsRNAs will depend on whether the SATs form base-pairs before or after splicing and regulate the transcription of SAT genes.

Identification and numbers of SATs in various organisms

The identification of large numbers of SATs by genome-wide analyses has raised the possibility that sense–antisense interactions are a common mechanism for regulating gene expression in various organisms. *In silico* analyses of SATs, as described below, have yielded substantial evidence that antisense transcripts are much more widespread than was previously thought. The total number of SATs may increase when more data, for example, such as generated by next-generation sequencers become available.

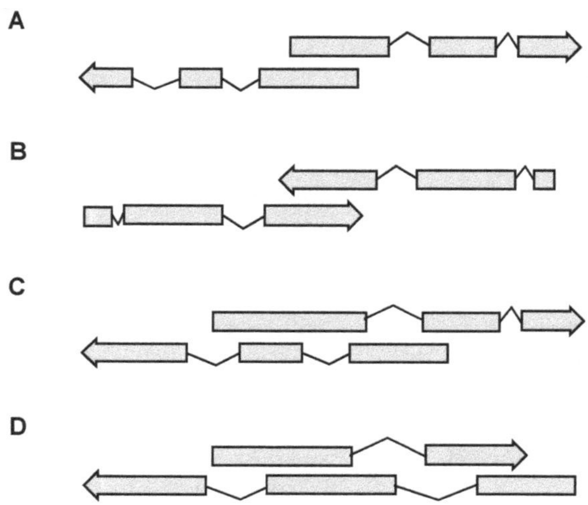

Figure 1.2 Classification of SATs by Lehner *et al.* (2002). Classification of SATs by Lehner *et al.* in terms of genomic arrangement is presented. (A) Head-to-head (5′ UTR and 5′ UTR overlap, divergent). (B) Tail-to-tail (3′ UTR and 3′ UTR overlap, convergent). (C) One transcript starting in an intron of the other gene. (D) One transcript entirely contained within the other gene. Exons are depicted as boxes. Arrow-shaped boxes indicate the direction of transcription.

A
B
C
D
E
F

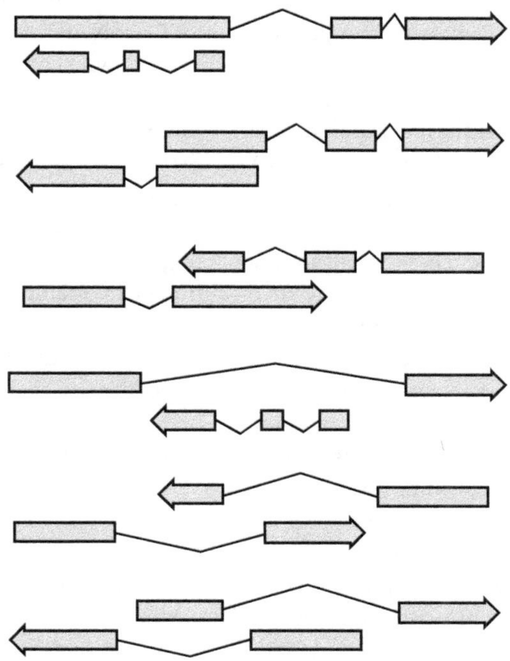

Figure 1.3 Classification of SATs and NABTs by Numata *et al.* (2007). Classification of SATs and NABTs by Numata *et al.* is presented. Overlaps of SATs and NABTs are categorized according to the following three patterns: (A) One transcription unit of the SAT pair is completely contained within an exon of the other. (B) Head-to-head (5′ UTR and 5′ UTR overlap, divergent). (C) Tail-to-tail (3′ UTR and 3′ UTR overlap, convergent). (D) One transcription unit of the NABT pair is completely contained within an intron of the other. (E) An NABT pair overlaps in convergent orientation. (F) An NABT pair overlaps in divergent orientation.

A
B
C

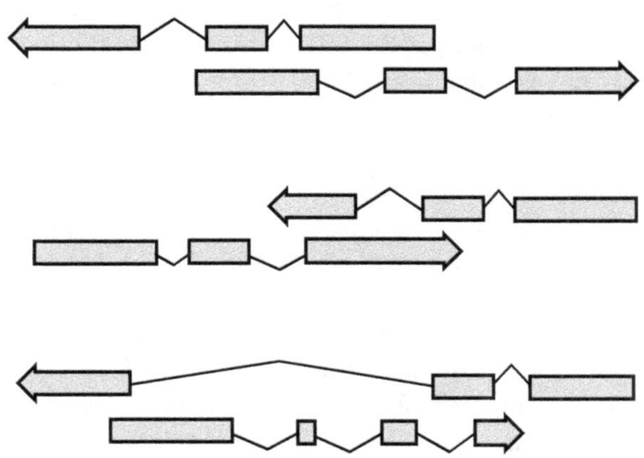

Figure 1.4 Basic classification of SATs. (A) Head-to-head (5′ UTR and 5′ UTR overlap, divergent). (B) Tail-to-tail (3′ UTR and 3′ UTR overlap, convergent). (C) Fully overlapping type (one transcriptional unit is completely covered by the other).

Humans and mice

The availability of the complete human genome sequence and the accumulation of data sets of millions of expressed sequences have made it possible to predict SATs in humans. The first genome-wide study identified 87 human SATs by using mRNA sequences in the public databases (Lehner *et al.*, 2002). These researchers started with a data set of only full-length mRNA sequences to ensure the reliability of the sequences. Therefore, the SAT pairs identified represented only a small portion of all the human SAT pairs, because a limited quantity of mRNA sequences was analysed. The next studies attempted to identify SATs by using expressed sequence tag (EST) sequences in addition to mRNA sequences. Shendure and Church (2002) found 144 human SAT pairs and 73 mouse SAT pairs by using ESTs with correct orientation. They also validated the expression of 26 of 33 candidate SAT pairs by orientation-specific RT-PCR (Shendure and Church, 2002). In 2003, Yelin *et al.* (2003) identified 2667 pairs of human SATs by using the entire set of EST sequences available in the public databases; by experimental validation based on microarray analysis they estimated that nearly 1600 of the 2667 SAT pairs were actually expressed. Their successful identification of larger numbers of SATs than were found by Shendure and Church was the result of comprehensive use of EST sequences. Chen *et al.* (2004) also paid more attention to the use of EST sequences, including sequences containing a poly(A) signal, and they used gene annotations to determine the correct orientations of the EST sequences. Their study revealed that nearly 22% of 26,741 transcriptional clusters form SAT pairs in humans. The SAT data sets published by the two research groups (Yelin *et al.*, 2003; Chen *et al.*, 2004) overlapped each other by only 38%, suggesting that the total number of human SAT pairs is even greater.

Use of the mouse full-length cDNA data from the FANTOM II project (led by RIKEN, Japan) made it possible to produce comprehensive SAT data without using EST sequences. We identified 2481 SAT pairs by using 60,770 mouse full-length cDNA sequences (FANTOM Consortium and RIKEN Genome Exploration Research Group Phase I and II Team, 2002; Kiyosawa *et al.*, 2003). Additionally, we compared the characteristics of SAT and NABT pairs. There was a substantial bias, in that SATs tended to be isolated from the same cDNA libraries, whereas NABTs did not. This bias towards one type of bidirectional pair implies that SATs found in the same cDNA library may exist in the same tissue or single cell. The simultaneous presence of both sense and antisense transcripts might be essential for antisense-mediated regulation in, for example, dsRNA formation.

Several more studies followed, and, from the results of cDNA sequence analyses, the numbers of SAT pairs in humans and mice have been updated to 6141 and 5248, respectively (Engström *et al.*, 2006).

Plants

In the plant kingdom, genome-wide SATs have been analysed in rice (*Oryza sativa*) and in *Arabidopsis thaliana*. Analysis of whole-genome tilling arrays in *Arabidopsis* has revealed nearly 7600 antisense transcripts, corresponding to roughly 25% of all annotated genes (Yamada *et al.*, 2003). These transcripts showed substantial tissue-specific antisense RNA expression, suggesting that antisense transcription is a general phenomenon in plant cells as well as in mammalian cells.

Several subsequent studies addressed the SATs in *Arabidopsis*. A computational analysis using a full-length cDNA data set found an estimated 1340 potential SATs; the expression of 957 of these was confirmed from massively parallel signature sequencing data (Wang *et al.*, 2005). Another study using The Arabidopsis Information Resource (TAIR) data set identified 1083 SAT pairs (2147 genes) (Jen *et al.*, 2005); microarray investigation of the expression of the convergent types (1866 genes) among these revealed that 63.8% to 67.9% of SATs were expressed in seedlings, roots, and flowers in 7-day suspension culture. Similarly, an *in silico* study using a TAIR data set identified 1126 SAT pairs in *Arabidopsis* (Henz *et al.*, 2007).

In the *japonica* rice subspecies, a genome-wide transcriptome analysis has been performed with 28,469 full-length cDNA sequences; 902 bidirectional transcript pairs (no distinction between SAT and NABT) were identified (Rice Full-Length cDNA Consortium, 2003). A subsequent study using the same data set, after removal of the redundant cDNA sequences, revealed 601 SAT pairs and 86 NABT pairs – up to about 7% of all cDNA clusters in rice (Osato *et al.*, 2003). They confirmed the expression of roughly 68% of bidirectional pairs by microarray analysis. Li *et al.* (2006a) also performed expression analyses in *indica* rice subspecies by using tilling arrays, showing that 23.8% of annotated genes (10,452 genes) had substantial antisense transcription.

Yeasts, flies, worms and other organisms

The yeast *Saccharomyces cerevisiae* has a relatively small eukaryotic genome and was the first eukaryote to have its genome sequenced. Up to 1555 of 5654 genes showed antisense expression, and in a high-resolution study using a tiling array 1153 of these had poly(A)-minus RNA (David *et al.*, 2006). Poly(A)-minus RNAs are also more abundant than poly(A)-plus RNAs in antisense transcripts of the fission yeast *Schizosaccharomyces pombe* (Dutrow *et al.*, 2008).

In *Drosophila melanogaster*, it was estimated by a computational approach that there were 1027 SAT pairs out of 13,474 gene loci, accounting for about 15% of annotated genes (Misra *et al.*, 2002). Similar numbers of SAT pairs have been identified by Sun M., *et al.* (2006), Zhang *et al.* (2006), and Numata *et al.* (2007). Strikingly, the number of SAT genes in the nematode *Caenorhabditis elegans* is remarkably low, at only 2.8% (Zhang *et al.*, 2006) of all genes (approximately 19,000; The *C. elegans* Sequencing Consortium, 1998). The numbers of transcript data used for the *C. elegans* analysis were comparable to those used in *Drosophila*, excluding the possibility that the paucity of SATs in the nematode was due to a shortage of transcript data. Zhang *et al.* (2006) hypothesized that the operon structure in the nematode genome may partly explain the low frequency of SATs in this organism.

SATs have been identified in other eukaryotic organisms, including rats, chickens, cows, malarial mosquitoes, frogs, zebrafish and sea squirts (Gunasekera *et al.*, 2004; M. Sun *et al.*, 2006; Zhang *et al.*, 2006). The results of all of these studies confirm that SATs are frequently observable in the animal, plant, and fungal kingdoms.

Characteristic features of SAT

Genome-wide analysis of SATs in various organisms by bioinformatics, as discussed above, has disclosed the very broad landscape of SATs. The experimental analyses have also revealed that SATs have unique features distinguishing them from ordinary transcripts. In this section, we discuss SATs in terms of their unique features, including their expression modes, expression patterns, and chromosomal distributions.

Absence of poly(A) tails and localization in nucleus

Mouse SATs are frequently identified as poly(A)-minus RNA transcripts by analysis using randomly primed CAGE (Cap Analysis Gene Expression) libraries, and also by Northern hybridization and microarray analyses (Kiyosawa *et al.*, 2005; RIKEN Genome Exploration Research Group *et al.*, 2005). In Northern analysis of mouse SATs, smear band patterns were detected with total RNA, but distinct bands were also detected with poly(A)-plus RNA (Kiyosawa *et al.*, 2005). These results suggest that the SAT gene loci tend to produce multiple-sized transcripts that are poly(A)-minus RNAs. In addition, these poly(A)-minus transcripts tended to be nuclear localized (Kiyosawa *et al.*, 2005). These features (absence of poly(A) tails, existence of various-sized RNAs, nuclear localization) are in agreement with those revealed in antisense RNAs discovered before the genome era, as described in the previous section. Poly(A)-minus RNAs have also been observed by Northern analysis of selected *Arabidopsis* SATs (Kiyosawa *et al.*, 2005). The biological function of these poly(A)-minus SATs remains to be elucidated. However, the conservation of this characteristic in animals and plants implies that it is of common importance to their existence.

Relationship between sense and antisense expression

A critical issue in evaluating the role of antisense transcripts in regulating different cellular phenomena is the expression patterns of sense and antisense transcripts within the same SAT pairs. Co-expression of SATs makes it possible to form dsRNA that can lead to antisense regulation.

Inverse expression between sense and antisense transcripts might be a good indicator of reciprocal regulation between sense and antisense genes, such as transcriptional interference.

The proportion of SAT pairs that are co-expressed and/or inversely expressed is significantly more frequent than would be expected by chance in humans and mice (Chen *et al.*, 2005; M. Sun *et al.*, 2006). This suggests that co-expression and inverse expression are properties of SATs and are related to antisense regulation (Chen *et al.*, 2005; M. Sun *et al.*, 2006). SAT pairs that exhibit co-expression patterns between sense and antisense genes are more frequent than those that exhibit inversely expressed patterns (Chen *et al.*, 2005; RIKEN Genome Exploration Research Group *et al.*, 2005; Engström *et al.*, 2006; Oeder *et al.*, 2007; Numata *et al.*, 2009; Watanabe *et al.*, 2010). Moreover, the sense–antisense expression ratios of many SATs in mice vary among different tissues and cell types, implying that SATs play a role in the maintenance of tissue-specific gene expression and cell differentiation (Kiyosawa *et al.*, 2005; Okada *et al.*, 2008; Watanabe *et al.*, 2010).

Although the exact mechanism of antisense regulation remains unclear, it is probable that there is a balance between the expression of sense and antisense transcripts under certain conditions; this balance is probably modulated by antisense regulation to adapt to different developmental and physiological stages or conditions.

SAT presence on mammalian X chromosomes

In humans and mice, SAT gene pairs are almost evenly distributed across the genome, except on the sex chromosomes (Kiyosawa and Abe, 2002; Kiyosawa *et al.*, 2003; Chen *et al.*, 2004; Zhang *et al.*, 2006). In both of these species the number of SATs is significantly lower (about half to one-third) on the X-chromosome than on the autosomes, whereas NABTs are evenly distributed among the X chromosome and autosomes (Fig. 1.5, revised from Kiyosawa and Abe, 2002; Table 1.2). This irregular distribution of SATs on the X chromosome promotes speculation about the putative mechanism of gene regulation by antisense transcripts as described below.

One of the characteristic differences between SAT and NABT pairs is the ability of the former to form dsRNA. In mammalian female nuclei, one of the two homologous X chromosomes is randomly inactivated, by which process mammals achieve dosage compensation of X-linked genes. This situation suggests that two active chromosomes are needed to achieve gene expression regulation via the formation of dsRNA between SATs, if any such regulation exists. It appears unlikely that bidirectional transcription events at a SAT locus on a single chromosome proceed simultaneously because of transcriptional collision. Therefore, in order for a cell to produce both sense and antisense transcripts in its single nucleus, random monoallelic expression of sense and antisense genes might be necessary (i.e. sense transcripts from one chromosome only, and antisense transcripts from the other chromosome only). This situation is impossible if there is only one active chromosome in the nucleus. If SATs need to perform gene expression regulation by such a mechanism, this might be why they are poorly represented on the X chromosome. This argument is supported by the observation that, in humans, certain SAT pairs on the X chromosome are accumulated in regions that escape X-inactivation, whereas NABT pairs are not (Kiyosawa and Abe, 2002).

Cross-species conservation of SAT pairs

Cross-species comparative analysis of large-scale data on nucleotide sequences, genomic structure, and gene expression is an effective approach to the determination of functionally important elements (Boffelli *et al.*, 2004). Isolation of the important factors that are related to SAT regulation and are conserved across various species may shed light on antisense-mediated gene regulation.

SATs conserved in human and mouse

Although the numbers of human–mouse conserved SATs reported so far are discrepant, ranging from a few hundred (M. Sun *et al.*, 2006; Zhang *et al.*, 2006; Galante *et al.*, 2007; Numata *et al.*, 2007; Okada *et al.*, 2008) to almost a thousand (Engström *et al.*, 2006), depending on the strategy and data sets used, these observations are

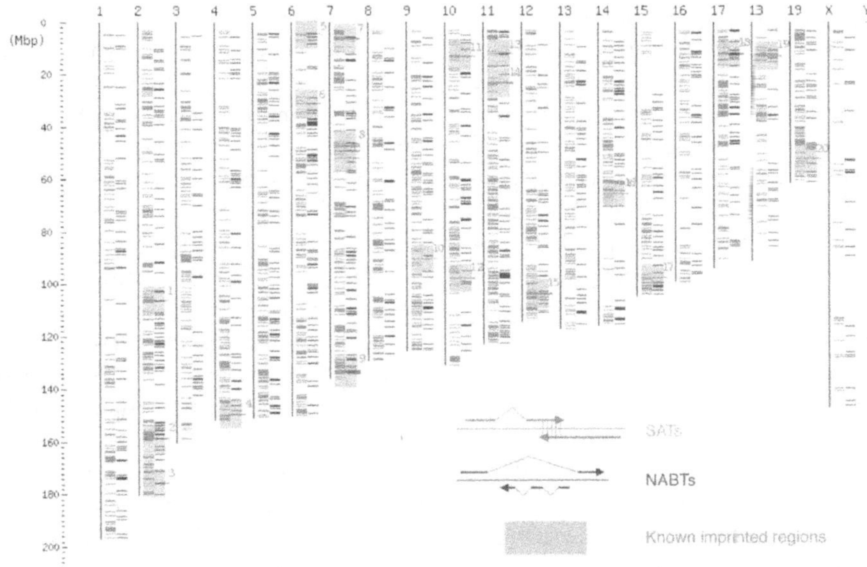

Figure 1.5 Distribution of SATs and NABTs on mouse chromosomes. The locations of mouse SATs and NABTs on chromosomes are presented. Dark grey lines (left-hand column to the right of each chromosome) indicate SATs that could potentially form natural dsRNA. Light grey lines (right-hand column to the right of each chromosome) indicate NABTs. Known imprinted regions are shown as grey boxes and numbered as follows: *Wt1* (1); *Nnat* (2); *Gnas*, *Gnasxl*, *Nespas* (3); *p73* (4); *Sgce* (5); *Copg2*, *Mit1/lb9* (6); *Peg3/Pw1*, *Usp29*, *Zim1*, *Zim3*, *Zfp264* (7); *Frat3*, *GABRA5*, *GABRB3*, *GABRG3*, *Magel2*, *Mkrn3/Zfp127*, *Ndn*, *Snrpn*, *Snurf*, *Ube3a* (8); *H19*, *Igf2*, *Igf2as*, *Ins2*, *Ipl/Tssc3*, *Kvlqt1*, *Mash2*, *Nap1l4*, *Obph1*, *p57KIP2/Cdkn1c*, *Slc22a1l*, *Tapa1/Cd81*, *Tssc4* (9); *Rasgrf1* (10); *Zac1* (11); *Dcn* (12); *Meg1/Grb10* (13); *U2af1-rs1 region1*, *U2af1-rs1 region2* (14); *Dlk*, *Meg3/Gtl2* (15); *Htr2a* (16); *Ata3* (17); *Igf2r*, *Mas1*, *Slc22a2*, *Slc22a3* (18); *Impact* (19); *Ins1* (20) (modified from Kiyosawa and Abe, 2002).

Table 1.2 Numbers of SATs and NABTs on X chromosome and autosomes (revised from Kiyosawa and Abe, 2002)

	SATs				NABTs			
	No. of genes[1]	Average[2]	Highest[3]	Lowest[4]	No. of genes[1]	Average[2]	Highest[3]	Lowest[4]
Autosomes	2448	15.1%	chr 19; 19.7%	chr 1; 12.5%	877	5.4%	chr 6; 6.9%	chr 3; 3.6%
X chromosome	33	6.3%	–	–	22	5.3%	–	–

[1]Total numbers of SAT or NABT genes mapped on autosomes or X chromosome.

[2]Average percentages of SAT or NABT genes on autosomes, or percentages of SAT or NABT genes on X chromosome.

[3]Highest percentage SAT or NABT gene appearance among autosomes.

[4]Lowest percentage SAT or NABT gene appearance among autosomes.

consistent in their support of the functional relevance of overlaps, such as overlapping patterns and coding potential. Most human–mouse conserved SATs are of the convergent overlapping type (Sun *et al.*, 2005; Engström *et al.*, 2006; Galante *et al.*, 2007; Numata *et al.*, 2007). There may be regulatory elements on the 3′ UTR for antisense regulation.

Among human–mouse conserved SATs, 95.6% are annotated as protein-coding gene–protein-coding gene pairs. In contrast, 42% of SATs among SAT pairs that are not conserved between humans and mice are protein-coding gene–non-coding gene pairs or non-coding gene–non-coding gene pairs (Numata *et al.*, 2007). This observation suggests that most ncRNA-associated SAT pairs emerge from species-specific transcripts, whereas most of the conserved SAT transcripts are protein-coding genes.

Further expression analysis by using microarrays and Northern hybridization has revealed the distinct nature of expression patterns between human and mouse SAT pairs (Okada *et al.*, 2008). Of the 291 human–mouse conserved SATs found, only 33% show similar expression in the two species (Okada *et al.*, 2008). Intriguingly, transcripts of human–mouse conserved SAT genes tend to be poly(A)-minus RNAs only in mice (Okada *et al.*, 2008). Moreover, in mice these poly(A)-minus RNAs displayed smear hybridization patterns, whereas those in humans showed specific band patterns (Okada *et al.*, 2008). This difference indicates that regulation of transcripts from SAT genes may occur in a species-specific manner, despite their cross-species conservation on genomic sequences (Okada *et al.*, 2008).

Comparison of overlap patterns of SATs in various organisms

Comparative analysis across multiple species has shown that a large number of conserved SATs maintain the same overlapping patterns and some SAT pairs are conserved between non-mammalian vertebrates and mammalian species (Zhang *et al.*, 2006). In humans, mice, *Drosophila*, and *Arabidopsis*, the number of convergent types of SATs is larger than that of divergent types, suggesting that regulation through 3′ regions is functionally important in SATs (Galante *et al.*, 2007; Numata *et al.*, 2007).

In contrast, the number of NABTs (i.e. bidirectional pairs but not exon-overlapping) is high in animals but not in plants (Galante *et al.*, 2007; Numata *et al.*, 2007). The most frequent pattern in NABTs is one in which a gene is completely embedded in an intron of its partner gene on the opposite strand (Galante *et al.*, 2007; Numata *et al.*, 2007). Expression analysis has shown that many intronic regions are transcriptionally active, and some antisense intronic transcription may be involved in tumour differentiation and developmental regulation (Reis *et al.*, 2004; Stolc *et al.*, 2004). Possible regulatory mechanisms that may act through intronic antisense transcription include transcriptional interference and splicing regulation of specific exons. Because of their genomic organization (i.e. overlap of intronic regions), the putative role of these intronic transcripts would have to be restricted to the nucleus in order to regulate transcription of the other transcripts.

It is important to note that many SATs remain to be discovered in species for which there are insufficient EST and full-length cDNA data. Moreover, expression analysis by DNA microarray has revealed more SATs that could not be identified by cDNA resources (Numata *et al.*, 2009). Therefore, the data sets of conserved SATs are still far from being complete. Next-generation high-throughput transcriptome sequencing definitely helps to gain a detailed view of SATs.

Proposed gene regulation mechanisms mediated by antisense transcripts/transcription

The finding that numerous SAT loci are harboured within the eukaryotic genome has raised further questions about the functional role of SATs. Below, we describe several proposed mechanisms of gene regulation by SATs through transcription, dsRNA formation, and modification of chromosome state. A summary of various gene expression mechanisms starting with antisense transcription is shown in Fig. 1.6.

Gene expression regulation mechanisms by antisense transcription

Transcriptional collision

RNA polymerase II binds to upstream regions of the gene and synthesizes mRNAs, moving towards the 3′ end of the gene. When transcription occurs at the SAT gene locus, an RNA

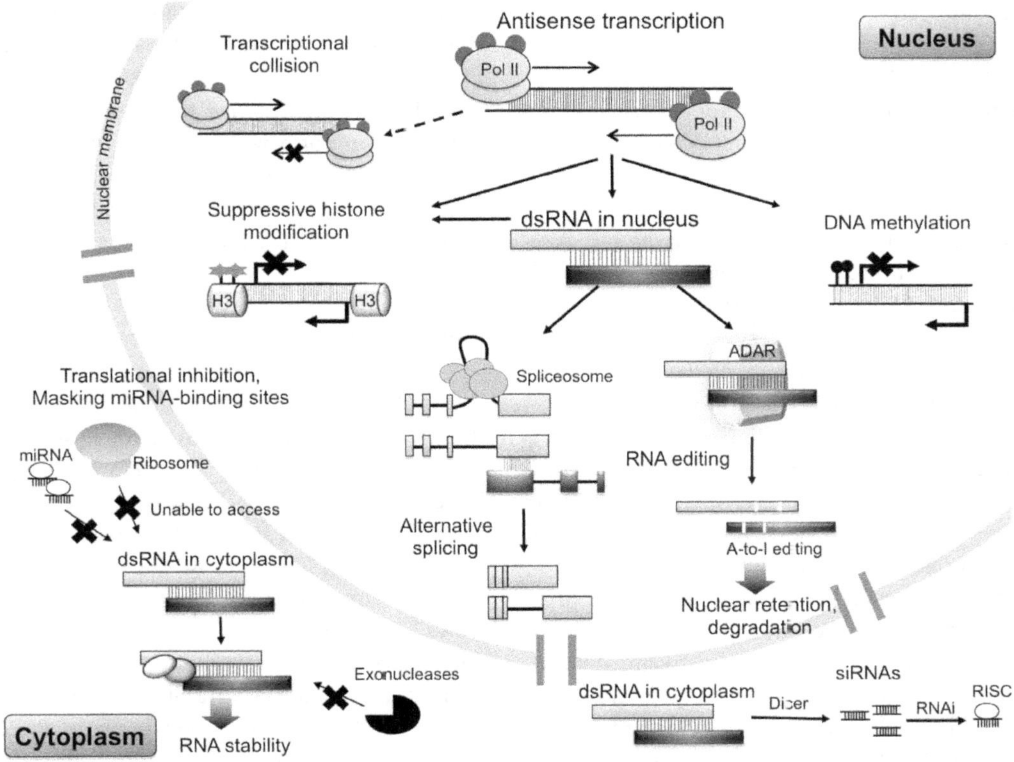

Figure 1.6 Summary of types of gene expression regulation starting with antisense transcription in the cell. Antisense transcription from SAT genes can lead to transcriptional collision, suppressive histone modification and methylation of DNA on the opposite strand. Antisense transcripts generate dsRNAs in the nucleus and cytoplasm. Some nuclear dsRNAs trigger alternative splicing and RNA editing. dsRNA production in the cytoplasm can lead to RNA stability by translational silencing, an impediment to the binding of miRNAs and exonucleases. The RNAi machinery produces siRNAs from dsRNAs. Pol II, RNA polymerase II; ADAR, adenosine deaminase acting on RNA; H3, histone 3; RISC, RNA-induced silencing complex.

polymerase bound to a strand of the genome should collide with the RNA polymerase bound to the opposite strand during the transcription of both strands of the gene. This leads to inhibition of transcription, which has been proposed as the transcriptional collision model (see review by Shearwin *et al.*, 2005). In this model, antisense RNA is not involved in interfering with transcription of the sense RNA. Instead, transcription in an antisense direction decreases the synthesis of sense RNA.

Collision of RNA polymerase II has been observed in *Escherichia coli* by using atomic force microscopy (Crampton *et al.*, 2006). Transcriptional collision has also been observed in *S. cerevisiae,* in which transcription elongation is restricted on the convergent type of SAT gene (Prescott and Proudfoot, 2002). A bioinformatics study of SATs has revealed that, in humans and mice, the expression levels of SAT pairs decrease as the length of the overlapping regions of SAT genes increases (Osato *et al.*, 2007). This observation suggests that collision of RNA polymerases occurs and indicates that the frequency of collision of RNA polymerases probably increases when the overlapping regions increase in length (Osato *et al.*, 2007). Transcriptional collision may explain the inverse expression levels observed between sense and antisense genes.

Genomic rearrangements

Immunoglobulin in B lymphocytes and T-cell receptor molecules in T lymphocytes are generated from the transcription of variable regions of immunoglobulin and T-cell receptor genes, respectively. In order to generate variability in response to antigens, T and B lymphocytes perform V(D)J recombination. Antisense transcription at this locus increases the frequency of recombination events in both B and T lymphocytes (Bolland *et al.*, 2007; Giallourakis *et al.*, 2010).

Additional processes that enhance variability and enable effector functions to be performed are known to occur in B lymphocytes. These processes include somatic hypermutation (SHM) in the exons of the variable regions of immunoglobulin and class-switch recombination (CSR), which promotes replacement of the constant region exons with a set of downstream constant region exons. Activation-induced cytosine deaminase (AID), which deaminates cytosine to uracil in single-strand DNA (ssDNA), is essential for both SHM and CSR. For DNA recombination, DNA must uncoil locally and become single stranded. It has been suggested that antisense transcription in the variable regions makes the ssDNA accessible to AID (Ronai *et al.*, 2007; Perlot *et al.*, 2008). Antisense transcription opens the transcription bubble and exposes the ssDNA to AID, or otherwise antisense transcription is involved in remodelling the chromatin structure, thus enabling access by recombination machinery (Ronai *et al.*, 2007; Perlot *et al.*, 2008). A recent study has also revealed that AID interacts broadly with promoter-proximal sequences, not just the immunoglobulin gene locus; this broad action is responsible for the low frequencies of hypermutation across the B-lymphocyte genome (Yamane *et al.*, 2011).

Regulation by dsRNAs in nucleus

Alternative splicing and termination

Antisense RNA may also bind to sense RNA, masking the splice sites and thereby changing the balances between splice variants. The thyroid hormone receptor-α gene, *erbAa*, encodes two isoform proteins, TRα1 and TRα2, which arise from the alternatively processed mRNAs *erbAa1* and *erbAa2* (Lazar *et al.*, 1989). The transcript of the antisense gene *RevErb* is complementary to the last exon of *erbAa2* mRNA but not to *erbAa1* mRNA (Fig. 1.7). A clear correlation between *RevErb* expression and *erbAa1*/*erbAa2* mRNA levels has been observed, indicating that, by RNA masking, the *RevErb* transcript prevents the *erbAa* primary transcript from splicing into *erbAa2* (Munroe and Lazar, 1991; Hastings *et al.*, 1997). Splicing inhibition by the SAT is specific and efficient and could thus lead to alternative splicing and termination.

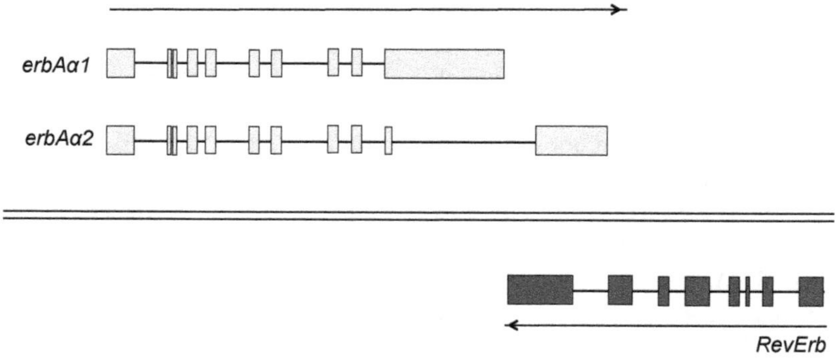

Figure 1.7 Structural organization of the *erbAa* locus in the rat. *erbAa* encodes the two alternatively spliced *erbAa* mRNAs, *erbAa1* and *erbAa2*. Boxes represent exons and lines indicate introns. The 3′ portion of the last exon of *RevErb* overlaps with the *erbAa2*-specific 3′ portion of the last exon, but not with that of *erbAa1*.

RNA editing

dsRNAs might also serve as substrates for RNA-editing adenosine deaminases that act on dsRNA to deaminate adenosine to inosine (A- to I-editing). Hyper-editing of long, perfect dsRNAs can result in their nuclear retention (Zhang and Carmichael, 2001) or cytoplasmic degradation (Scadden and Smith, 2006). RNA editing leads to altered coding potential in mRNAs, with the expression of specifically altered proteins. Basic fibroblast growth factor (*bFGF*) in *Xenopus* oocytes is an example in which the region overlapping with the *bFGF* antisense transcript is edited (Kimelman and Kirschner, 1989). The *Drosophila melanogaster* gene *Rnp4f* and its antisense transcript *Sas10* also induce editing in the overlapping regions of *Rnp4f* mRNA (Peters *et al.*, 2003). It was recently shown in both humans and mice that the level of RNA editing in SAT transcripts is negligible, except in the regions residing within Alu repeat elements (Neeman *et al.*, 2005). Therefore, regulation through RNA editing may not be one of the main mechanisms of SATs in humans and mice.

Regulation by dsRNAs in cytoplasm

RNA stability

dsRNA formation in the cytoplasm can change sense mRNA stability. The overlapping region might affect mRNA stability by reducing mRNA decay. Antisense transcripts of inducible nitric oxide synthase (*iNOS*) increase the stability of *iNOS* mRNA (Matsui *et al.*, 2008). *iNOS* gene expression is important in inflammatory diseases; an *iNOS* antisense transcript without a poly(A) tail is transcribed in response to inflammatory stimuli (Matsui *et al.*, 2008). In the 3′ UTR of *iNOS* mRNAs in humans, rats, and mice there are several AU-rich elements (AREs). *iNOS* mRNA and the *iNOS* antisense transcript have been proposed to form an RNA-binding protein complex

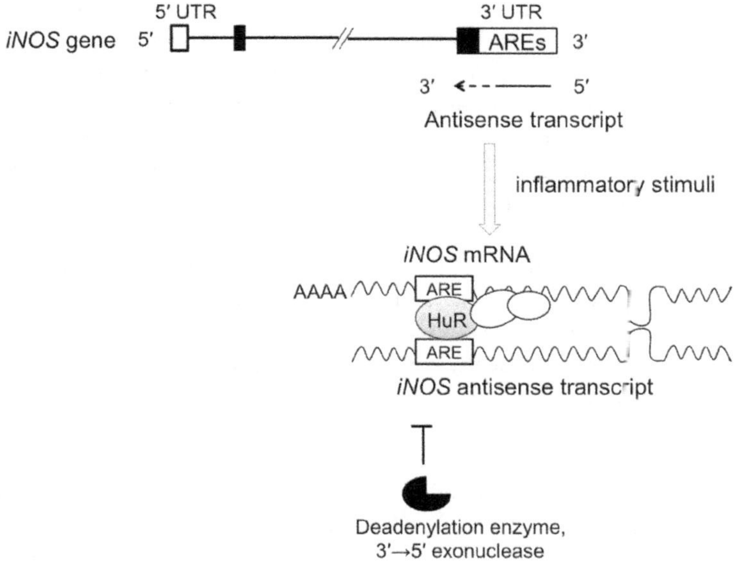

Figure 1.8 Model of stabilization of *iNOS* mRNA by interaction with *iNOS* antisense transcript. When the *iNOS* antisense transcript is expressed in response to inflammatory stimuli, it stabilizes *iNOS* mRNA by interaction with the *iNOS* mRNA 3′ UTR through the AU-rich elements (ARE)-binding human antigen R (HuR). *iNOS* mRNA may cross-hybridize with the *iNOS* antisense transcript at the loops that have sequence complementarity with each other. It is speculated that heterogeneous nuclear ribonucleoprotein L and polypyrimidine tract binding protein are involved in the protein complex (white circles) (Matsui *et al.*, 2008). The resulting protein complex may inhibit access by a deadenylation enzyme and 3′→5′ exonuclease (Matsui *et al.*, 2008). White and black boxes indicate untranslated and protein-coding regions, respectively.

with the ARE-binding human antigen R (HuR) protein for *iNOS* mRNA stability (Matsui *et al.*, 2008). Formation of this complex may help to inhibit access by deadenylation enzyme and exonucleases (Fig. 1.8) (Matsui *et al.*, 2008).

Translational inhibition

Translational inhibition can be mediated not only by miRNAs but also by long RNAs of SATs. In the case of the transcription factor *PU.1*, antisense RNA is non-coding and polyadenylated and present at low concentrations, but it has a longer half-life than *PU.1* mRNA (Ebralidze *et al.*, 2008). Processed antisense RNA in the cytoplasm can bind to the sense transcript and stall translation between the initiation and elongation steps (Ebralidze *et al.*, 2008). Naturally occurring small interfering RNAs (siRNAs) have not been detected at the *PU.1* gene locus (Ebralidze *et al.*, 2008). Therefore, it is probable that antisense RNA mediates the translational inhibition of sense transcripts by dsRNA formation (Ebralidze *et al.*, 2008).

Masking of miRNA-binding sites

SATs can cover miRNA-binding sites following cytosolic dsRNA formation. A non-coding antisense transcript of beta-secretase-1 (*BACE1*), *BACE1-AS*, forms dsRNA with the sense *BACE1* RNA and increases the stability of *BACE1* RNA (Faghihi *et al.*, 2008). This is owing to covering of the miRNA-binding sites of *BACE1* RNA (Faghihi *et al.*, 2010). *BACE1-AS* might eliminate miRNA-induced translational repression and sense *BACE1* RNA decay (Faghihi *et al.*, 2010). Because the 3′ UTRs of mRNAs frequently contain target sites for miRNAs (Arteaga-Vazquez *et al.*, 2006), dsRNA formation in the cytoplasm can inhibit the interaction between miRNAs and transcripts; this explains the enhancement of the stability of SAT transcripts.

Short RNAs associated with protein-coding-gene loci

Two classes of short RNA (<200 nt) localized to around the 5′ and 3′ termini of the annotated gene loci in both sense and antisense directions have been found in human protein-coding genes; these classes are promoter-associated short RNAs (PASRs) and termini-associated short RNAs (TASRs) (Kapranov *et al.*, 2007). The expression of genes for which PASRs or TASRs overlap is correlated with the density of these short RNAs (Kapranov *et al.*, 2007). It is possible that short RNAs often mark the boundaries of genes and regulate the expression of the genes associated with them (Kapranov *et al.*, 2007). Recently, it has been revealed that in human cells subsets of short RNAs in PASRs and TASRs have a poly(U) tail at their 5′ ends (Kapranov *et al.*, 2010). These short RNAs are probably generated by RdRp-mediated copying in which mRNA molecules are used as templates in synthesis (Kapranov *et al.*, 2010). It is possible that 5′ poly(U) PASRs and TASRs function either as primers or incomplete antisense products of this RNA copying, although the biological significance of 5′ poly(U) RNAs remains unclear (Kapranov *et al.*, 2010).

Transcription start site–associated RNAs (TSSa-RNAs) (20–90 nt) have also been identified by using deep sequencing (Core *et al.*, 2008; Seila *et al.*, 2008). TSSa-RNAs flank active promoters in both the sense and antisense direction with regard to downstream genes (Core *et al.*, 2008; Seila *et al.*, 2008). Divergent promoter activity, which is a non-overlapping transcription initiation event and is thought to produce TSSa-RNAs, is observed with more than half of human and mouse genes (Core *et al.*, 2008; Seila *et al.*, 2008). Similarly, promoter upstream transcripts (PROMPTs) have been identified as short and unstable transcripts upstream of genes in both the sense and the antisense direction (Preker *et al.*, 2008). The transcribed regions of TSSa-RNAs and PROMPTs overlap with RNA polymerase II occupancy and active chromatin marks, suggesting that TSSa-RNAs and PROMPTs play a role in maintaining a chromatin structure that permits transcription initiation (Core *et al.*, 2008; Preker *et al.*, 2008; Siela *et al.*, 2008). Interestingly, generation of these TSSa-RNAs is independent of Dicer, and there is no evidence of dsRNA formation or the corresponding natural antisense transcripts, suggesting that these small RNAs are not mediated through a common RNA interference (RNAi) (Seila *et al.*, 2008).

The abundance of short RNA molecules at transcription initiation sites and termination sites suggests the possibility of involvement of these RNA transcripts in the regulation of protein-coding gene expression by epigenetic regulation processes such as chromatin remodelling, which is explained in the following subsection.

Epigenetic regulation

Many natural antisense transcripts are involved in epigenetic regulation, in association with methylation and/or chromatin modification, as described below. Some of the epigenetic changes are independent of the RNAi pathway mediated by dsRNA formation. Instead, antisense transcripts accumulate on the chromosomes and trigger DNA or chromatin modifications. These modifications affect the expression of a single gene or a gene cluster, or even (occasionally) of entire chromosomes. In this subsection, we introduce several examples of antisense-mediated epigenetic regulation.

Silencing of tumour-suppressor genes through chromatin modification

In a variety of cancers, many tumour-suppressor genes are silenced epigenetically (Jones and Laird, 1999). An imbalance in SAT expression can potentially lead to this situation. Here, we give some well-documented examples, namely of *p15AS* and *p21AS*, which are the antisense RNAs for *p15* and *P21*, respectively.

p15AS

p15 encodes a cyclin-dependent kinase inhibitor that acts as a tumour suppressor (Nobori *et al.*, 1994). In many cancers, the *p15* gene is silenced (Lübbert, 2003). Yu *et al.* (2008) found a naturally occurring *p15* antisense RNA (*p15AS*) that is 34.8 kb long. The level of *p15AS* was up-regulated in leukaemia cell lines, whereas the *p15* mRNA level was reduced (Yu *et al.*, 2008). Additionally, heterochromatin marks of silencing – a decrease in H3K4 dimethylation and an increase in H3K9 dimethylation – were observed on the *p15* gene promoter in leukaemia cell lines (Yu *et al.*, 2008). When *p15AS* was overexpressed in cells, the *p15* gene was silenced and the silencing was maintained even after expression of *p15AS* was eliminated (Yu *et al.*, 2008). This suggests that *p15AS* is involved in initiating the silencing of *p15*, but not in maintaining the silencing (Yu *et al.*, 2008). However, an increase in DNA methylation in the *p15* promoter was not observed, and the silencing of *p15* was Dicer independent, suggesting that *p15* silencing is due to the formation of heterochromatin mediated by *p15AS* (Yu *et al.*, 2008).

p21AS

Small RNA-directed transcriptional activation has been reported in human cells in the case of *p21*, a tumour-suppressor gene (Li *et al.*, 2006b). Transcriptional activation of *p21* gene expression results from the post-transcriptional silencing of *p21* antisense transcript (*p21AS*); this is mediated by a small RNA that is homologous to *p21AS* (Morris *et al.*, 2008). Following the suppression of *p21AS*, loss of H3K27 trimethylation at the *p21* promoter is observed (Morris *et al.*, 2008). In contrast, when *p21* mRNA is suppressed, an increase in *p21AS* expression at the *p21* promoter is observed (Morris *et al.*, 2008). However, when the expression of *p21* mRNA and Argonaute (AGO)-1 is suppressed, there is a loss of enrichment of *p21AS* at the *p21* promoter and an increase in *p21* RNA expression at the *p21* promoter (Morris *et al.*, 2008). These data suggest that *p21AS* recruits the regulatory complex that induces suppressive epigenetic modifications in the *p21* promoter in an AGO-1 dependent manner, resulting in transcriptional gene silencing of *p21* (Morris *et al.*, 2008). Thus, the balance between *p21* and *p21AS* transcription can be maintained by *p21AS* RNA, which modulates chromatin at the *p21* promoter (Morris *et al.*, 2008).

In addition to the case with *p21*, gene regulation of activation and repression by small RNAs coupled with antisense RNA is also reported in the case of the human progesterone receptor (*PR*) gene (Schwartz *et al.*, 2008). AGO proteins are involved in this process and, in the case of gene silencing, HP1γ (heterochromatin protein gamma) association has been observed, suggesting that epigenetic regulation is also involved (Schwartz *et al.*, 2008).

Establishment of DNA methylation at gene promoters

Natural antisense transcripts can also change the DNA methylation status of the promoter regions of genes. For example, an antisense RNA for the human *α2-globin gene (HBA2)* can induce DNA methylation at the *HBA2* promoter, thus silencing *HBA2* expression (Tufarelli *et al.*, 2003). Tufarelli *et al.* (2003) reported that deletion of a region downstream of *HBA2* containing the poly(A) addition site of a neighbouring gene transcribed from the opposite strand gave rise to an accidental antisense transcript that overlapped with both the coding and the promoter region of *HB2A*. This genome rearrangement leads to methylation of a CpG island in the *HBA2* promoter and maintains the transcriptional silencing of *HBA2 in cis* (Fig. 1.9) (Tufarelli *et al.*, 2003).

Tufarelli *et al.* (2003) discussed the possibility that abnormal gene expression in malignant cells causes similar epigenetic phenomena at a genome-wide level. Because a large region of the genome becomes hypomethylated in malignant cells, transcription of SINEs, LINEs, and retroviral sequences on the chromosomes is de-repressed, resulting in aberrant antisense transcriptions (Tufarelli *et al.*, 2003). Some of these antisense transcripts may randomly initiate methylation of the CpG islands of key oncogenes and tumour-suppressor genes (Tufarelli *et al.*, 2003).

In the case of rat sphingosine kinase1 (*Sphk1*), overexpression of the antisense transcript for *Sphk1* induces demethylation of CpG sites and methylation of non-CpG sites in the tissue-dependent differentially methylated region (T-DMR) associated with *Sphk1* (Imamura *et al.*, 2004). In the case of the human *HBA2* locus, the effect of the antisense RNA is *in cis*, whereas in rat *Sphk1* the antisense RNA influence occurs *in trans*.

Altering chromatin modifications *in trans*

HOX gene clusters play an important role in embryonic development. The human *HOX* genes are contained within four clusters (*HOXA* to *-D*) on four independent chromosomes. *HOTAIR* is a 2.2-kb ncRNA made from the *HOXC* locus on chromosome 12. *HOTAIR* acts in *trans* to silence transcription across the 40-kb *HOXD* locus on chromosome 2 (Rinn *et al.*, 2007; Gupta *et al.*, 2010). *HOTAIR* transcribed from the *HOXC* locus recruits Polycomb repressive complex 2

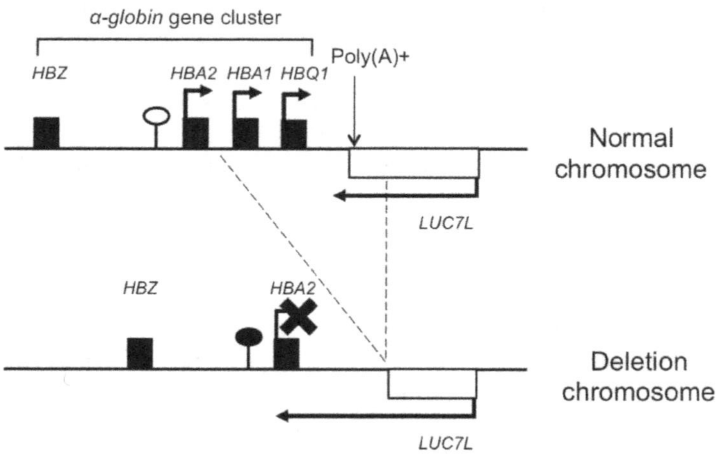

Figure 1.9 Schematic illustration showing how DNA methylation at the *HBA2* promoter occurs by deletion of a poly(A) addition site on the gene of the opposite DNA strand. Positions of genes of the *α-globin* gene cluster (*HBZ*, *HBA2*, *HBA1*, *HBQ1*) are schematically shown, together with that of the neighbouring gene *LUC7L* transcribed from the opposite strand. On the normal chromosome, a CpG island in the *HBA2* promoter is unmethylated (shown as a white circle). On the deletion chromosome, *HBA1* and *HBQ1* are removed, along with the last three exons of *LUC7L*, including its poly(A) addition site, but *HBA2* is intact. However, *HBA2* is silenced and its promoter CpG island is methylated (black circle). This methylation pattern is associated with the presence of antisense transcripts from the truncated *LUC7L* gene running through into the *HBA2* promoter.

(PRC2) to the *HOXD* locus, leading to repressive chromatin modification (Fig. 1.10) (Rinn *et al.*, 2007; Gupta *et al.*, 2010). This antisense transcript is unique because it acts in *trans* to silence different chromosomal domains, not the *HOX* gene cluster that encodes it.

X inactivation

In mammalian females, one of the two X chromosomes in a single nucleus is transcriptionally silenced to equalize the X-linked gene dosage relative to that in males. This X chromosome inactivation (XCI) is achieved through the heterochromatinization of an entire X chromosome. One important component in XCI is the long ncRNA, *Xist* (Borsani *et al.*, 1991). In female embryos, *Xist* is up-regulated on the X chromosome to be inactivated in the future, and it coats the entire X chromosome. *Xist* initiates XCI by facilitating the recruitment of silencing factors such as Polycomb group proteins that modify histones (Plath *et al.*, 2003; Silva *et al.*, 2003). As a result, following *Xist* expression, chromatin modifications spread across the entire X chromosome from which *Xist* was transcribed and lead to heterochromatin formation (Heard and Disteche, 2006).

Tsix is an antisense transcript of *Xist*, and its expression and that of *Xist* on the X chromosome are mutually exclusive (Lee *et al.*, 1999). *Tsix* expression protects the active X chromosome from inactivation by *Xist* (Heard and Disteche, 2006). *Tsix* RNA expression persists on the X chromosome that stays active, whereas expression ceases on the X chromosome that is inactivated.

Tsix modulates the chromatin status at the *Xist* promoter on the X chromosome, making the chromatin active (B. Sun *et al.*, 2006). Transcription of *Tsix* through the *Xist* promoter in the antisense direction is required for down-regulation of *Xist*, following the recruitment of a histone-modifying protein complex in the promoter region (Ohhata *et al.*, 2008).

Genomic imprinting

Mammalian cells have two alleles, one from each homologous chromosome: one is inherited from the mother and the other from the father. A small subset of genes exhibits monoallelic expression, in which a gene is expressed from only the maternal

Figure 1.10 Model of regulation of chromatin domains via histone-modification enzymes with *HOTAIR*. *HOTAIR* transcribed from the *HOXC* locus recruits PRC2 to the *HOXD* locus *in trans*. PRC2 recruitment leads to H3K27 methylation and transcriptional silencing of neighbouring *HOX* genes.

or paternal allele. This phenomenon is called genomic imprinting and is established by epigenetic mechanisms. Approximately 100 imprinted genes have been identified in humans and mice, and many of them are positioned in clusters on the chromosomes. Some of the genes in an imprinted cluster express antisense/ncRNAs. Natural antisense transcripts are associated with up to 81% of imprinted genes (RIKEN Genome Exploration Research Group *et al.*, 2005).

The role of antisense RNA in genomic imprinting has been experimentally explored in several clusters in detail. However, the molecular mechanisms by which antisense RNAs mediate silencing at a single allele or in clusters seem to differ at each imprinted gene locus. Below we discuss two well-characterized examples of the relationship between genomic imprinting and antisense RNA.

Air RNA

Air RNA is an antisense transcript of the insulin-like growth factor receptor type 2 (*Igf2r*) gene on mouse chromosome 17. It is paternally expressed. In mice, *Igf2r* is maternally expressed in most tissues except the brain (Barlow *et al.*, 1991). *Air* RNA is transcribed from intron 2 of *Igf2r* and is 108 kb long (Lyle *et al.*, 2000). This antisense transcript is reported to be polyadenylated but unspliced in the nucleus (Seidl *et al.*, 2006). *Air* transcription influences imprinted genes spanning a 400-kb region. Targeted deletion of the paternal *Air* promoter or truncation of the paternal *Air* transcript results in loss of silencing of not only paternal expression of *Igf2r*, but also of *Slc22a2* and *Slc22a3*, which do not overlap with the *Air* transcription unit (Zwart *et al.*, 2001; Sleutels *et al.*, 2002, 2003). Thus, transcription of *Air* represses the expression of *Igf2r*/*Slc22a2*/*Slc22a3 in cis*.

A recent study showed that, in the mouse placenta, *Air* RNA localizes to the silenced, paternal *Slc22a3* promoter region and recruits histone lysine methyltransferase; this leads to H3K9 methylation and allele-specific gene silencing by chromatin modelling (Nagano *et al.*, 2008). Although several studies have indirectly indicated that *Air* transcription plays a functional role in *Igf2r* silencing (Yamazaki *et al.*, 2005; Latos *et al.*, 2009), *Air* RNA accumulation has not been observed in the *Igf2r* promoter – at least in the placenta (Nagano *et al.*, 2008). This evidence suggests that *Air* uses different mechanisms to silence overlapping and non-overlapping genes (Fig. 1.11).

Kcnq1ot1 RNA

The maternally expressed gene *Kcnq1* (potassium voltage-gated channel, subfamily Q member 1) on mouse chromosome 7 is also regulated by its antisense RNA, *Kcnq1ot1*. *Kcnq1ot1* RNA is paternally expressed and is transcribed from intron 10 of the *Kcnq1* gene in an antisense direction (Mancini-DiNardo *et al.*, 2003). Mouse *Kcnq1ot1* RNA is 91 kb long and polyadenylated, but not spliced, and is localized in the nucleus (Pandey *et*

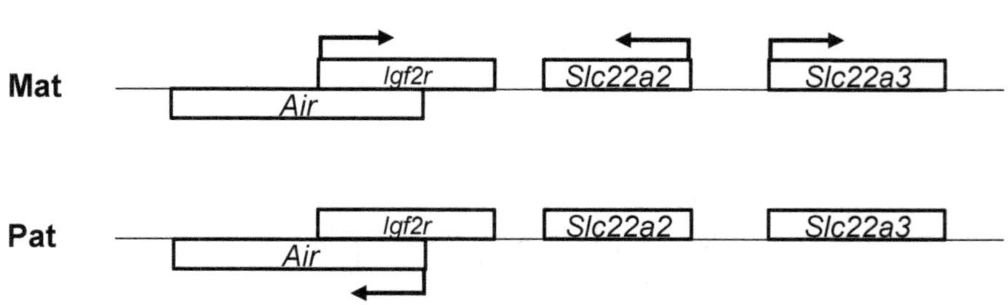

Figure 1.11 Model of antisense RNA–mediated regulation of imprinted genes at the *Igf2r* domain. The imprinted gene cluster containing *Igf2r* also contains the *Air* ncRNA gene. The *Air* promoter lies in an antisense direction in the *Igf2r* intron region. On the paternal chromosome, *Air* RNA is expressed and silencing of *Igf2r*, *Slc22a2*, and *Slc22a3* is observed. Accumulation of *Air* RNA and histone lysine methyltransferase is observed only at the *Slc22a3* promoter (Nagano *et al.*, 2008). On the maternal chromosome, *Air* expression is suppressed and other imprinted genes can be expressed only from this chromosome. Arrows indicate the direction of transcription. Mat, maternal chromosome; Pat, paternal chromosome.

al., 2008). Production of *Kcnq1ot1* RNA on the paternal allele is linked to the silencing of eight to ten protein-coding genes spread over a 1-Mb region (Fitzpatrick *et al.*, 2002; Mancini-DiNardo *et al.*, 2006).

Kcnq1ot1 RNA interacts with both PRC2 and histone lysine methyltransferase, resulting in the enrichment of repressive histone modifications in the promoter regions of some flanking genes (Pandey *et al.*, 2008; Terranova *et al.*, 2008). A recent study has suggested that *Kncq1ot1* recruits DNA methyltransferase 1 (Dnmt1) to silence other imprinted genes (Mohammad *et al.*, 2010). Therefore, *Kcnq1ot1* may help to mediate domain-wide silencing by repressive histone modification and DNA methylation.

Endogenous siRNA formation

Transcription of both DNA strands at the same locus can lead to the formation of dsRNA. One of the pathways involving dsRNA is RNAi. RNAi is a sequence-specific gene regulation conserved among various eukaryotes. The sequence specificity of RNAi is determined by small regulatory RNAs. One of the major classes of endogenous small RNAs is that of siRNAs. siRNAs are generated from long dsRNA and are involved mainly in defence against viruses, transposons, and transgenes. SAT transcripts can trigger the formation of endogenous siRNAs. In this section, we give examples of endogenous siRNA formation in animals and plants.

Plants

Endogenous siRNAs derived from the regions of overlap between 1-pyrroline-5-carboxylate dehydrogenase (*P5CDH*) and *SRO5* were first reported in *Arabidopsis* (Borsani *et al.*, 2005). *SRO5* RNA expression is induced by salt stress and triggers siRNA production, leading to the degradation of *P5CDH* mRNA. Down-regulation of *P5CDH* leads to proline accumulation, resulting in an increase in the salt stress tolerance of *Arabidopsis* (Borsani *et al.*, 2005). In *Arabidopsis*, the expression of siRNAs derived from the SAT genes *ATGB2/PPRL* and *SRRLK/AtRAP* is induced by bacterial infection (Katiyar-Agarwal *et al.*, 2006, 2007). Different types of DICER-like proteins are involved in different types of endogenous siRNA production (Borsani *et al.*, 2005; Katiyar-Agarwal *et al.*, 2006, 2007). The other protein components involved in this process, including AGO proteins, also vary, and the sizes of the siRNAs produced differ slightly in each case (21 nt in *P5CDH/SRO5*; 22 nt in *ATGB2/PPRL*; 39 to 41 nt in *SRRLK/AtRAP*) (Borsani *et al.*, 2005; Katiyar-Agarwal *et al.*, 2006, 2007). 64% of protein-coding SAT pairs (1008 pairs) may produce the endogenous siRNAs in *Arabidopsis* (Jin *et al.*, 2008), suggesting that gene expression regulation mediated by endogenous siRNAs from SATs is a common mechanism in *Arabidopsis*.

Animals

The endogenous small RNAs, Piwi-interacting RNA (piRNA) and siRNA, originating from mouse oocytes have been identified in recent studies that have used high-throughput sequencing (Watanabe *et al.*, 2006, 2008). piRNAs interact with a subset of AGO proteins and are generated by a Dicer-independent mechanism (Klattenhoff and Theurkauf, 2008), whereas siRNAs are processed from the dsRNAs of SAT pairs by Dicer endonucleases. Watanabe *et al.* (2008) identified 17 SAT gene loci that potentially produce endogenous siRNAs through dsRNA formation. In one example of a head-to-head type of SAT gene pair, at the *Pdzd11/Kif4* locus, the level of siRNAs from this SAT declined with the loss of Dicer activity, suggesting that these siRNAs were produced by a Dicer-dependent pathway. Endogenous siRNAs were derived not only from SAT genes, but also from expressed pseudogenes and retrotransposons that were capable of forming dsRNAs (Watanabe *et al.*, 2008). These findings suggest that mammalian cells lacking the RdRps found in lower eukaryotes can produce endogenous siRNAs from naturally occurring dsRNAs (Watanabe *et al.*, 2008). Carlile *et al.* (2009) also identified siRNAs generated from the *Slc34a1/Pfn3* locus in mouse kidney and testis. Intriguingly, the orientation of siRNAs differs in the kidney and testis, suggesting that there are different regulatory pathways for the generation of siRNAs in different tissues (Carlile *et al.*, 2009).

Conclusion

The existence of endogenous antisense RNA has attracted attention since RNAi-based gene knockdown methods utilizing synthetic siRNA became widely available in various model organisms. Antisense transcription is now regarded as a universal phenomenon in many organisms. Especially in plants such as *Arabidopsis*, there is now much evidence that the RNAi pathway functions by siRNAs produced from endogenous sense–antisense transcripts via dsRNA formation. In humans and mice, however, with few exceptions, RNAi utilizing endogenous siRNAs has not been reported, although thousands of SAT pairs have been identified by bioinformatics approaches. Most SAT pairs in humans and mice exhibit coordinated expression of sense and antisense genes, not inverse expression. These SATs may have a role in maintaining the balance of sense and antisense gene expression via dsRNA formation, rather than by transcriptional interference. Extremely long ncRNAs (as large as several hundred kilobases) have been reported in the imprinted regions of mice. 'Endogenous antisense RNAs' are apparently extremely heterogeneous entities and cannot be explained by one principle. As we have stated in our review, SATs are involved in spatial and temporal gene expression regulation, which utilizes various biological mechanisms in a highly species-specific manner.

'Epigenome' is one of the keywords of the post-genome era. Many examples have begun to be documented in which the transcription of antisense RNA and/or antisense RNA is associated with epigenetic gene regulation. With advances in computing and sequencing technologies, such as next-next generation sequencers, not only the production but also the integration of transcriptome and epigenome data is expected. The hidden roles of antisense RNA/transcription have just begun to be revealed.

Acknowledgement

We warmly thank Dr Koji Numata for his comments on, and improvements to, our manuscript.

References

Adelman, J.P., Bond, C.T., Douglass, J., and Herbert, E. (1987). Two mammalian genes transcribed from opposite strands of the [illegible] 1514–1517.

Arteaga-Vazquez, M., Caballero-Perez, J., and Vielle-Calzada, J.P. (2006). A family of microRNAs present in plants and animals. Plant Cell *18*, 3355–3369.

Barlow, D.P., Stöger, R., Herrmann, B.G., Saito, K., and Schweifer, N. (1991). The mouse insulin-like growth factor type-2 receptor is imprinted and closely linked to the Tme locus. Nature *349*, 84–87.

Bartsch, H., Voigtsberger, S., Baumann, G., Morano, I., and Luther, H.P. (2004). Detection of a novel sense–antisense RNA-hybrid structure by RACE experiments on endogenous troponin I antisense RNA. RNA *10*, 1215–1224.

Batshake, B., and Sundelin, J. (1996). The mouse genes for the EP1 prostanoid receptor and the PKN protein kinase overlap. Biochem. Biophys. Res. Commun. *227*, 70–76.

Bedford, M., Arman, E., Orr-Urtreger, A., and Lonai, P. (1995). Analysis of the Hoxd-3 gene: structure and localization of its sense and natural antisense transcripts. DNA Cell Biol. *14*, 295–304.

Belhumeur, P., Lussier, M., and Skup, D. (1988). Expression of naturally occurring RNA molecules complementary to the murine L27′ ribosomal protein mRNA. Gene 72, 277–285.

Bender, T.P., Thompson, C.B., and Kuehl, W.M. (1987). Differential expression of c-myb mRNA in murine B lymphomas by a block to transcription elongation. Science *237*, 1473–1476.

Boffelli, D., Nobrega, M.A., and Rubin E.M. (2004). Comparative genomics at the vertebrate extremes. Nat. Rev. Genet. *5*, 456–465.

Bolland, D.J., Wood, A.L., Afshar, R., Featherstone, K., Oltz, E.M., and Corcoran, A.E. (2007). Antisense intergenic transcription precedes Igh D-to-J recombination and is controlled by the intronic enhancer Emu. Mol. Cell. Biol. *27*, 5523–5533.

Bond, V.C., and Wold, B. (1993). Nucleolar localization of *myc* transcripts. Mol. Cell. Biol. *13*, 3221–3230.

Borsani, G., Tonlorenzi, R., Simmler, M.C., Dandolo, L., Arnaud, D., Capra, V., Grompe, M., Pizzuti, A., Muzny, D., Lawrence, C., *et al.* (1991). Characterization of a murine gene expressed from the inactive X chromosome. Nature *351*, 325–329.

Borsani, O., Zhu, J., Verslues, P.E., Sunkar, R., and Zhu J.K. (2005). Endogenous siRNAs derived from a pair of natural *cis*-antisense transcripts regulate salt tolerance in *Arabidopsis*. Cell *123*, 1279–1291.

Carlile, M., Swan, D., Jackson, K, Preston-Fayers, K., Ballester, B., Flicek, P., and Werner, A. (2009). Strand selective generation of endo-siRNAs from the Na/phosphate transporter gene Slc34a1 in murine tissues. Nucleic Acids Res. *37*, 2274–2282.

Chen, J., Sun, M., Kent, W.J., Huang, X., Xie H., Wang, W., Zhou, G., Shi, R.Z., and Rowley, J.D. (2004). Over 20% of human transcripts might form sense–antisense pairs. Nucleic Acids Res. *32*, 4812–4820.

Chen, J., Sun, M., Hurst, L.D., Carmichael, G.G., and Rowley, J.D. (2005). Genome-wide analysis of coordinate expression and evolution of human cis-encoded sense–antisense transcripts. Trends Genet. *21*, 326–329.

Cheng, J., Kapranov, P., Drenkow, J., Dike, S., Brubaker, S., Patel, S., Long, J., Stern, D., Tammana, H., Helt, G., *et al.* (2005). Transcriptional maps of 10 human chromosomes at 5-nucleotide resolution. Science *308*, 1149–1154.

Coker, R.K., Laurent, G.J., Dabbagh, K., Dawson, J., and McAnulty, R.J. (1998). A novel transforming growth factor beta2 antisense transcript in mammalian lung. Biochem. J. *332*, 297–301.

Core, L.J., Waterfall, J.J., and Lis J.T. (2008). Nascent RNA sequencing reveals widespread pausing and divergent initiation at human promoters. Science *322*, 1845–1848.

Crampton, N., Bonass, W.A., Kirkham, J., Rivetti, C., and Thomson N.H. (2006). Collision events between RNA polymerases in convergent transcription studied by atomic force microscopy. Nucleic Acids Res. *34*, 5416–5425.

David, L., Huber, W., Granovskaia, M., Toedling, J., Palm, C.J., Bofkin, L., Jones, T., Davis, R.W., and Steinmetz, L.M. (2006). A high-resolution map of transcription in the yeast genome. Proc. Natl. Acad. Sci. U.S.A. *103*, 5320–5325.

Dolnick, B.J. (1993). Cloning and characterization of a naturally occurring antisense RNA to human thymidylate synthase mRNA. Nucleic Acids Res. *21*, 1747–1752.

van Duin, M., van Den Tol, J., Hoeijmakers, J.H., Bootsma, D., Rupp, I.P., Reynolds, P., Prakash, L., and Prakash, S. (1989). Conserved pattern of antisense overlapping transcription in the homologous human *ERCC-1* and yeast *RAD10* DNA repair gene regions. Mol. Cell. Biol. *9*, 1794–1798.

Dutrow, N., Nix, D.A., Holt, D., Milash, B., Dalley, B., Westbroek, E., Parnell, T.J., and Cairns, B.R. (2008). Dynamic transcriptome of *Schizosaccharomyces pombe* shown by RNA–DNA hybrid mapping. Nat. Genet. *40*, 977–986.

Ebralidze, A.K., Guibal, F.C., Steidl, U., Zhang, P., Lee, S., Bartholdy, B., Jorda, M.A., Petkova, V., Rosenbauer, F., Huang, G., *et al.* (2008). PU.1 expression is modulated by the balance of functional sense and antisense RNAs regulated by a shared cis-regulatory element. Genes Dev. *22*, 2085–2092.

Eccles, M.R., Grubb, G., Ogawa, O., Szeto, J., and Reeve, A.E. (1994). Cloning of novel Wilms tumor gene (WT1) cDNAs; evidence for antisense transcription of WT1. Oncogene *9*, 2059–2063.

Engström, P.G., Suzuki, H., Ninomiya, N., Akalin, A., Sessa, L., Lavorgna, G., Brozzi, A., Luzi, L., Tan, S.L., Yang, L., *et al.* (2006). Complex loci in human and mouse genomes. PLoS Genet. *2*, e47.

Faghihi, M.A., Modarresi, F., Khalil, A.M., Wood, D.E., Sahagan, B.G., Morgan, T.E., Finch, C.E., St Laurent, G., 3rd, Kenny, P.J., and Wahlestedt, C. (2008). Expression of a noncoding RNA is elevated in Alzheimer's disease and drives rapid feed-forward regulation of beta-secretase. Nat. Med. *14*, 723–730.

Faghihi, M.A., Zhang, M., Huang, J., Modarresi, F., Van der Brug, M.P., Nalls, M A., Cookson, M.R., St Laurent, G., 3rd, and Wahlestedt, C. (2010). Evidence for natural antisense transcript-mediated inhibition of microRNA function. Genome Biol. *11*, R56.

FANTOM Consortium and RIKEN Genome Exploration Research Group Phase I & II Team. (2002). Analysis of the mouse transcriptome based on functional annotation of 60,770 full-length cDNAs. Nature *420*, 563–573.

Farnham, P.J., Abrams, J.M., and Schimke, R.T. (1985). Opposite-strand RNAs from the 5′ flanking region of the mouse dihydrofolate reductase gene. Proc. Natl. Acad. Sci. U.S.A. *82*, 3978–3982.

Fitzpatrick, G.V., Soloway, P.D., and Higgins, M.J. (2002). Regional loss of imprinting and growth deficiency in mice with a targeted deletion of KvDMR1. Nat. Genet. *32*, 426–431.

Fu, X.D., and Maniatis, T. (1992). Isolation of a complementary DNA that encodes the mammalian splicing factor SC35. Science *256*, 535–538.

Gagneur, J., Sinha, H., Perocchi, F., Bourgon, R., Huber, W., and Steinmetz, L.M. (2009). Genome-wide allele- and strand-specific expression profiling. Mol. Syst. Biol. *5*, 274.

Galante, P.A., Vidal, D.O., de Souza, J.E., Camargo, A.A., and de Souza, S.J. (2007). Sense–antisense pairs in mammals: functional and evolutionary considerations. Genome Biol. *8*, R40.

Ge, X., Wu, Q., Jung, Y.C., Chen, J., and Wang, S.M. (2006). A large quantity of novel human antisense transcripts detected by LongSAGE. Bioinformatics *22*, 2475–2479.

Ge, X., Rubinstein, W.S., Jung, Y.C., and Wu, Q. (2008). Genome-wide analysis of antisense transcription with Affymetrix exon array. BMC Genomics *9*, 27.

Giallourakis, C.C., Franklin, A., Guo, C., Cheng, H.L., Yoon, H.S., Gallagher, M., Perlot, T., Andzelm, M., Murphy, A.J., Macdonald, L.E., *et al.* (2010). Elements between the IgH variable (V) and diversity (D) clusters influence antisense transcription and lineage-specific V(D)J recombination. Proc. Natl. Acad. Sci. U.S.A. *107*, 22207–22212.

Gunasekera, A.M., Patankar, S., Schug, J., Eisen, G., Kissinger, J., Roos, D., and Wirth, D.F. (2004). Widespread distribution of antisense transcripts in the *Plasmodium falciparum* genome. Mol. Biochem. Parasitol. *136*, 35–42.

Gupta, R.A., Shah, N., Wang, K.C., Kim, J., Horlings, H.M., Wong, D.J., Tsai, M.C., Hung, T., Argani, P., Rinn, J.L., *et al.* (2010). Long non-coding RNA *HOTAIR* reprograms chromatin state to promote cancer metastasis. Nature *464*, 1071–1076.

Hastings, M.L., Milcarek, C., Martincic, K., Peterson, M.L., and Munroe, S.H. (1997). Expression of the thyroid hormone receptor gene, *erbAalpha*, in B lymphocytes: alternative mRNA processing is independent of differentiation but correlates with antisense RNA levels. Nucleic Acids Res. *25*, 4296–4300.

He, Y., Vogelstein, B., Velculescu, V.E., Papadopoulos, N., and Kinzler, K.W. (2008). The antisense transcriptomes of human cells. Science *322*, 1855–1857.

Heard, E., and Disteche, C.M. (2006). Dosage compensation in mammals: fine-tuning the expression of the X chromosome. Genes Dev. *20*, 1848–1867.

Henz, S.R., Cumbie, J.S., Kasschau, K.D., Lohmann, J.U., Carrington, J.C., Weigel, D., and Schmid, M. (2007). Distinct expression patterns of natural antisense transcripts in *Arabidopsis*. Plant Physiol. *144*, 1247–1255.

Hervieu, G., and Nahon, J.L. (1995). Pro-melanin concentrating hormone messenger ribonucleic acid and peptides expression in peripheral tissues of the rat. Neuroendocrinology *61*, 348–364.

Hsieh-Li, H.M., Witte, D.P., Weinstein, M., Branford, W., Li, H., Small, K., and Potter, S.S. (1995). Hoxa 11 structure, extensive antisense transcription, and function in male and female fertility. Development *121*, 1373–1385.

Imamura, T., Yamamoto, S., Ohgane, J., Hattori, N., Tanaka, S., and Shiota, K. (2004). Non-coding RNA directed DNA demethylation of *Sphk1* CpG island. Biochem. Biophys. Res. Commun. *322*, 593–600.

Imanishi, T., Itoh, T., Suzuki, Y., O'Donovan, C., Fukuchi, S., Koyanagi, K.O., Barrero, R.A., Tamura, T., Yamaguchi-Kabata, Y., Tanino, M., *et al.* (2004). Integrative annotation of *21*,037 human genes validated by full-length cDNA clones. PLoS Biol. *2*, e162.

International Human Genome Sequencing Consortium (2001). Initial sequencing and analysis of the human genome. Nature *409*, 860–921.

Jen, C.H., Michalopoulos, I., Westhead, D.R., and Meyer, P. (2005). Natural antisense transcripts with coding capacity in *Arabidopsis* may have a regulatory role that is not linked to double-stranded RNA degradation. Genome Biol. *6*, R51.

Jin, H., Vacic, V., Girke, T., Lonardi, S., and Zhu, J.K. (2008). Small RNAs and the regulation of *cis*-natural antisense transcripts in *Arabidopsis*. BMC Mol. Biol. *9*, 6.

Jones, P.A., and Laird, P.W. (1999). Cancer epigenetics comes of age. Nat. Genet. *21*, 163–167.

Kampa, D., Cheng, J., Kapranov, P., Yamanaka, M., Brubaker, S., Cawley, S., Drenkow, J., Piccolboni, A., Bekiranov, S., Helt, G., *et al.* (2004). Novel RNAs identified from an in-depth analysis of the transcriptome of human chromosomes 21 and 22. Genome Res. *14*, 331–342.

Kapranov, P., Cawley, S.E., Drenkow, J., Bekiranov, S., Strausberg, R.L., Fodor, S.P., and Gingeras, T.R. (2002). Large-scale transcriptional activity in chromosomes 21 and 22. Science *296*, 916–919.

Kapranov, P., Cheng, J., Dike, S., Nix, D.A., Duttagupta, R., Willingham, A.T., Stadler, P.F., Hertel, J., Hackermüller, J., Hofacker, I.L., *et al.* (2007). RNA maps reveal new RNA classes and a possible function for pervasive transcription. Science *316*, 1484–1488.

Kapranov, P., Ozsolak, F., Kim, S.W., Foissac, S., Lipson, D., Hart, C., Roels, S., Borel, C., Antonarakis, S.E., Monaghan, A.P., John, B., and Milos, P.M. (2010). New class of gene-termini-associated human RNAs suggests a novel RNA copying mechanism. Nature *466*, 642–646.

Katiyar-Agarwal, S., [illegible] O., Villegas, A. Jr., Zhu, J.K., Staskawicz, B.J., and Jin, H. (2006). A pathogen-inducible endogenous siRNA in plant immunity. Proc. Natl. Acad. Sci. U.S.A. *103*, 18002–18007.

Katiyar-Agarwal, S., Gao, S., Vivian-Smith, A., and Jin, H. (2007). A novel class of bacteria-induced small RNAs in *Arabidopsis*. Genes Dev. *21*, 3123–3134.

Khochbin, S., and Lawrence, J.J. (1989). An antisense RNA involved in p53 mRNA maturation in murine erythroleukemia cells induced to differentiate. EMBO J. *8*, 4107–4114.

Kimelman, D., and Kirschner, M.W. (1989). An antisense mRNA directs the covalent modification of the transcript encoding fibroblast growth factor in *Xenopus* oocytes. Cell *59*, 687–696.

Kiyosawa, H., and Abe, K. (2002). Speculations on the role of natural antisense transcripts in mammalian X chromosome evolution. Cytogenet. Genome Res. *99*, 151–156.

Kiyosawa, H., Yamanaka, I., Osato, N., Kondo, S., RIKEN GER Group and GSL Members, and Hayashizaki, Y. (2003). Antisense transcripts with FANTOM2 clone set and their implications for gene regulation. Genome Res. *13*, 1324–1334.

Kiyosawa, H., Mise, N., Iwase, S., Hayashizaki, Y., and Abe, K. (2005). Disclosing hidden transcripts: mouse natural sense–antisense transcripts tend to be poly(A) negative and nuclear localized. Genome Res. *15*, 463–474.

Klattenhoff, C., and Theurkauf, W. (2008). Biogenesis and germline functions of piRNAs. Development *135*, 3–9.

Krinke, L., and Wulff, D.L. (1987). OOP RNA, produced from multicopy plasmids, inhibits lambda cII gene expression through an RNase III-dependent mechanism. Genes Dev. *1*, 1005–1013.

Krystal, G.W., Armstrong, B.C., and Battey, J.F. (1990). N-myc mRNA forms an RNA-RNA duplex with endogenous antisense transcripts. Mol. Cell. Biol. *10*, 4180–4191.

Laabi, Y., Gras, M.P., Brouet, J.C., Berger, R., Larsen, C.J., and Tsapis, A. (1994). The BCMA gene, preferentially expressed during B lymphoid maturation, is bidirectionally transcribed. Nucleic Acids Res. *22*, 1147–1154.

Lapidot, M., and Pilpel, Y. (2006). Genome-wide natural antisense transcription: coupling its regulation to its different regulatory mechanisms. EMBO Rep. *7*, 1216–1222.

Latos, P.A., Stricker, S.H., Steenpass, L., Pauler, F.M., Huang, R., Senergin, B.H., Regha, K., Koerner, M.V., Warczok, K.E., Unger, C., and Barlow, D.P. (2009). An *in vitro* ES cell imprinting model shows that imprinted expression of the *Igf2r* gene arises from an allele-specific expression bias. Development *136*, 437–448.

Lazar, M.A., Hodin, R.A., Darling, D.S., and Chin, W.W. (1989). A novel member of the thyroid/steroid hormone receptor family is encoded by the opposite strand of the rat *c-erbA* alpha transcriptional unit. Mol. Cell. Biol. *9*, 1128–1136.

Lee, J.T., Davidow, L.S., and Warshawsky, D. (1999). *Tsix*, a gene antisense to *Xist* at the X-inactivation centre. Nat. Genet. *21*, 400–404.

Lehner, B., Williams, G., Campbell, R.D., and Sanderson, C.M. (2002). Antisense transcripts in the human genome. Trends Genet. *18*, 63–65.

Lerner, A., D'Adamio, L., Diener, A.C., Clayton, L.K., and Reinherz, E.L. (1993). CD3 *zeta/eta/theta* locus is colinear with and transcribed antisense to the gene encoding the transcription factor Oct-1. J. Immunol. *151*, 3152–3162.

Li, L., Wang, X., Stolc, V., Li, X., Zhang, D., Su, N., Tongprasit, W., Li, S., Cheng, Z., Wang, J., and Deng, X.W. (2006a). Genome-wide transcription analyses in rice using tiling microarrays. Nat. Genet. *38*, 124–129.

Li, L.C., Okino, S.T., Zhao, H., Pookot, D., Place, R.F., Urakami, S., Enokida, H., and Dahiya, R. (2006b). Small dsRNAs induce transcriptional activation in human cells. Proc. Natl. Acad. Sci. U.S.A. *103*, 17337–17342.

Liao, S.M., Wu, T.H., Chiang, C.H., Susskind, M.M., and McClure, W.R. (1987). Control of gene expression in bacteriophage P22 by a small antisense RNA. I. Characterization *in vitro* of the *Psar* promoter and the *sar* RNA transcript. Genes Dev. *1*, 197–203.

Lübbert, M. (2003). Gene silencing of the *p15/INK4B* cell-cycle inhibitor by hypermethylation: an early or later epigenetic alteration in myelodysplastic syndromes? Leukemia *17*, 1762–1764.

Luther, H.P., Haase, H., Hohaus, A., Beckmann, G., Reich, J., and Morano, I. (1998). Characterization of naturally occurring myosin heavy chain antisense mRNA in rat heart. J. Cell. Biochem. *70*, 110–120.

Lyle, R., Watanabe, D., te Vruchte, D., Lerchner, W., Smrzka, O.W., Wutz, A., Schageman, J., Hahner, L., Davies, C., and Barlow, D.P. (2000). The imprinted antisense RNA at the *Igf2r* locus overlaps but does not imprint *Mas1*. Nat Genet. *25*, 19–21.

Mammalian Gene Collection Program Team (2002). Generation and initial analysis of more than *15*,000 full-length human and mouse cDNA sequences. Proc. Natl. Acad. Sci. U.S.A. *99*, 16899–16903.

Mancini-DiNardo, D., Steele, S.J., Ingram, R.S., and Tilghman, S.M. (2003). A differentially methylated region within the gene *Kcnq1* functions as an imprinted promoter and silencer. Hum. Mol. Genet. *12*, 283–294.

Mancini-DiNardo, D., Steele, S.J., Levorse, J.M., Ingram, R.S., and Tilghman, S.M. (2006). Elongation of the *Kcnq1ot1* transcript is required for genomic imprinting of neighboring genes. Genes Dev. *20*, 1268–1282.

Matsui, K., Nishizawa, M., Ozaki, T., Kimura, T., Hashimoto, I., Yamada, M., Kaibori, M., Kamiyama, Y., Ito, S., and Okumura, T. (2008). Natural antisense transcript stabilizes inducible nitric oxide synthase messenger RNA in rat hepatocytes. Hepatology *47*, 686–697.

Misra, S., Crosby, M.A., Mungall, C.J., Matthews. B.B., Campbell, K.S., Hradecky, P., Huang, Y., Kaminker, J.S., Millburn, G.H., Prochnik, S.E., *et al.* (2002). Annotation of the *Drosophila melanogaster* euchromatic genome: a systematic review. Genome Biol. 3, RESEARCH0083.

Mohammad, F., Mondal, T., Guseva, N., Pandey, GK., and Kanduri, C. (2010). *Kcnq1ot1* noncoding RNA mediates transcriptional gene silencing by interacting with *Dnmt1*. Development *137*, 2493–2499.

Morris, K.V., Santoso, S., Turner, A.M., Pastori, C., and Hawkins, P.G. (2008). Bidirectional transcription directs both transcriptional gene activation and suppression in human cells. PLoS Genet. *4*, e1000258.

Mouse Genome Sequencing Consortium (2002). Initial sequencing and comparative analysis of the mouse genome. Nature *420*, 520–562.

Munroe, S.H. (2004). Diversity of antisense regulation in eukaryotes: multiple mechanisms, emerging patterns. J. Cell Biochem. *93*, 664–671.

Munroe, S.H., and Lazar, M.A. (1991). Inhibition of *c-erbA* mRNA splicing by a naturally occurring antisense RNA. J. Biol. Chem. *266*, 22083–22086.

Murashov, A.K., and Wolgemuth, D.J. (1996). Distinct transcripts are recognized by sense and antisense riboprobes for a member of the murine *HSP70* gene family, *HSP70.2*, in various reproductive tissues. Mol. Reprod. Dev. *43*, 17–24.

Murphy, P.R., and Knee, R.S. (1994). Identification and characterization of an antisense RNA transcript (*gfg*) from the human basic fibroblast growth factor gene. Mol. Endocrinol. *8*, 852–859.

Nagano, T., Mitchell, J.A., Sanz, L.A., Pauler, F.M., Ferguson-Smith, A.C., and Feil, R. (2008). The *Air* noncoding RNA epigenetically silences transcription by targeting G9a to chromatin. Science *322*, 1717–1720.

Neeman, Y., Dahary, D., Levanon, E.Y., Sorek, R., and Eisenberg, E. (2005). Is there any sense in antisense editing? Trends Genet. *21*,544–547.

Nepveu, A., and Marcu, K.B. (1986). Intragenic pausing and anti-sense transcription within the murine *c-myc* locus. EMBO J. *5*, 2859–2865.

Nobori, T., Miura, K., Wu, D.J., Lois, A., Takabayashi, K., and Carson, D.A. (1994). Deletions of the cyclin-dependent kinase-4 inhibitor gene in multiple human cancers. Nature *368*, 753–756.

Numata, K., Okada, Y., Saito, R., Kiyosawa, H., Kanai, A., and Tomita, M. (2007). Comparative analysis of *cis*-encoded antisense RNAs in eukaryotes. Gene *392*, 134–41.

Numata, K., Osada, Y., Okada, Y., Saito, R., Hiraiwa, N., Nakaoka, H., Yamamoto, N., Watanabe, K., Okubo, K., Kohama, C., *et al.* (2009). Identification of novel endogenous antisense transcripts by DNA microarray analysis targeting complementary strand of annotated genes. BMC Genomics *10*, 392.

Oeder, S., Mages, J., Flicek, P., and Lang, R. (2007). Uncovering information on expression of natural antisense transcripts in Affymetrix MOE430 data sets. BMC Genomics *8*, 200.

Ohhata, T., Hoki, Y., Sasaki, H., and Sado, T. (2008). Crucial role of antisense transcription across the *Xist* promoter in *Tsix*-mediated *Xist* chromatin modification. Development *135*, 227–235.

Okada, Y., Tashiro, C., Numata, K., Watanabe, K., Nakaoka, H., Yamamoto, N., Okubo, K., Ikeda, R., Saito, R., Kanai, A., *et al.* (2008). Comparative expression analysis uncovers novel features of endogenous antisense transcription. Hum. Mol. Genet. *17*, 1631–1640.

Osato, N., Yamada, H., Satoh, K., Ooka, H., Yamamoto, M., Suzuki, K., Kawai, J., Carninci, P., Ohtomo, Y., Murakami, K., *et al.* (2003). Antisense transcripts with rice full-length cDNAs. Genome Biol. *5*, R5.

Osato, N., Suzuki, Y., Ikeo, K., and Gojobori, T. (2007). Transcriptional interferences in *cis* natural antisense transcripts of humans and mice. Genetics *176*, 1299–1306.

Ota, T., Suzuki, Y., Nishikawa, T., Otsuki, T., Sugiyama, T., Irie, R., Wakamatsu, A., Hayashi, K., Sato, H., Nagai, K., *et al.* (2004). Complete sequencing and characterization of 21,243 full-length human cDNAs. Nat. Genet. *36*, 40–45.

Pandey, R.R., Mondal, T., Mohammad, F., Enroth, S., Redrup, L., Komorowski, J., Nagano, T., Mancini-DiNardo, D., and Kanduri, C. (2008). Kcnq1ot1 antisense noncoding RNA mediates lineage-specific transcriptional silencing through chromatin-level regulation. Mol. Cell. *32*, 232–246.

Perlot, T., Li, G., and Alt, F.W. (2008). Antisense transcripts from immunoglobulin heavy-chain locus V(D)J and switch regions. Proc. Natl. Acad. Sci. U.S.A. *105*, 3843–3848.

Peters, N.T., Rohrbach, J.A., Zalewski, B.A., Byrkett, C.M., and Vaughn, J.C. (2003). RNA editing and regulation of Drosophila 4f-rnp expression by sas-10 antisense readthrough mRNA transcripts. RNA *9*, 698–6710.

Plath, K., Fang, J., Mlynarczyk-Evans, S.K., Cao, R., Worringer, K.A., Wang, H., de la Cruz, C.C., Otte, A.P., Panning, B., and Zhang, Y. (2003). Role of histone H3 lysine 27 methylation in X inactivation. Science *300*, 131–135.

Podlowski, S., Bramlage, P., Baumann, G., Morano, I., and Luther, H.P. (2002). Cardiac troponin I sense–antisense RNA duplexes in the myocardium. J. Cell. Biochem. *85*, 198–207.

Preker, P., Nielsen, J., Kammler, S., Lykke-Andersen, S., Christensen, M.S., Mapendano, C.K., Schierup, M.H., and Jensen, T.H. (2008). RNA exosome depletion reveals transcription upstream of active human promoters. Science *322*, 1851–1854.

Prescott, E.M., and Proudfoot, N.J. (2002). Transcriptional collision between convergent genes in budding yeast. Proc. Natl. Acad. Sci. U.S.A. *99*, 8796–8801.

Reis, E.M., Nakaya, H.I., Louro, R., Canavez, F.C., Flatschart, A.V., Almeida, G.T., Egidio, C.M., Paquola, A.C., Machado, A.A., Festa, F., *et al.* (2004). Antisense intronic non-coding RNA levels correlate to the degree of tumor differentiation in prostate cancer. Oncogene *23*, 6684–6692.

Rice Full-Length cDNA Consortium; National Institute of Agrobiological Sciences Rice Full-Length cDNA Project Team, Kikuchi, S., Satoh, K., Nagata, T., Kawagashira, N., Doi, K., Kishimoto, N., Yazaki, J., Ishikawa, M., *et al.* (2003). Collection, mapping, and annotation of over 28,000 cDNA clones from japonica rice. Science *301*, 376–379.

RIKEN Genome Exploration Research Group Phase II Team and the FANTOM Consortium. (2001). Functional annotation of a full-length mouse cDNA collection. Nature *409*, 685–690.

RIKEN Genome Exploration Research Group and Genome Science Group (Genome Network Project Core Group) and the FANTOM Consortium (2005). Antisense transcription in the mammalian transcriptome. Science *309*, 1564–1566.

Rinn, J.L., Kertesz, M., Wang, J.K., Squazzo, S.L., Xu, X., Brugmann, S.A., Goodnough, L.H., Helms, J.A., Farnham, P.J., Segal, E., *et al.* (2007). Functional demarcation of active and silent chromatin domains in human *HOX* loci by noncoding RNAs. Cell *129*, 1311–1323.

Ronai, D., Iglesias-Ussel, M.D., Fan, M., Li, Z., Martin, A., and Scharff, M.D. (2007). Detection of chromatin-associated single-stranded DNA in regions targeted for somatic hypermutation. J. Exp. Med. *204*, 181–190.

Scadden, A.D., and Smith, C.W. (2006). Specific cleavage of hyper-edited dsRNAs. EMBO J. *20*, 4243–4252.

Schwartz, J.C., Younger, S.T., Nguyen, N.B., Hardy, D.B., Monia, B.P., Corey, D.R., and Janowski, B.A. (2008). Antisense transcripts are targets for activating small RNAs. Nat. Struct. Mol. Biol. *15*, 842–848.

Seidl, C.I., Stricker, S.H., and Barlow, D.P. (2006). The imprinted *Air* ncRNA is an atypical RNAPII transcript that evades splicing and escapes nuclear export. EMBO J. *25*, 3565–3575.

Seila, A.C., Calabrese, J.M., Levine, S.S., Yeo, G.W., Rahl, P.B., Flynn, R.A., Young, R.A., and Sharp, P.A. (2008). Divergent transcription from active promoters. Science *322*, 1849–1851.

Shearwin, K.E., Callen, B.P., and Egan, J.B. (2005). Transcriptional interference – a crash course. Trends Genet. *21*, 339–345.

Shendure, J., and Church, G.M. (2002). Computational discovery of sense–antisense transcription in the human and mouse genomes. Genome Biol. *3*, RESEARCH0044.

Shi, M., Yan, X., Ryan, D.H., and Harris, R.B. (2000). Identification of urocortin mRNA antisense transcripts in rat tissue. Brain Res. Bull. *53*, 317–324.

Silva, J., Mak, W., Zvetkova, I., Appanah, R., Nesterova, T.B., Webster, Z., Peters, A.H., Jenuwein, T., Otte, A.P., and Brockdorff, N. (2003). Establishment of histone h3 methylation on the inactive X chromosome requires transient recruitment of *Eed-Enx1* polycomb group complexes. Dev. Cell *4*, 481–495.

Silverman, T.A., Noguchi, M., and Safer, B. (1992). Role of sequences within the first intron in the regulation of expression of eukaryotic initiation factor 2 alpha. J. Biol. Chem. *267*, 9738–9742.

Simons, R.W., and Kleckner, N. (1983). Translational control of IS10 transposition. Cell *34*, 683–691.

Sleutels, F., Zwart, R., and Barlow, D.P. (2002). The non-coding *Air* RNA is required for silencing autosomal imprinted genes. Nature *415*, 810–813.

Sleutels, F., Tjon, G., Ludwig, T., and Barlow, D.P. (2003). Imprinted silencing of *Slc22a2* and *Slc22a3* does not need transcriptional overlap between *Igf2r* and *Air*. EMBO J. *22*, 3696–3704.

Stolc, V., Gauhar, Z., Mason, C., Halasz, G., van Batenburg, M.F., Rifkin, S.A., Hua, S., Herreman, T., Tongprasit, W., Barbano, P.E., *et al.* (2004). A gene expression map for the euchromatic genome of *Drosophila melanogaster.* Science *306*, 655–660.

Stougaard, P., Molin, S., and Nordström, K. (1981). RNAs involved in copy-number control and incompatibility of plasmid R1. Proc. Natl. Acad. Sci. U.S.A. *78*, 6008–6012.

Sun, B.K., Deaton, A.M., and Lee, J.T. (2006). A transient heterochromatic state in *Xist* preempts X inactivation choice without RNA stabilization. Mol. Cell *21*, 617–628.

Sun, M., Hurst, L.D., Carmichael, G.G., and Chen, J. (2005). Evidence for a preferential targeting of 3′-UTRs by cis-encoded natural antisense transcripts. Nucleic Acids Res. *33*, 5533–5543.

Sun, M., Hurst, L.D., Carmichael, G.G., and Chen, J. (2006). Evidence for variation in abundance of antisense transcripts between multicellular animals but no relationship between antisense transcription and organismic complexity. Genome Res. *16*, 922–933.

Terranova, R., Yokobayashi, S., Stadler, M.B., Otte, A.P., van Lohuizen, M., Orkin, S.H., and Peters, A.H. (2008). Polycomb group proteins Ezh2 and Rnf2 direct genomic contraction and imprinted repression in early mouse embryos. Dev. Cell *15*, 668–679.

The *C. elegans* Sequencing Consortium. (1998). Genome sequence of the nematode *C. elegans*: a platform for investigating biology. Science *282*, 2012–2018.

The Mammalian Gene Collection Project Team. (2004). The status, quality, and expansion of the NIH full-length cDNA project: the Mammalian Gene Collection (MGC). Genome Res. *14*, 2121–2127.

Thrash-Bingham, C.A., and Tartof, K.D. (1999). *aHIF*: a natural antisense transcript overexpressed in human renal cancer and during hypoxia. J. Natl. Cancer Inst. *91*, 143–151.

Tomizawa, J., and Itoh, T. (1981). Plasmid ColE1 incompatibility determined by interaction of RNA I with primer transcript. Proc. Natl. Acad. Sci. U.S.A. *78*, 6096–6100.

Tufarelli, C., Stanley, J.A., Garrick, D., Sharpe, J.A., Ayyub, H., Wood, W.G., and Higgs, D.R. (2003). Transcription of antisense RNA leading to gene silencing and methylation as a novel cause of human genetic disease. Nat. Genet. *34*, 157–165.

Volloch, V. (1986). Cytoplasmic synthesis of globin RNA in differentiated murine erythroleukemia cells: possible involvement of RNA-dependent RNA polymerase. Proc. Natl. Acad. Sci. U.S.A. *83*, 1208–1212.

Wang, X.J., Gaasterland, T., and Chua, N.H. (2005). Genome-wide prediction and identification of *cis*-natural antisense transcripts in *Arabidopsis thaliana.* Genome Biol. *6*, R30.

Watanabe, T., Takeda, A., Tsukiyama, T., Mise, K., Okuno, T., Sasaki, H., Minami, N., and Imai, H. (2006). Identification and characterization of two novel classes of small RNAs in the mouse germline: retrotransposon-derived siRNAs in oocytes and germline small RNAs in testes. Genes Dev. *20*, 1732–1743.

Watanabe, T., Totoki, Y., Toyoda, A., Kaneda, M., Kuramochi-Miyagawa, S., Obata, Y., Chiba, H., Kohara, Y., Kono, T., Nakano, T., *et al.* (2008). Endogenous siRNAs from naturally formed dsRNAs regulate transcripts in mouse oocytes. Nature *453*, 539–543.

Watanabe, Y., Numata, K., Murata, S., Osada, Y., Saito, R., Nakaoka, H., Yamamoto, N., Watanabe, K., Kato, H., Abe, K., *et al.* (2010). Genome-wide analysis of expression modes and DNA methylation status at sense–antisense transcript loci in mouse. Genomics *96*, 333–341.

Werner, A., Carlile, M., and Swan, D. (2009). What do natural antisense transcripts regulate? RNA Biol. *6*, 43–48.

Williams, T., and Fried, M. (1986). A mouse locus at which transcription from both DNA strands produces mRNAs complementary at their 3′ ends. Nature *322*, 275–279.

Yamada, K., Lim, J., Dale, J.M., Chen, H., Shinn, P., Palm, C.J., Southwick, A.M., Wu, H.C., Kim, C., Nguyen, M., *et al.* (2003). Empirical analysis of transcriptional activity in the *Arabidopsis* genome. Science *302*, 842–846.

Yamane, A., Resch, W., Kuo, N., Kuchen, S., Li, Z., Sun, H.W., Robbiani, D.F., McBride, K., Nussenzweig, M.C., and Casellas, R. (2011). Deep-sequencing identification of the genomic targets of the cytidine deaminase AID and its cofactor RPA in B lymphocytes. Nat. Immunol. *12*, 62–69.

Yamasaki, Y., Kayashima, T., Soejima, H., Kinoshita, A., Yoshiura, K., and Matsumoto, N. (2005). Neuron-specific relaxation of *Igf2r* imprinting is associated with neuron-specific histone modifications and lack of its antisense transcript *Air.* Hum. Mol. Genet. *14*, 2511–2520.

Yelin, R., Dahary, D., Sorek, R., Levanon, E.Y., Goldstein, O., Shoshan, A., Diber, A., Biton, S., Tamir, Y., Khosravi, R., *et al.* (2003). Widespread occurrence of antisense transcription in the human genome. Nat. Biotechnol. *21*, 379–386.

Yu, W., Gius, D., Onyango, P., Muldoon-Jacobs, K., Karp, J., Feinberg, A.P., and Cui, H. (2008). Epigenetic silencing of tumour-suppressor gene *p15* by its antisense RNA. Nature *451*, 202–206.

Zavadil, J., Svoboda, P., Liang, H., Kottickal, L.V., and Nagarajan, L. (1999). An antisense transcript to *SMAD5* expressed in fetal and tumor tissues. Biochem. Biophys. Res. Commun. *255*, 668–672.

Zhang, Y., Liu, X.S., Liu, Q.R., and Wei, L. (2006). Genome-wide *in silico* identification and analysis of *cis* natural antisense transcripts (*cis*-NATs) in ten species. Nucleic Acids Res. *34*, 3465–3475.

Zhang, Z., and Carmichael, G.G. (2001). The fate of dsRNA in the nucleus: a *p54*(*nrb*)–containing complex mediates the nuclear retention of promiscuously A-to-I edited RNAs. Cell *106*, 465–475.

Zwart, R., Sleutels, F., Wutz, A., Schinkel, A.H., and Barlow, D.P. (2001). Bidirectional action of the *Igf2r* imprint control element on upstream and downstream imprinted genes. Genes Dev. *15*, 2361–2366.

Processing and Regulatory Impact of Endogenous siRNAs in Animals

2

Andreas Werner

Abstract

The complexity of an organism is driven by a positive balance between creative and destructive forces in the process of evolution. Small non-protein-coding RNAs play an instrumental role in both regulating gene expression (creative influence) as well as suppressing selfish genetic elements (defensive role). There are three main groups of small RNAs including microRNAs, endogenous siRNAs and piRNAs. MicroRNAs represent the most important gene regulatory small RNAs and act by tuning the expression level of eukaryotic mRNAs. Interestingly, endogenous siRNAs as well as piRNAs apparently serve both regulatory and defensive purposes. They suppress the expression of repetitive DNA elements but also influence the expression of protein coding genes. For endo-siRNAs, intriguing roles in epigenetic regulation are emerging. The multiple tasks of small RNAs are in line with a role as drivers of organismal complexity.

Introduction

The research output on small RNAs and the related process of RNA interference has virtually exploded. In the recent years a myriad of cellular functions for small RNAs in health and disease have been reported. Small RNAs are increasingly used as novel therapeutic agents and are widely used in biological research to knock down specific genes. Technical progress in nucleic acid sequencing has helped to identify novel classes of small RNAs and to describe their biological roles (Marguerat and Bahler, 2009). There are three main classes of small RNAs that can be separated with respect to their origin, length and chemical modification, including microRNAs, piRNAs and endogenous siRNAs (Kim *et al.*, 2009). The amount of relevant information on all small RNAs exceeds the scope of this chapter by far. This report will only focus on the latest member of the trio, endogenous siRNAs or endo-siRNAs. Endo-siRNAs, microRNAs and piRNAs, however, show overlaps with respect to structure, synthesis or apparent biological role that make it difficult to discuss one class of short RNAs completely isolated from the other two. Consequently, two sections are included in this chapter giving a brief introduction to microRNAs and piRNAs.

The production and biological actions of small RNAs include two obligatory steps. The first step involves one or two nucleases that cut a RNA hybrid into small fragments of 20–30 bases. MicroRNAs are produced from RNA stem–loop structures with imperfect complementarity and require the sequential action of two type III RNases called 'Drosha' and 'Dcr'. Endo-siRNAs derive from perfectly complementary RNA hybrids and only involve the action of Dcr (Kim *et al.*, 2009). The enzyme responsible for the primary processing of piRNAs is still unknown, but with the phosphodiesterase 'Courgette' a strong candidate has recently been identified (Haase *et al.*, 2010). The second step prominently involves a multifunctional protein of the Ago protein family. First, Agos bind to one strand of the double-stranded small RNAs. Thereafter, the two strands are separated and the guide strand remains attached to the Ago protein. The unbound passenger strand becomes quickly degraded. The guide strand will then direct the RNA–Ago complex to target RNA molecules with high sequence complementarity (Czech and Hannon, 2010). This principle is common to all three classes of RNAs. Ago proteins exist in various types with essential functional differences. For example, certain Ago isoforms cut the RNA complementary to the guide strand with high specificity whereas others do not. In addition, individual Agos interact with various other proteins to assemble specific protein machines which influence the future of the complementary RNA (Hock *et al.*, 2007). These features suggest that Agos exhibit a double function as effector and adaptor proteins.

Phylogenetically, Dcr and Ago proteins derive from large protein families. Interestingly, the repertoires of the two proteins show species-specific characteristics that suggest adaptive evolution of small RNA mediated pathways. For example, plants express four different Dcr proteins: One isoform is responsible for miRNA synthesis whereas the other isoforms mediate antiviral defence (Ding and Voinnet, 2007). A clear division between miRNA- and endo-siRNA synthesis is also observed in flies. Each pathway involves a specific enzymatic machinery with only limited substrate overlap. MiRNA synthesis involves Dcr1 and Ago1 whereas siRNAs are predominantly produced by Dcr2 and Ago2 (Bartel, 2004). In contrast, human and mice have retained a single Dcr isoform but express four Agos with only Ago2 showing enzymatic (endonuclease) activity. Ago3 has an established role in piRNA synthesis but the redundancy of the other Ago isoforms is not yet understood (Ender and Meister, 2010). The specific cellular roles of Ago proteins merit scientific attention and may help to explain the raison d'être of RNA interference in animals (Okamura and Lai, 2008).

In addition to the three key proteins Drosha/Dcr and Ago a number of proteins have been identified which contribute to the biological impact of small RNAs. Notably, RNA-dependent RNA polymerases (RDRs) are essential components of a system that provides protection against viral infection in plants. This 'immune system' or 'RNA mediated plant surveillance mechanism' is crucially based on small RNAs and is thought to be the main biological purpose of RNA interference. RDRs amplify virus-derived transcripts and lead to silencing of viral genes but also affect endogenous genes that share sequence homology with the initial RNA trigger (Wassenegger and Krczal, 2006). However, RDRs are only expressed in plants, fungi and *C. elegans* but not in *Drosophila* and vertebrates. Because the presence of a small RNA amplification mechanism skews the repertoire – and probably influences the biological role of endo-siRNAs – only model system that do not express RDRs will be discussed here.

To summarize, this chapter will give a brief introduction to microRNAs and piRNAs with the main focus being on endo-siRNAs. The biological role of this class of small RNAs will be discussed in the context of coding transcription and repetitive DNA elements, or, in other words, elaborate the 'constructive' and 'protective' impact of endo-siRNAs to organismal complexity.

MicroRNAs

MicroRNAs are the best characterized and generally the most abundant class of small RNAs. They serve a well established function in repressing gene expression at a post-transcriptional level. There are currently more than 1000 human microRNAs listed in the miRNA database (www.mirbase.org). Reflecting their widespread and important regulatory functions, microRNAs are ubiquitously expressed with a cell specific pattern. Single microRNAs have rather subtle tuning effects but may act cooperatively on the same target mRNA to reinforce inhibition (Hafner *et al.*, 2010).

MicroRNAs are synthesized from stem loops with imperfect complementarity in the stem structure (Fig. 2.1). The majority of loci encoding microRNAs contain clusters of several hairpin structures. The primary transcripts originate about equally from non-coding loci and from genic regions. In the latter case the stem loops are predominantly found in introns and non-protein-coding regions but are also found in exons in rare cases. The factors that regulate transcriptional activation of miRNA genes are still poorly understood. The precursors are generally synthesized as a single transcript by RNA polymerase II and are capped and polyadenylated. The so-called

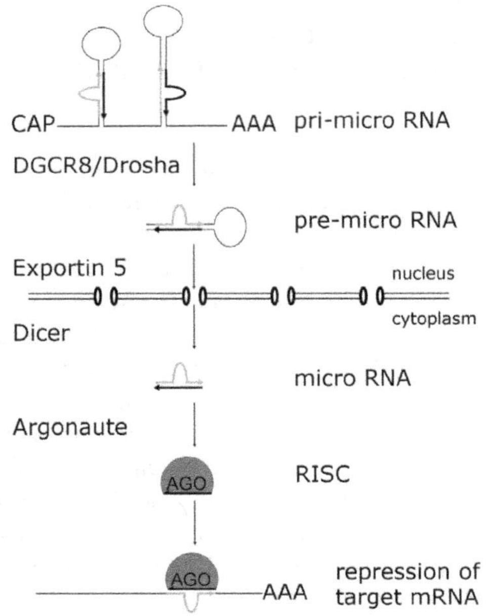

Figure 2.1 Schematic representation of miRNA processing. Stem loop structures in the pri-miRNA transcript are recognized and cleaved by DGCR8/Drosha to form the pre-miRNA structures. The stem loop is recognized by Exportin 5 and shuttled to the cytoplasm. Dicer removes the stem loop to generate microRNAs that can enter the pre-RISC complex. Strand displacement results in mature RISC formation and eventual repression of the target mRNA.

pri-miRNA hairpin structures form within the primary transcript and microRNA processing initiates co-transcriptionally (Kim *et al.*, 2009).

As a first step in microRNA processing, a single stranded RNA binding protein, DGCR8 (DiGeorge syndrome region gene 8 protein), associates with the foot of the stem loop. DGCR8 then guides Drosha to cleave the stem about 11 bp from the foot. The resulting pre-miRNAs are bound to Exportin 5 and exported to the cytoplasm with Ran-GTP acting as a cofactor. In the cytoplasm, Dcr trims the pre-miRNA to form the double-stranded oligonucleotides. This canonical pathway applies to the vast majority of RNAs; however, substrates that fulfil the structural requirements of a miRNA intermediate can feed into the canonical pathway at a later stage. For example, a short intron that forms a hairpin structure after splicing can bind to Exportin 5 and end up as a mature miRNA without the input of DCCR8/Drosha (Bartel, 2004; Czech and Hanno, 2011).

In the next step, the double-stranded short RNAs are loaded onto Ago proteins to initiate the formation of the so-called RNA-induced silencing complex (RISC). During this process, the biologically active strand of the RNA duplex is selected resulting in a 'guide-strand' bound to Ago and a 'passenger strand' that becomes quickly degraded. Theoretically, the asymmetry in the original pre-miRNA could ensure the selection of a specific strand. However, this does not seem to be the case in vertebrates because there are examples where both strands of the stem structure result in mature miRNAs; it is rather the thermodynamic properties of the RNA hybrid and the availability of a potential target sequence that play a role in guide-strand selection (Hu *et al.*, 2004). Interestingly, the binding of the RNA duplex to the Ago proteins is ATP dependent whereas the dissociation of the passenger strand does not require ATP (Kawamata *et al.*, 2009).

RISC essentially contains an Ago protein and a single-stranded RNA molecule of 22 nucleotides.

The sequence of miRNA defines the nature of the RNA structure targeted by the specific RISC. A cell does not usually express perfectly complementary sequences to miRNAs and consequently, the RISC complex is guided to targets with few mismatches. This ambiguity complicates target prediction for specific miRNAs considerably and candidate interactions need stringent experimental scrutiny (Huang *et al.*, 2010). Despite these challenges, an increasingly accurate atlas of miRNA targets is currently being established (Hafner *et al.*, 2010; Thomas *et al.*, 2010). The widespread functions of targeted mRNAs suggest that most cellular pathways are influenced by miRNAs.

Once a specific RISC has found a target sequence the translation of the complementary mRNA is repressed, with a member of the GW182 protein family playing an essential role in this step. GW182 are multidomain proteins that bind directly to Ago proteins of RISC and promote deadenylation of the target mRNA. Subsequently the mRNA decays or the complex is moved to P-bodies (RNA processing bodies). Whether the different outcomes are subject to biological regulation is as yet unknown (Eulalio *et al.*, 2009); however, eventually RNA degradation prevails (Guo *et al.*, 2010).

The impact of specific miRNAs is linked to the cellular function of the protein encoded by the target mRNAs. However, every miRNA has multiple targets as shown by knock-down or overexpression approaches in human or mouse cells: In the range of 30 proteins were found to be specifically up- or down-regulated by ≥ 50% in response to knockdown or overexpression of a single miRNA (Baek *et al.*, 2008). This amazing complexity of targets and the underlying regulatory networks are the focus of intense research. The emerging links between miRNA regulation, physiological processes and the development of disease have opened exiting new perspectives for pharmaceutical interventions.

A more general impact of miRNA mediated gene regulation becomes tangible when components of the miRNA machinery are reduced or completely ablated by conditional knock-out strategies. There is a flurry of papers indicating that virtually any cell shows pathophysiological traits if miRNA synthesis is interrupted, however, cell proliferation is possible even in the absence of miRNAs (Wang *et al.*, 2007). Accordingly, Dcr disruption leads to embryonic lethality around day 7.5 (Bernstein *et al.*, 2003). Dcr is also essential for zebrafish embryonic development but mutants can be rescued by injecting one specific miRNA species (mir-430) into fertilized eggs (Giraldez *et al.*, 2005). These results confirm that, despite near omnipresence, the influence of miRNAs on gene regulation is rather subtle. Interestingly, cells have established ways of avoiding the regulatory input of miRNAs by expressing short 3′ untranslated regions lacking the recognition sequences. Short transcripts prevail during rapid proliferation and whereas alternative polyadenylation sites are used in phases of cell differentiation (Ji *et al.*, 2009; Sandberg *et al.*, 2008).

piRNAs

In higher eukaryotes, a large proportion of the genome consists of various repetitive and transposable elements. These sequences do not contain gene coding potential and were long thought to be transcriptionally inert. However, with increasingly powerful analytical tools to monitor RNA expression it became clear, that large proportions of these repetitive sequences give rise to low level transcripts. Recently, a novel family of small RNAs has been discovered that is crucially involved in silencing the spurious transcription and reducing transposon mobility. With 26–33 nucleotides, these short RNAs are marginally longer than miRNAs and siRNAs. They also bind to a different clade of Agos, the so-called Piwi proteins. Piwi proteins are only expressed in animal gonads and are closely linked to the emergence of specialized germ cells (Aravin *et al.*, 2007).

The synthesis of piRNAs occurs in two steps that are separated between the nucleus and the cytoplasm and involve specific 3 Piwi isoforms (Piwi, Ago3, Aubergine in *Drosophila* or Miwi2, Miwi and Mili in mouse; Fig 2.2). The model sketched here is still rather speculative and alternative mechanisms may complement the picture. In a first step, primary piRNAs are produced in the nucleus involving the phosphodiesterase 'Courgette' and Piwi. Interestingly, this step

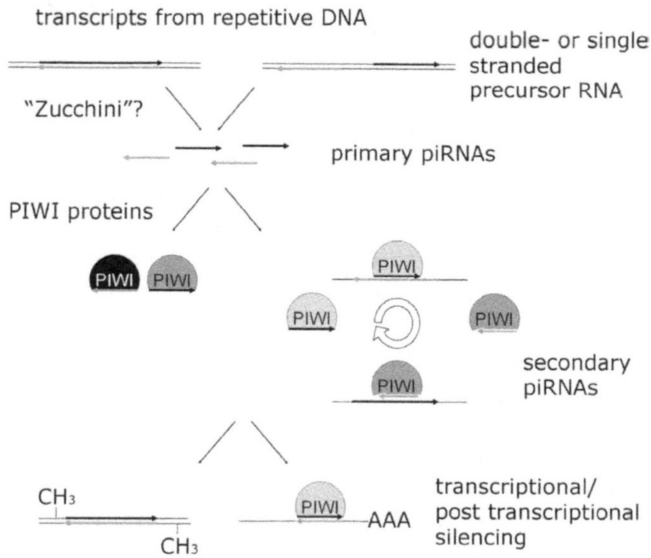

Figure 2.2 Schematic summary of piRNA formation. Transcripts from repetitive sequences, piRNA clusters, are transcribed and processed into primary piRNAs. It is yet unclear if RNA hybrids are involved. In a hypothesized ping-pong mechanisms piRNAs are amplified and target transcripts are degraded in the process to form secondary piRNAs. piRNAs are then involved in transcriptional silencing of repetitive elements. Recently, a gene regulatory role in selective degradation of maternal mRNA has been established (Rouget *et al.*, 2010). In each step specific isoforms of PIWI proteins are involved, as indicated by grey tones.

takes predominantly place in either nurse cells or somatic follicle cells. The resulting so-called primary piRNAs show a strong antisense strand bias and derive from few defined genomic loci (Haase *et al.*, 2010); Malone *et al.*, 2009). The next step in piRNA synthesis takes place in germ cells and involves the amplification of primary piRNAs. How these are transported from the surrounding somatic cells to the cytoplasm of the germ cell is unknown. Alternatively, germ cells could produce their own primary piRNAs at low levels and the somatic primary piRNAs could serve another biological purpose. Amplification of piRNAs involves Ago3 and Aubergine. A specific sequence pattern within piRNAs, a uridine at the 5′ end and an adenosine at position 10, suggests a ping-pong mechanism between sense-piRNA-bound Ago3 and antisense-piRNA-bound Aubergine. Through endonucleic cleavage and exonucleic processing novel piRNAs are generated and the target transcript is degraded (Saito and Siomi, 2010). Interestingly, Mili and Miwi2 deficient mice show reduced *de novo* methylation of repetitive DNA suggesting that piRNAs also contribute to transcriptional silencing (Aravin *et al.*, 2008; Kuramochi-Miyagawa *et al.*, 2008).

piRNAs provide a powerful line of defence against selfish transposable elements in the germ line. Disruption of piRNA synthesis leads to infertility and increased transposon activity (Vagin *et al.*, 2006). Protection against DNA transposition may become particularly important in the context of transcription and DNA compactation during spermatogenesis. The second meiotic division and the formation of haploid spermatids coincide with a wave of general transcription without mRNA translation. It is plausible that under conditions of widespread transcriptional stimulation transposons become co activated and jeopardize genome integrity (Shima *et al.*, 2004).

In addition to the clearly defensive task in transposon control a regulatory role of piRNAs recently emerged. A significant number of mRNAs expressed in the germline of flies and mice give rise to primary piRNAs that match to sequences in their 3′ untranslated region. The synthesis of these potentially regulatory piRNAs depends on key enzymatic components such as Piwi and Courgette (Robine *et al.*, 2009). Another example of piRNA-mediated gene regulation was reported in fly embryos where they induce the degradation of maternal mRNA. The ping-pong components of the (secondary) piRNA synthesis pathway are reported to be involved in the repression and eventual degradation of maternal Nanos encoding mRNA. The patterning of fly embryos depends on a gradient of the morphogen Nanos that is formed by posterior stabilization of corresponding mRNA. The decay of maternal Nanos mRNA in the rest of the embryo is promoted by piRNAs. Accordingly, mutants with impaired secondary piRNA formation show defects in head development (Rouget *et al.*, 2010).

Endo-siRNAs

Since the detection of RNA interference (RNAi) it is known that small RNAs act as exquisitely specific tags to direct RNA–protein complexes to complementary nucleotide sequences. This paradigm is universally adopted in eukaryotic cells. With the regulatory and defensive roles convincingly established for miRNAs and piRNAs, respectively, the biological purpose of endo-siRNAs is rather unclear (Okamura and Lai, 2008). Species dependent variations in the enzymatic make-up responsible for endo-siRNA production suggest that the biological raison d'être for endo-siRNAs may have changed during evolution. Early work in plants suggested that siRNA tags complementary to viral RNA provided protection against virus infection. The hypothesis is strongly supported by the finding that plant cells produce siRNAs complementary to a viral genome upon infection with the virus. Indeed, comparative studies suggest an arms race unfolded between viruses and plant cells involving inhibitors and new variations of the RNAi machinery (Ding and Voinnet, 2007). This scenario, however, seems not to apply to animals. Here, viruses and exogenous double-stranded RNA both trigger an immune response and eventually apoptosis, indicating that the immune system may have taken over the defence against exogenous RNA. Nevertheless, animal cells can be transfected with exogenous siRNAs and complementary transcripts become efficiently silenced (Elbashir *et al.*, 2001) demonstrating functional RNA interference. The occurrence of siRNAs from endogenous sources that would justify evolutionary conservation of an RNAi

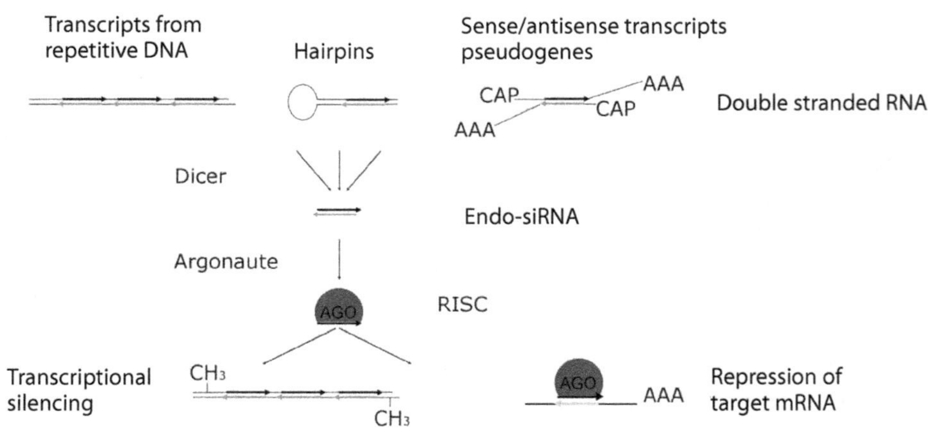

Figure 2.3 Endo-siRNA processing. Double-stranded sequences derive from repetitive elements, from hairpin structures or from co-expressed sense and antisense transcripts. Antisense transcripts can also originate form pseudogenes. Dicer processes the RNA hybrids into siRNAs which are loaded onto Argonaute proteins from a pre-RISC complex. Strand replacement results in RISC formation. Both transcriptional silencing and post transcriptional effects are attributed to endo-siRNA loaded RISC by indirect evidence.

machinery was long assumed to be dependent on the enzyme RNA-dependent RNA polymerase (RdRP). Without amplification by RdRP it was argued that siRNAs would not reach critical numbers to make a regulatory impact. The fact that intrinsic substrates for the RNAi machinery remained elusive until very recently reinforced the dilemma and it is still unclear why animals express components reminiscent to the RNAi machinery in almost every cell.

Then, in 2008, a series of papers reported the discovery of siRNAs mapping to endogenous DNA at a large scale in *Drosophila* (Czech *et al.*, 2008; Ghildiyal *et al.*, 2008; Okamura *et al.*, 2008) and mouse (Tam *et al.*, 2008; Watanabe *et al.*, 2008) as well as in zebrafish (Carlile *et al.*, 2008) and the term endo-siRNA was coined. The oligonucleotides originated from hairpin structures or complementary RNAs from sense–antisense transcription. The endo-siRNAs mapped predominantly to transposable element and, to a much lesser extent, to pseudogenes or genic sense–antisense transcripts.

The identification of endo-siRNAs in various species has opened exciting new perspectives for RNA-mediated regulation of gene expression. This section will discuss both regulatory and defensive tasks of endo-siRNAs. Emphasis will also be drawn to technical challenges that may hamper progress in the field.

endo-siRNA, an elusive species

Two strategies have eventually led to the discovery of endo-siRNAs in animal cells: Next generation sequencing and the identification of a gene that potentially produces complementary transcripts followed by the identification of specific endo-siRNAs by Northern blotting. Both approaches contain pitfalls that potentially skew the significance of the entire experiment, as will be explained below. The experimental difficulties associated with endo-siRNA detection also provide a plausible account for the delayed discovery of this RNA species in animal systems.

Next generation sequencing stands for three related platforms that apply parallel sequencing technology to deliver vast amounts of sequencing data. Differences exist with respect to the number and the length of the reads generated. One specific platform (Illumina Genome Analyser/Solexa Sequencing, Illumina, San Diego, CA, USA) has proven particularly powerful for the analysis of short RNA transcriptomes because it produces short reads (<40 bases) in high numbers. The entire repertoire of small RNAs in a cell population becomes accessible in one single experiment (RNA-Seq). The profile of small RNAs is highly complex and for practical reasons either the input RNA is pre-selected or the raw results are parsed. For example, immunoprecipitation of a specific Ago protein prior to RNA-seq will limit the complexity of the data set and focus on a selected type of small RNA. Alternatively, the data can be parsed with respect to expression level and/or mapping criteria. Apart from making the raw data more amenable, these parsing steps also aim to reduce the impact of RNA break-down products and eliminate reverse transcription- and amplification artefacts (reviewed in (Costa *et al.*, 2010; Hawkins *et al.*, 2010; Marguerat and Bahler, 2009).

In a whole cell short RNA profile endo-siRNAs take up only a small proportion of the total. Usually most of the reads represent microRNAs and in gonads piRNAs and less than 10% are endo-siRNAs (Lau *et al.*, 2009). The majority of the latter reads map to transposable elements and only a negligible number derive from genic sense–antisense transcripts. The fact that genic endo-siRNAs cannot be distinguished from breakdown products of mRNAs makes these species highly challenging to track. Despite their elusive nature, endo-siRNAs were comprehensively described in 2008, including a few examples that map to protein coding genes (Okamura and Lai, 2008).

Three groups simultaneously reported the discovery of endo-siRNAs from *Drosophila* using next generation sequencing. Ghildiyal *et al.* focused on the short RNA population with 2′-O-methyl modification and thus selected against microRNAs (which are not modified). RNA from *Drosophila* heads and from the *Drosophila* Schneider 2 cell line (S2) was included in the analysis (Ghildiyal *et al.*, 2008). Czech *et al.* (2008) and Kawamura *et al.* (2008) immunoprecipitated tagged Ago2 from transgenic flies or S2 cells and deep-sequenced the associated RNA. All three studies reported comparable findings.

The majority of reads were clustered and *mapped to transposable* elements and also hairpin structures. Interestingly, immunoprecipitation with Ago2 (endo-siRNA) and Ago3 (piRNA) included reads that mapped to the same cluster indicating that both pathways may contribute to the silencing of transposable elements (Czech *et al.*, 2008). This view is also supported by the increased transposon activity in Ago2 and Dcr2 deficient fly strains. In addition to the most abundant species, endo-siRNAs originating from sense–antisense transcripts were found in considerable numbers. They mapped with high preference to convergently transcribed regions or hairpin structures that give potentially rise to double-stranded RNA (Okamura *et al.*, 2008). The detection of these genic endo-siRNAs opens the intriguing perspective of siRNA mediated gene regulation. Somewhat counterintuitively, however, Ago2 depletion only marginally influenced the level of endo-siRNAs from convergent genic transcripts.

Investigations into the protein machinery responsible for the processing of endo-siRNAs in *Drosophila* revealed a novel combination of enzymatic components. In the canonical pathways Dcr1 combines with the accessory protein 'loquacious' to mediate the synthesis of microRNAs (Saito and Siomi, 2010). On the other hand, Dcr2 binds to the protein 'R2D2' for the production of siRNA from exogenous sources (Liu *et al.*, 2003). The synthesis of endo-siRNA was shown to depend on a novel the combination of Dcr2 with a specific isoform of loquacious (Hartig *et al.*, 2009). The findings from fly are difficult to extrapolate to vertebrates because these animals express only one isoform of Dcr.

The short RNA transcriptome of mouse oocytes and testis was at the focus of a conventional short RNA cloning/sequencing project two years before next generation sequencing became widely available. 395 sequences were obtained from oocyte libraries and 1111 clones were sequenced from three testis libraries. More than 75% of the sequences derived from ribosomal and transfer RNAs or could not be mapped to the mouse genome. The remaining sequences were micro-RNAs (11%), repeat-associated siRNAs (rasiRNAs, 10%) or mapped to mRNAs or non-protein-coding RNAs (2.3%). The testis libraries contained 1111 clones and contained comparable proportions of short RNAs: piRNAs were most abundant (34%), followed by micro RNAs (24%), rRNAs and tRNAs (19%) as well as rasiRNA/mRNA-related reads that would now be categorized as endo-siRNAs (Watanabe *et al.*, 2006). These results established a pattern of small RNAs that was later confirmed by much deeper RNA-seq analysis. Importantly, the high number of reads in the reports by Watanabe *et al.* (2008) and Tam *et al.* (2008) underpinned a biological relevance of endo-siRNAs that was at best speculative before publication of their work. The prevalence of repeat associated sequences is striking in all these reports and it was suggested that endo-siRNAs are being generated to keep transposons at bay. Endo-siRNAs mapping to piRNA clusters confirm this view and abundant reads from hairpin structures indicate that this pathway deals efficiently with double-stranded RNA in the nucleus. Accordingly, Dcr or Ago2 mutants show deregulation of transposable elements (Tam *et al.*, 2008). RNA hybrids can also arise from convergent genic transcription or, alternatively, involve antisense transcripts from pseudogenes. The hypothesis of a pseudogene regulating the expression of its protein-coding homologue seemed very attractive; however, specific biological examples of such a regulatory interplay are rare. The possibility of a regulatory interaction between a protein coding transcript with its corresponding antisense RNA seems more intuitive but attracted little attention in the original papers. This aspect but will be discussed in detail below.

Apart from cDNA sequencing, the occurrence of endo-siRNAs can be detected by Northern blotting if regions producing complementary transcripts can be identified. This strategy was pursued by two groups, one focusing on repetitive elements (Yang and Kazazian, 2006), and the other one on a specific bidirectionally transcribed gene (Carlile *et al.*, 2008). Yang and Kazazian argued that RNAi-deficient *C. elegans* shows increased transposon mobility and tested the impact of RNAi on long interspersed element 1 (LINE-1) in human cultured cells. After confirming bidirectional of transcription of LINE-1, Northern blotting detected LINE-1 related sense- and antisense siRNAs. Interestingly, knockdown of Dcr1 significantly increased the mobility of Dcr-1 retrotransposons (Yang and Kazazian, 2006).

The second report describes an attempt to identify endo-siRNAs from a locus that produces complementary sense and antisense transcripts. The authors of this report argued that RNA interference was the most probable mechanism involved in processing the astonishing number of co-expressed sense and antisense transcripts. The scale of these enigmatic naturally occurring antisense transcripts (NATs) was discovered during whole transcriptome analysis in 2003 (Kiyosawa *et al.*, 2003; Yelin *et al.*, 2003). Focusing on the expression pattern of a protein coding sense/non-protein coding antisense RNA pair during zebrafish embryonic development Carlile *et al.* found striking reciprocal expression of the two complementary transcripts with a short period of co-expression (Carlile *et al.*, 2008). During this time window gene specific endo-siRNAs were visualized using conventional Northern blotting. The detection of endo-siRNAs by Northern blotting harbours technical challenges that merit special considerations. Endo-siRNAs that derive from coding regions are generally expressed at a very low level (Werner *et al.*, 2007). As a consequence, hybridization probes that either carry multiple labels and/or allow for signal amplification are necessary for reliable detection. In addition, the expression of endo-siRNAs is both tissue and development dependent. These potential pitfalls make a serendipitous discovery of (genic) endo-siRNAs highly unlikely.

To summarize, endo-siRNAs have been discovered as a result of large-scale sequencing efforts and targeted experiments. They are predominantly expressed in germ cells but may also be present is somatic cells. Intuitively, endo-siRNAs are assumed to regulate the expression of the complementary transcripts and the encoded proteins. This assumption and other implications of endo-siRNA expression have only been experimentally tested in very limited detail. The following part focusing on the biological implications of endo-siRNAs is therefore largely based on indirect evidence and contains hypothetical elements.

Endo-siRNA synthesis

The synthesis of endo-siRNAs has been firmly linked to the occurrence of RNA duplexes in flies and mice and other animal model systems (Fig. 2.3). The double-stranded region was required to be perfectly complementary in *Xenopus* oocytes and needed to be at least 30 bp to warrant efficient processing (Carlile *et al.*, 2009). In flies and mice, on the other hand mismatches seem to be tolerated (Okamura *et al.*, 2008; Tam *et al.*, 2008). This difference may be the result of species specific variations of RNAi components or, alternatively, reflect experimental differences. Endo-siRNA production in oocytes was monitored after injection of perfect or imperfect RNA duplexes in the cell nucleus whereas the observation in flies and mice are based on the mapping of endo-siRNA sequencing reads to genomic DNA including a mismatch.

There is strong evidence that the initial processing of the RNA duplex by Dcr in vertebrates takes place in the nucleus. First, many of the precursor RNA from non-protein coding regions that give rise to endo-siRNAs are unlikely to be exported to the cytoplasm. Second, RNA duplexes were shown to be efficiently cleaved in the nucleus of *Xenopus* oocytes whereas they were stable for days in the cytoplasm (Carlile *et al.*, 2008). The RNA hybrids are protected by specific RNA-binding chaperones. Third, in most animal cells, cytoplasmic RNA duplexes trigger a strong protein kinase R/interferon mediated immune response (Wang and Carmichael, 2004). It is therefore unlikely that endo-siRNA precursors accumulate significantly in the cytoplasm. Whether the diced RNA molecules are exported from the nucleus is unclear. In *Xenopus* oocytes the endo-siRNAs do leave the nucleus since cytoplasmic RNA interference can be triggered by nuclear injection of double-stranded precursor RNA. However, this observation could be cell type specific.

In a next step, the double-stranded RNA molecules bind to an Ago protein under the expense of ATP resulting in the so-called pre-RISC complex. Unwinding of the double-stranded RNA leads to the formation of RISC. Interestingly, the latter step occurs also in the absence of ATP (Kawamata *et al.*, 2009). Ago isoforms that are catalytically active are more efficient in removing the passenger strand; however, RNA slicing is not essential for strand displacement (Matranga *et al.*, 2005). Both flies (2) and mice (4) express multiple Ago isoforms that potentially accommodate siRNAs from endogenous sources. Results from RNA-seq experiments on samples of Ago-bound RNA essentially confirmed the canonical pathways in flies: Dcr1/Ago1 mediate miRNA processing whereas Dcr2/Ago2 is involved in endo-siRNA synthesis (Czech *et al.*, 2008). The situation is less clear in mice because the contribution of Agos 1, 3 and 4 is unknown. Ago 2 was shown to complex endo-siRNAs, in addition, Ago 2 mutant mouse oocytes show increased levels of RNA targets complementary to endo-siRNAs (Watanabe *et al.*, 2008; Kaneda *et al.*, 2009). Ago 2, however, is not always the most abundant isoform and the contribution of the other isoforms may be biologically significant. In pachytene, round and elongated spermatids, for example, Ago4 mRNA is about twice as abundant as Ago2 mRNA. The proportion is usually more than 10 times in favour of Ago2 in somatic cells (Gonzalez-Gonzalez *et al.*, 2008). Interestingly, mouse testis is also a rich source of small RNAs including endo-siRNAs (Watanabe *et al.*, 2006). This coincidence may provide a lead towards investigating mechanisms of action and biological roles of endo-siRNAs.

Endo-siRNAs and gene regulation

The detection of small siRNAs from transposons and genic regions as well as the isolated observation of altered endo-siRNA expression during zebrafish development suggest that endo-siRNAs serve both defensive and regulatory roles.

Endo-siRNA and transposon silencing

Interference of endo-siRNA synthesis by knocking down either Dcr or Ago2 results in increased mobility of transposable elements (Saito and Siomi, 2010). The mechanism by which transposon silencing is mediated is not known. Mechanistic parallels may exist between transcriptional gene silencing induced by *exogenous* siRNAs (as discussed in Chapter 13) and the process triggered by endo-siRNAs. However, experimental scrutiny of this hypothesis is challenging and yet missing.

Pseudogene derived endo-siRNAs

Both RNA-seq studies that discovered endo-siRNAs at a large scale in mice found a small but significant number of reads from pseudogenes (Tam *et al.*, 2008; Watanabe *et al.*, 2008). These endo-siRNAs also mapped to the parent gene with no or only few mismatches. Consequently, the parent gene could potentially be regulated by the pseudogene derived endo-siRNAs. Pseudogenes constitute a significant part of the non-protein coding genome in vertebrates. They were long perceived as transcriptionally inert and non-functional. A counter indication to this view was reported in 1999 when Korneev *et al.* (1999) found an antisense transcript from a nitric oxide synthase (NOS) pseudogene in snail. The antisense RNA was demonstrated to bind to the sense transcript of the parent gene and reduce its expression. The concomitant decrease in NOS activity led to an altered feeding behaviour of the animal. It was not established whether the decrease in NOS mRNA involved endo-siRNAs. Interestingly, the same group detected a related NOS pseudogene in human and other primate genomes that gives rise to a potential trans- acting antisense transcript (Korneev *et al.*, 2008). Phylogenetic conservation of the unusual genomic arrangement offers strong support for a putative regulatory role of the pseudogene derived antisense transcript. On the other hand, there is little indication for a significant regulatory role of pseudogenes via sense–antisense RNA interactions and endo-siRNAs on a genome-wide scale. One example of a regulatory pseudogene represents a transcribed pseudogene which destabilizes Makorin-1 mRNA. Mice which lacked the regulatory input of the pseudogene developed polycystic kidneys (Hirotsune *et al.*, 2003). These conclusions, however, were rejected by another group who failed to reproduce the findings (Grey *et al.*, 2006).

The very limited examples of regulatory pseudogenes are an indication that they may not have a general impact on gene expression. Another observation supports this conclusion; despite the large number of pseudogenes – approximately 20,000 in human – only a very limited number of pseudogene-derived transcripts have been reported. Detection methods that rely on nucleic acid hybridization such as Northern blotting and *in situ* hybridization (with double-stranded probes) or reverse-transcription PCR tend to give false-positive results if transcripts with extended complementarity to the target mRNA are present. Whereas this methodological ambiguity resulted in the detection and characterization of large numbers of natural antisense transcripts, pseudogene-derived transcripts were hardly found.

Endo-siRNAs derived from natural antisense transcripts

A significant source of endo-siRNAs was found to be bi-directionally transcribed genes that produce sense and antisense transcripts. The fact that endo-siRNAs strictly mapped to complementary regions suggest that the two overlapping transcripts hybridize to become substrates for the endo-siRNA synthesis. Natural antisense transcripts (NATs) are a particularly intriguing and well researched family of non-protein-coding RNAs (Beiter *et al.*, 2008). NATs are discussed in detail in another chapter of this book; however, some special features that are relevant in the context of endo-siRNAs are highlighted here. A substantial number of NATs are synthesized from the opposite strand of protein coding genes. Estimates vary but more than 20% of human and mice protein-coding genes seem to express corresponding NATs. These numbers are based on large-scale sequencing approaches and bioinformatic studies that use hallmarks of mRNA processing such as cap, splicing and polyA tail to exclude experimental artefacts and breakdown products (Katayama *et al.*, 2005). Compared with the ever-increasing range of non-protein coding RNAs that accompany the synthesis of a protein coding transcript, the number of fully processed NATs may be comparably small. Of particular interest, however, this special group of NATs shows hallmarks that make them particularly suitable substrates for endo-siRNA synthesis. First, these NATs are rather stable thanks to an mRNA-like structure. Second, per definition they contain sequences that are complementary to the cognate sense mRNAs (Werner and Berdal, 2005; Werner *et al.*, 2009). Third, NATs show a strong tendency to be co-expressed – at least transiently – with the related sense transcripts (Kiyosawa *et al.*, 2003). A fundamental biological importance of NATs is suggested by their sheer prevalence in the mouse and human genomes (Katayama *et al.*, 2005). In addition, the regulatory potential of an increasing number of NATs has been experimentally confirmed and shown to impact on disease onset and progression as in cancer and Alzheimer's disease (Faghihi and Wahlestedt, 2009) as detailed in other chapters of this book. Whether these regulatory events involve endo-siRNAs has not been established, neither have the NATs-derived endo-siRNAs from large-scale screens been tested for function.

There is a single gene, Slc34a, encoding an epithelial sodium/phosphate (Na/P_i) cotransporter, where antisense transcription and endo-siRNA synthesis have been demonstrated in a broader epigenetic and physiological context (Werner and Sayer, 2009). The strategic aim behind this research was to fill the knowledge gap between the genomics of antisense transcription and physiological regulation Na/P_i transport. The Na/P_i transporter is crucially involved in maintaining vertebrate P_i homeostasis and the effector of numerous regulatory loops (Murer *et al.*, 2003). The expression pattern of the antisense RNA, however, does not support a direct involvement of the antisense transcript in adjusting P_i transport to physiological requirements. Could the Slc34a-related NAT help in establishing an epigenetic imprint at the Slc34a locus rather than becoming involved in the acute regulation P_i homeostasis? A link between antisense expression and transcriptional gene silencing has been found for individual genes and this hypothesis is also supported by early genomic findings focusing on the identification and mapping of NATs. Independent studies observed that NATs are significantly under-represented on the human and mouse X chromosome (Chen *et al.*, 2004; Kiyosawa *et al.*, 2003). A strong, collective bias against a heterogeneous group of bidirectionally transcribed genes is hardly compatible with a scenario where every NAT represents an exclusive regulator of its sense transcript. The potential of NATs to induce transcriptional gene silencing offers a conceivable explanation for the bias against antisense transcription on the X chromosome. Because only one copy of the X chromosome is usually active, NATs-induced silencing would trigger complete gene knock down with potentially fatal consequences for the affected cell. If a comparable incident happened on autosomes, the second allele would compensate for the loss of function. This argument is supported by the fact that NATs expressed from mammalian X chromosomes tend to be clustered in areas that escape inactivation (Kiyosawa *et al.*, 2003). In addition, flies and worms that adjust the output from the X chromosomes, but keep both alleles active, show equal distribution of NATs on the X chromosomes and autosomes (Zhang *et al.*, 2006). If antisense RNA induced gene silencing applied to autosomes one would expect a significant number of genes expressed from only one allele, in other words, monoallelic expression or random imprinting. Gimelbrant *et al.* (2007) reported widespread monoallelic expression of human genes, thus supporting the argument. Importantly, the monoallelically expressed genes show a strong tendency be transcribed in both directions (Carlile *et al.*, 2009).

The findings outlined above support a view that antisense transcription leads to epigenetic changes; however, it remains largely unclear whether endo-siRNAs are generated along this process. For various reasons, other cellular strategies to deal with RNA hybrids such as RNA editing, transcriptional interference or transcriptional interference represent unlikely alternatives to RNA interference to deal with sense/antisense duplexes (Werner and Berdal, 2005). Investigations focusing on the Slc34a gene in mice and zebrafish argue that antisense transcription is indeed closely linked to the occurrence of endo-siRNAs. This research also suggested that the decision which strand of the endo-siRNA duplex becomes the guide and which one becomes the passenger represents a key step in the biological function of endo-siRNAs.

In this respect, an intriguing observation was made in developing zebrafish. Slc34a sense and antisense transcripts show reciprocal expression with a window of significant co-expression at 48 hours post fertilization (hpf). It is during this period that the zebrafish kidney starts producing urine and the sense-encoded Na/P_i transporter becomes physiologically important. Before 48 hpf the antisense transcript is exclusively expressed and no endo-siRNAs are produced. Between 48 hpf and 72 hpf there is not only a switch from antisense to sense transcript production but also a reorientation of endo-siRNAs from antisense to sense (Carlile *et al.*, 2008). This indicates that a significant shift in transcript abundance also results in reversed 'guide' and 'passenger' roles for the two endo-siRNA strands. In other words, endo-siRNAs represent a molecular switch that is turned according to the dominant transcript; but, what would such a switch be good for?

A paradigm for endo-siRNA function?

A recently published series of articles aims to integrate endo-siRNAs into a comprehensive picture of epigenetic chromatin regulation and links aspects of natural NATs and RISC formation and transcriptional silencing (Werner *et al.*, 2009). The suggested model is based on a close link between endo-siRNAs and NATs which concurs with the expression pattern of both RNAi components and the shape of the antisense transcriptome. Both show prominent expression in the germ line and ubiquitous but significantly lower levels of NATs as well as an altered repertoire of Ago proteins in somatic cells. The attempt to integrate the latest findings in the field of endo-siRNAs resulted in an attractive, yet highly speculative, model that is relevant to the overarching hypothesis of this book (Fig 2.4).

The model takes into account that sense and antisense transcripts of bi-directionally transcribed genes tend to be co-expressed (Kiyosawa *et al.*, 2003). Co-expression may be linked to a transcriptional wave during sperm development or, in somatic cells, represent the consequence of 'transcriptional ripples' (Kimmins *et al.*, 2004; Ebisuya *et al.*, 2008), meaning that spurious transcription of neighbouring regions occurs upon activation of a gene. Transcripts in both directions will be processed into mRNAs that either hybridize and become processed into endo-siRNAs or, alternatively, persist as mRNAs. Thus, the mRNAs not only act as starting material for endo-siRNA synthesis but also represent fully complementary targets for both endo-siRNA strands.

The endo-siRNAs will then bind to Ago proteins, in testis likely to the non-slicing isoform Ago-4. Investigations into the assembly of the RISC complex indicate that double-stranded

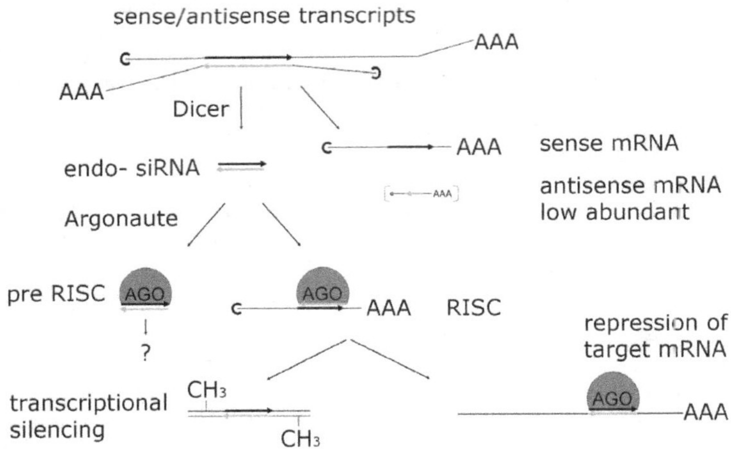

Figure 2.4 Hypothetical model of gene regulation by genic endo-siRNAs. A protein coding sense transcript and a low abundance antisense transcript are produced. Both transcripts are either processed into endo-siRNAs or persist as mRNAs. The endo-siRNAs are loaded onto a non-slicing Argonaute protein (most probably Ago4 or Ago1) in both orientations (pre-RISC). The availability of a target sequence will drive RISC formation. How the resulting complexes contribute to gene regulation is unclear, however both transcriptional silencing and mRNA repression are conceivable.

endo-siRNAs may bind in random orientation to form the 'pre-RISC' (Kawamata *et al.*, 2009). ATP-dependent loading of the endo-siRNA onto Ago is then followed by the ATP-independent unwinding of the siRNA duplex. The availability of complementary target sequences will therefore influence the efficiency of 'unwinding' or rather 'strand replacement'. The accessibility of complementary sequences as found in the remaining full length sense or antisense transcripts, has a direct influence on strand selection. The process will result in an Ago-guide-mRNA RISC complex if a relevant target is accessible or, alternatively, in an Ago-passenger pre-RISC seeking for complementary sequences. The two resulting complexes will show different half-life times, different cellular distributions and different biological fates. Importantly, any mutation that influences transcription efficiency, RNA processing or folding, for example, and skews the proportion between the sense and the antisense transcript may lead to a change of 'guide' and 'passenger' function. It is conceivable that, in the absence of a target mRNA, the Ago-passenger pre-RISC binds to nascent transcripts in the nucleus and initiates transcriptional silencing. In support of this assumption, siRNAs have the potential to induce transcriptional silencing in mammalian cultured cells; moreover, NATs are associated with gene silencing genome-wide.

Are endo-siRNAs driving organismal complexity?

The most prominent expression of NATs occurs in testis, in haploid spermatocytes (Carlile *et al.*, 2009). In haploid cells, mutations that trigger a reversal of endo-siRNA strand selection and redirect gene silencing to the opposite strand will cause a complete gene knock-down and reduce the cell's fitness. As a consequence of the reduced survival of mutated cells a positively selected sperm population will emerge. A screening process that monitors the genetic output at RNA level and avoids the build-up of an entire organism seems perfectly in line with enhancing complexity.

References

Aravin, A.A., Hannon, G.J., and Brennecke, J. (2007). The Piwi-piRNA pathway provides an adaptive defense in the transposon arms race. Science *318*, 761–764.

Aravin, A.A., Sachidanandam, R., Bourc'his, D., Schaefer, C., Pezic, D., Toth, K.F., Bestor, T., and Hannon, G.J. (2008). A piRNA pathway primed by individual transposons is linked to *de novo* DNA methylation in mice. Mol. Cell *31*, 785–799.

Baek, D., Villen, J., Shin, C., Camargo, F.D., Gygi, S.P., and Bartel, D.P. (2008). The impact of microRNAs on protein output. Nature *455*, 64–71.

Bartel, D.P. (2004). MicroRNAs: genomics, biogenesis, mechanism, and function. Cell *116*, 281–297.

Beiter, T., Reich, E., Williams, R.W., and Simon, P. (2008). Antisense transcription: A critical look in both directions. Cell. Mol. Life Sci. *66*, 94–112.

Bernstein, E., Kim, S.Y., Carmell, M.A., Murchison, E.P., Alcorn, H., Li, M.Z., Mills, A.A., Elledge, S.J., Anderson, K.V., and Hannon, G.J. (2003). Dicer is essential for mouse development. Nat. Gen. *35*, 215–217.

Carlile, M., Nalbant, P., Preston-Fayers, K., McHaffie, G.S., and Werner, A. (2008). Processing of naturally occurring sense/antisense transcripts of the vertebrate Slc34a gene into short RNAs. Phys. Gen. *34*, 95–100.

Carlile, M., Swan, D., Jackson, K., Preston-Fayers, K., Ballester, B., Flicek, P., and Werner, A. (2009). Strand selective generation of endo-siRNAs from the Na/phosphate transporter gene Slc34a1 in murine tissues. Nucleic Acids Res. *37*, 2274–2282.

Chen, J., Sun, M., Kent, W.J., Huang, X., Xie, H., Wang, W., Zhou, G., Shi, R.Z., and Rowley, J.D. (2004). Over 20% of human transcripts might form sense–antisense pairs. Nucleic Acids Res. *32*, 4812–4820.

Costa, V., Angelini, C., De Feis, I., and Ciccodicola, A. (2010). Uncovering the complexity of transcriptomes with RNA-Seq. J. Biomed. Biotechnol. Epub: 853916.

Czech, B., and Hannon, G.J. (2011). Small RNA sorting: matchmaking for Argonautes. Nat. Rev. *12*, 19–31.

Czech, B., Malone, C.D., Zhou, R., Stark, A., Schlingeheyde, C., Dus, M., Perrimon, N., Kellis, M., Wohlschlegel, J.A., Sachidanandam, R., *et al.* (2008). An endogenous small interfering RNA pathway in Drosophila. Nature *453*, 798–802.

Ding, S.W., and Voinnet, O. (2007). Antiviral immunity directed by small RNAs. Cell *130*, 413–426.

Ebisuya, M., Yamamoto, T., Nakajima, M., and Nishida, E. (2008). Ripples from neighbouring transcription. Nat. Cell. Biol. *10*, 1106–1113.

Elbashir, S.M., Harborth, J., Lendeckel, W., Yalcin, A., Weber, K., and Tuschl, T. (2001). Duplexes of 21-nucleotide RNAs mediate RNA interference in cultured mammalian cells. Nature *411*, 494–498.

Ender, C., and Meister, G. (2010). Argonaute proteins at a glance. J. Cell. Sci. *123*, 1819–1823.

Eulalio, A., Tritschler, F., and Izaurralde, E. (2009). The GW182 protein family in animal cells: new insights into domains required for miRNA-mediated gene silencing. RNA *15*, 1433–1442.

Faghihi, M.A., and Wahlestedt, C. (2009). Regulatory roles of natural antisense transcripts. Nat. Rev. Mol. Cell. Biol. *10*, 637–643.

Ghildiyal, M., Seitz, H., Horwich, M.D., Li, C., Du, T., Lee, S., Xu, J., Kittler, E.L., Zapp, M.L., Weng, Z., *et al.* (2008). Endogenous siRNAs derived from transposons and mRNAs in *Drosophila* somatic cells. Science *320*, 1077–1081.

Gimelbrant, A., Hutchinson, J.N., Thompson, B.R., and Chess, A. (2007). Widespread monoallelic expression on human autosomes. Science *318*, 1136–1140.

Giraldez, A.J., Cinalli, R.M., Glasner, M.E., Enright, A.J., Thomson, J.M., Baskerville, S., Hammond, S.M., Bartel, D.P., and Schier, A.F. (2005). MicroRNAs regulate brain morphogenesis in zebrafish. Science *308*, 833–838.

Gonzalez-Gonzalez, E., Lopez-Casas, P.P., and del Mazo, J. (2008). The expression patterns of genes involved in the RNAi pathways are tissue-dependent and differ in the germ and somatic cells of mouse testis. Biochim. Biophys. Acta *1779*, 306–311.

Gray, T.A., Wilson, A., Fortin, P.J., and Nicholls, R.D. (2006). The putatively functional Mkrn1-p1 pseudogene is neither expressed nor imprinted, nor does it regulate its source gene in trans. Proc. Natl. Acad. Sci. U.S.A. *103*, 12039–12044.

Guo, H., Ingolia, N.T., Weissman, J.S., and Bartel, D.P. (2010). Mammalian microRNAs predominantly act to decrease target mRNA levels. Nature *466*, 835–840.

Haase, A.D., Fenoglio, S., Muerdter, F., Guzzardo, P.M., Czech, B., Pap (2010). Probing the initiation and effector phases of the somatic piRNA pathway in Drosophila. Genes Dev. *24*, 2499–2504.

Hafner, M., Landthaler, M., Burger, L., Khorshid, M., Hausser, J., Berninger, P., Rothballer, A., Ascano, M., Jr., Jungkamp, A.C., Munschauer, M., *et al.* (2010). Transcriptome-wide identification of RNA-binding protein and microRNA target sites by PAR-CLIP. Cell *141*, 129–141.

Hartig, J.V., Esslinger, S., Bottcher, R., Saito, K., and Forstemann, K. (2009). Endo-siRNAs depend on a new isoform of loquacious and target artificially introduced, high-copy sequences. EMBO J. *28*, 2932–2944.

Hawkins, R.D., Hon, G.C., and Ren, B. (2010). Next-generation genomics: an integrative approach. Nat. Rev. *11*, 476–486.

Hirotsune, S., Yoshida, N., Chen, A., Garrett, L., Sugiyama, F., Takahashi, S., Yagami, K., Wynshaw-Boris, A., and Yoshiki, A. (2003). An expressed pseudogene regulates the messenger-RNA stability of its homologous coding gene. Nature *423*, 91–96.

Hock, J., Weinmann, L., Ender, C., Rudel, S., Kremmer, E., Raabe, M., Urlaub, H., and Meister, G. (2007). Proteomic and functional analysis of Argonaute-containing mRNA–protein complexes in human cells. EMBO Rep. *8*, 1052–1060.

Hu, X., Hipolito, S., Lynn, R., Abraham, V., Ramos, S., and Wong-Staal, F. (2004). Relative gene-silencing efficiencies of small interfering RNAs targeting sense and antisense transcripts from the same genetic locus. Nucleic Acids Res. *32*, 4609–4617.

Huang, Y., Zou, Q., Song, H., Song, F., Wang, L., Zhang, G., and Shen, X. (2010). A study of miRNAs targets prediction and experimental validation. Protein Cell *1*, 979–986.

Ji, Z., Lee, J.Y., Pan, Z., Jiang, B., and Tian, B. (2009). Progressive lengthening of 3′ untranslated regions of mRNAs by alternative polyadenylation during mouse embryonic development. Proc. Natl. Acad. Sci. U.S.A. *106*, 7028–7033.

Kaneda, M., Tang, F., O'Carroll, D., Lao, K., and Surani, M.A. (2009). Essential role for Argonaute2 protein in mouse oogenesis. Epigen. Chromatin 2, 9.

Katayama, S., Tomaru, Y., Kasukawa, T., Waki, K., Nakanishi, M., Nakamura, M., Nishida, H., Yap, C.C., Suzuki, M., Kawai, J., *et al.* (2005). Antisense transcription in the mammalian transcriptome. Science *309*, 1564–1566.

Kawamata, T., Seitz, H., and Tomari, Y. (2009). Structural determinants of miRNAs for RISC loading and slicer-independent unwinding. Nat. Struct. Mol. Biol. *16*, 953–960.

Kawamura, Y., Saito, K., Kin, T., Ono, Y., Asai, K., Sunohara, T., Okada, T.N., Siomi, M.C., and Siomi, H. (2008). Drosophila endogenous small RNAs bind to Argonaute 2 in somatic cells. Nature *453*, 793–797.

Kim, V.N., Han, J., and Siomi, M.C. (2009). Biogenesis of small RNAs in animals. Nat. Rev. Mol. Cell. Biol. *10*, 126–139.

Kimmins, S., Kotaja, N., Davidson, I., and Sassone-Corsi, P. (2004). Testis-specific transcription mechanisms promoting male germ-cell differentiation. Reproduction *128*, 5–12.

Kiyosawa, H., Yamanaka, I., Osato, N., Kondo, S., and Hayashizaki, Y. (2003). Antisense transcripts with FANTOM2 clone set and their implications for gene regulation. Genome Res. *13*, 1324–1334.

Korneev, S.A., Park, J.H., and O'Shea, M. (1999). Neuronal expression of neural nitric oxide synthase (nNOS) protein is suppressed by an antisense RNA transcribed from an NOS pseudogene. J Neurosci *19*, 7711–7720.

Korneev, S.A., Korneeva, E.I., Lagarkova, M.A., Kiselev, S.L., Critchley, G., and O'Shea, M. (2008). Novel noncoding antisense RNA transcribed from human anti-NOS2A locus is differentially regulated during neuronal differentiation of embryonic stem cells. RNA *14*, 2030–2037.

Kuramochi-Miyagawa, S., Watanabe, T., Gotoh, K., Totoki, Y., Toyoda, A., Ikawa, M., Asada, N., Kojima, K., Yamaguchi, Y., Ijiri, T.W., *et al.* (2008). DNA methylation of retrotransposon genes is regulated by Piwi family members MILI and MIWI2 in murine fetal testes. Genes Dev. *22*, 908–917.

Lau, N.C., Robine, N., Martin, R., Chung, W.J., Niki, Y., Berezikov, E., and Lai, E.C. (2009). Abundant primary piRNAs, endo-siRNAs, and microRNAs in a Drosophila ovary cell line. Genome Res. *19*, 1776–1785.

Liu, Q., Rand, T.A., Kalidas, S., Du, F., Kim, H.E., Smith, D.P., and Wang, X. (2003). R2D2, a bridge between the initiation and effector steps of the *Drosophila* RNAi pathway. Science *301*, 1921–1925.

Malone, C.D., Brennecke, J., Dus, M., Stark, A., McCombie, W.R., Sachidanandam, R., and Hannon, G.J. (2009). Specialized piRNA pathways act in germline and somatic tissues of the Drosophila ovary. Cell *137*, 522–535.

Marguerat, S., and Bahler, J. (2009). RNA-seq: from technology to biology. Cell. Mol. Life Sci. *67*, 569–579.

Matranga, C., Tomari, Y., Shin, C., Bartel, D.P., and Zamore, P.D. (2005). Passenger-strand cleavage facilitates assembly of siRNA into Ago2-containing RNAi enzyme complexes. Cell *123*, 607–620.

Murer, H., Hernando, N., Forster, I., and Biber, J. (2003). Regulation of Na/Pi transporter in the proximal tubule. Ann. Rev. Physiol. *65*, 531–542.

Okamura, K., and Lai, E.C. (2008). Endogenous small interfering RNAs in animals. Nat. Rev. Mol. Cell. Biol. *9*, 673–678.

Okamura, K., Chung, W.J., Ruby, J.G., Guo, H., Bartel, D.P., and Lai, E.C. (2008). The *Drosophila* hairpin RNA pathway generates endogenous short interfering RNAs. Nature *453*, 803–806.

Robine, N., Lau, N.C., Balla, S., Jin, Z., Okamura, K., Kuramochi-Miyagawa, S., Blower, M.D., and Lai, E.C. (2009). A broadly conserved pathway generates 3'UTR-directed primary piRNAs. Curr. Biol. *19*, 2066–2076.

Rouget, C., Papin, C., Boureux, A., Meunier, A.C., Franco, B., Robine, N., Lai, E.C., Pelisson, A., and Simonelig, M. (2010). Maternal mRNA deadenylation and decay by the piRNA pathway in the early Drosophila embryo. Nature *467*, 1128–1132.

Saito, K., and Siomi, M.C. (2010). Small RNA-mediated quiescence of transposable elements in animals. Dev. Cell *19*, 687–697.

Sandberg, R., Neilson, J.R., Sarma, A., Sharp, P.A., and Burge, C.B. (2008). Proliferating cells express mRNAs with shortened 3′ untranslated regions and fewer microRNA target sites. Science *320*, 1643–1647.

Shima, J.E., McLean, D.J., McCarrey, J.R., and Griswold, M.D. (2004). The murine testicular transcriptome: characterizing gene expression in the testis during the progression of spermatogenesis. Biol. Reprod. *71*, 319–330.

Tam, O.H., Aravin, A.A., Stein, P., Girard, A., Murchison, E.P., Cheloufi, S., Hodges, E., Anger, M., Sachidanandam, R., Schultz, R.M., *et al.* (2008). Pseudogene-derived small interfering RNAs regulate gene expression in mouse oocytes. Nature *453*, 534–538.

Thomas, M., Lieberman, J., and Lal, A. (2010). Desperately seeking microRNA targets. Nat. Struct. Mol. Biol. *17*, 1169–1174.

Vagin, V.V., Sigova, A., Li, C., Seitz, H., Gvozdev, V., and Zamore, P.D. (2006). A distinct small RNA pathway silences selfish genetic elements in the germline. Science *313*, 320–324.

Wang, Q., and Carmichael, G.G. (2004). Effects of length and location on the cellular response to double-stranded RNA. Microbiol. Mol. Biol. Rev. *68*, 432–452.

Wang, Y., Medvid, R., Melton, C., Jaenisch, R., and Blelloch, R. (2007). DGCR8 is essential for microRNA biogenesis and silencing of embryonic stem cell self-renewal. Nat. Gen. *39*, 380–385.

Wassenegger, M., and Krczal, G. (2006). Nomenclature and functions of RNA-directed RNA polymerases. Trends Plant Sci. *11*, 142–151.

Watanabe, T., Takeda, A., Tsukiyama, T., Mise, K., Okuno, T., Sasaki, H., Minami, N., and Imai, H. (2006). Identification and characterization of two novel classes of small RNAs in the mouse germline: retrotransposon-derived siRNAs in oocytes and germline small RNAs in testes. Genes Dev. *20*, 1732–1743.

Watanabe, T., Totoki, Y., Toyoda, A., Kaneda, M., Kuramochi-Miyagawa, S., Obata, Y., Chiba, H., Kohara, Y., Kono, T., Nakano, T., *et al.* (2008). Endogenous siRNAs from naturally formed dsRNAs regulate transcripts in mouse oocytes. Nature *453*, 539–543.

Werner, A., and Berdal, A. (2005). Natural antisense transcripts: sound or silence? Phys. Gen. *23*, 125–131.

Werner, A., Carlile, M., and Swan, D. (2009). What do natural antisense transcripts regulate? RNA Biol. *6*, 43–48.

Werner, A., and Sayer, J.A. (2009). Naturally occurring antisense RNA: function and mechanisms of action. Curr. Opin. Nephrol. Hypertens. *18*, 343–349.

Werner, A., Schmutzler, G., Carlile, M., Miles, C.G., and Peters, H. (2007). Expression profiling of antisense transcripts on DNA arrays. Phys. Gen. *28*, 294–300.

Yang, N., and Kazazian, H.H., Jr. (2006). L1 retrotransposition is suppressed by endogenously encoded small interfering RNAs in human cultured cells. Nat. Struct. Mol. Biol. *13*, 763–771.

Yelin, R., Dahary, D., Sorek, R., Levanon, E.Y., Goldstein, O., Shoshan, A., Diber, A., Biton, S., Tamir, Y., Khosravi, R., *et al.* (2003). Widespread occurrence of antisense transcription in the human genome. Nat. Biotechnol. *21*, 379–386.

Zhang, Y., Liu, X.S., Liu, Q.R., and Wei, L. (2006). Genome-wide *in silico* identification and analysis of cis natural antisense transcripts (cis-NATs) in ten species. Nucleic Acids Res. *34*, 3465–3475.

Imprinted Small Non-coding RNA Genes: Time to Decipher their Physiological Functions

3

Stéphane Labialle, Patrice Vitali and Jérôme Cavaillé

Abstract

Genomic imprinting is a developmentally controlled form of epigenetic regulation that triggers parent-of-origin specific expression of a few mammalian genes. That is, for a given gene, only one of the two parental alleles is transcriptionally active. Over the last several years, many imprinted small regulatory non-coding RNAs (ncRNAs, including microRNAs and box C/D small nucleolar RNAs) have been described at four evolutionarily distinct imprinted loci: the Snurf-Snrpn/Prader–Willi syndrome and Dlk1-Dio3 chromosomal domains, and more recently at the C19MC locus and the Sfmbt2 gene. Remarkably, many imprinted small ncRNA loci are clustered within large arrays formed by tandemly repeated genes of related sequences and processed from long non-coding transcripts extending over hundreds of kilobases. Imprinted gene loci therefore give rise to unique opportunities to address both the impact of small and long non-coding RNA genes on the evolution, expression and function of mammalian genomes. In this chapter, we survey our current understanding of the functions of imprinted ncRNAs, with a particular attention to their suspected involvement in the Prader–Willi disease and higher-brain functions, as well as to their hypothetical contribution in the evolution and/or control of genomic imprinting.

Introduction

Recent advances support the view that most, if not all, of the nucleotides of mammalian genomes are transcribed, at least in a cell- or developmental stage-specific manner (Carninci *et al.*, 2005; Birney *et al.*, 2007; Mattick, 2009). Such widespread transcription, largely unsuspected a decade ago, produces a myriad of non-protein coding RNAs (npcRNAs), also referred to as non-coding RNAs (ncRNAs), whose complexity is at least of the same order of magnitude, if not greater, than that of the repertoire of protein-coding genes. Although our detailed knowledge of their biological function is still extremely limited, the study of ncRNAs is now revolutionizing how we consider genome organization, expression and evolution.

Depending upon their size, ncRNAs are generally ranked into two classes: (i) the long ncRNAs (from 300 to several hundreds of thousands nucleotides) and (ii) the small ncRNAs (~20–300 nt). Long ncRNAs represent an extremely heterogeneous class of molecules (Prasanth and Spector, 2007; Mercer *et al.*, 2009; Ponting *et al.*, 2009) whereas, in contrast, some (but not all) small ncRNAs can be subdivided into families sharing conserved sequences, specific secondary structure motifs, genomic organization or related molecular functions. This is the case for box C/D small nucleolar RNAs (or C/D snoRNAs) and microRNAs (miRNAs) that are the main focus of this chapter.

C/D snoRNAs and miRNAs are part of ribonucleoparticle complexes in which they serve as RNA guides to target other cellular RNAs through specific Watson–Crick base-pairing (Fig. 3.1). Through the formation of canonical RNA duplexes, the vast majority of known mammalian C/D snoRNAs direct site-specific 2′-O-methylation at selected nucleotides onto ribosomal

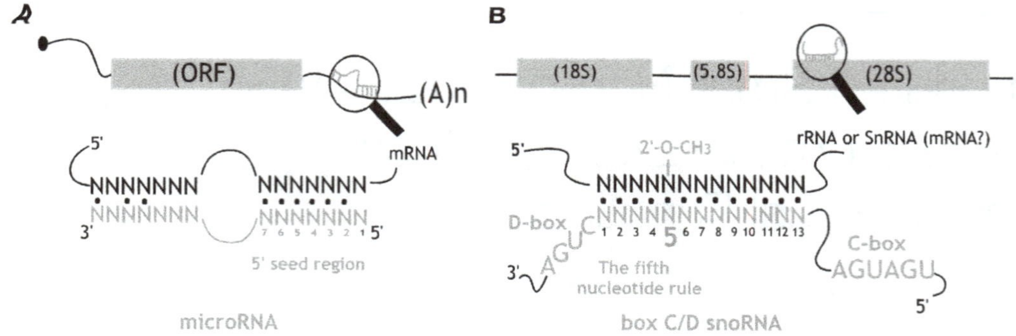

Figure 3.1 miRNAs and box C/D snoRNAs: two classes of RNA guides that fine-tune gene expression at the post-transcriptional level. Most miRNAs (A) and C/D snoRNAs (B) incorporate into ribonucleoparticles and act as antisense RNAs by base-pairing with their cellular RNA targets. In animals, the vast majority of miRNAs recognize mRNAs through imperfect base-pairing that preferentially occurs within the 3′ untranslated region of mRNAs. Note that perfect base-pairing of miRNA nucleotides 2–8 (or 2–7) – the so-called 'seed region' – is required for faithful mRNA targeting and gene silencing. Most animal C/D snoRNAs function as 2′-O-methylation RNA guides through the formation of a perfect ~10–20-nt RNA duplex at each modification site. According to the 'fifth-nucleotide rule,' the position to be sugar-methylated is systematically paired to the fifth nucleotide upstream of the conserved D-box motif (5′-CUGA-3′). Classical RNA targets are mature ribosomal RNA sequences (18S, 5.8S and 28S) and spliceosomal U6 small RNAs, and perhaps mRNA as well (see text for further details).

RNAs and spliceosomal U6 small nuclear RNAs, while a minority also plays some roles in pre-ribosomal RNA processing (Bachellerie *et al.*, 2002; Kiss, 2002; Henras *et al.*, 2004; Reichow *et al.*, 2007) (see also http://www-snorna.biotoul.fr/). miRNAs function in post-transcriptional gene silencing, either by accelerating degradation and/or by inhibiting translation of the targeted mRNAs with which they associate through partial complementarities (Filipowicz *et al.*, 2008) (Fig. 3.1). Mammalian genomes express hundreds (if not thousands) of miRNA genes (http://www.mirbase.org/) thought to target a huge number, perhaps as many as 30–50% of genes. As a consequence, microRNAs play regulatory roles in diverse biological and pathological contexts (Bushati and Cohen, 2007; Garzon *et al.*, 2010).

The functional relationship between ncRNAs and epigenetic control of chromatin or transcriptional regulation is now becoming clearer in mammals. This is particularly well studied in the case of dosage compensation where Xist, a 17 kb-long ncRNA, triggers gene silencing of nearly all genes encoded by one of the two X chromosomes in mammalian females (Lee, 2010). Other examples of large ncRNAs modulating gene expression at the transcriptional level, including through *trans*-acting mechanisms, have been described (or strongly suspected) for many other mammalian gene loci (Nagano and Fraser, 2009; Lee, 2010; Wang *et al.*, 2010; Morris, 2011). A role for small ncRNAs in epigenetic regulation has been attested, principally in yeast and plants where they target nascent transcripts to recruit DNA methylation complexes or chromatin modifying complexes at specific loci (Moazed, 2009). However, our understanding of their roles in mammalian epigenetics is not very well documented yet, even though piwi-interacting RNAs (piRNAs) are very probably players in the *de novo* DNA methylation of transposons in germ cells (Aravin and Bourc'his, 2008) and transcriptional gene silencing (or activation) by small RNAs targeting promoter regions has also been reported (Turner and Morris, 2010). Finally, short RNA species have recently been discovered at gene borders: promoter-associated short RNAs (PASR), transcription initiation RNAs (tiRNAs), promoter upstream transcripts (PROMPTs) and termini-associated short RNAs (TASR). It is still a matter of debate whether these novel species represent functional molecules or simply reflect transcriptional noise (Dinger *et al.*, 2009; Jacquier, 2009).

Specific epigenetic regulatory pathways can also impact on the expression of ncRNA genes, as exemplified by genomic imprinting that confers parent-specific mono-allelic expression at a subset of mammalian gene loci. In other words, for a given gene only one of the two parental alleles is transcriptionally active, while the other one remains in an epigenetic silent state (Reik and Walter, 2001; Bartolomei, 2009). This is undoubtedly one of the most fascinating examples of epigenetic regulation since the paternal and maternal alleles are present in the same nucleus, sometimes genetically identical, yet only one is competent for transcription. This phenomenon concerns mostly protein-coding genes but also ncRNA genes such as C/D snoRNAs, miRNAs and long ncRNAs can also be similarly affected. As a case in point, the maternally expressed H19 gene was one of the very first mammalian sequence encoding an mRNA-like transcript that was initially described as an 'unusual gene whose product may be an RNA molecule' (Brannan *et al.*, 1990). Interestingly, other long imprinted ncRNA genes, also referred to as macroRNAs, such as Airn or Kcnq1ot1, are not only regulated by genomic imprinting (only expressed from the paternally inherited chromosome), but also play a direct role in the control of genomic imprinting by inducing local silencing of some flanking protein-coding genes (Koerner *et al.*, 2009; Kanduri, 2011). Historically, their study has pioneered the functional analysis of long ncRNAs that behave as *cis*-acting silencers over large distances (Sleutels *et al.*, 2002).

Aside from the excellent opportunity it offers to apprehend and decipher the molecular basis that underlies epigenetic regulation, genomic imprinting also raises numerous intriguing questions in terms of mammalian genome organization and evolution. The characterization of Mbp-long domains that contain both maternally and paternally expressed genes shows that imprinted genes are not randomly dispersed but instead occur predominantly in large clusters (Verona *et al.*, 2003); however, isolated imprinted genes may be more prevalent than previously thought, and form imprinted micro-domains (Cowley and Oakey, 2010). At the molecular level, genomic imprinting implies the differential marking of maternal and paternal genomes, notably through DNA methylation and chromatin modification patterns established during germline differentiation and maintained after fertilization. These allele-specific, germline-derived epigenetic marks occur at specific genomic regions, the so-called imprinting centres (ICs), that regulate genomic imprinting throughout the entire chromosomal domain they control (Arnaud, 2010). The evolution of genomic imprinting is a dynamic and probably ongoing process, as imprinted genes did not arise at a single point during therian evolution (Renfree *et al.*, 2009). This mono-allelic mode of expression is rather enigmatic: what could be the evolutionarily advantages of repressing one allele in diploid organisms and thus leading to the risk that the sole remaining expressed allele is a dysfunctional one? The most prevalent explanation to date is the genomic conflict hypothesis, which proposes that genomic imprinting is selected by the different interests of maternal and paternal genomes for the transfer of resources from the mother to its progeny. In short, the paternally inherited genome will favour allocation of maternal resources to its progeny, while the maternally inherited genome will tend to preserve these resources in order to distribute them 'more equally' to the several litters that females might bear during their reproductive life (Moore and Haig, 1991).

This chapter will now describe the two main classes of imprinted small ncRNA genes, the C/D snoRNAs and miRNAs (Royo *et al.*, 2006; Royo and Cavaille, 2008). Recent years have been marked by the unexpected discovery that, in placental mammals, an important number of C/D snoRNAs and miRNAs are produced from imprinted domains, with most of them being encoded at four evolutionarily distinct loci: the Snurf-Snrpn (Prader–Willi syndrome, PWS) domain, the Dlk1-Dio3 domain, the C19MC locus and the Sfmbt2 gene. Perhaps more surprising is the general trend of their genomic organization as large arrays of tandemly repeated small-RNA genes of related sequences most strongly expressed in placenta and brain. These features deserve further consideration and we will review the most recent advances in our knowledge of the biogenesis and function of imprinted small-RNA genes, including in a pathological context such as the Prader–Willi syndrome, and also perhaps in the evolution of genomic imprinting. In addition, we will focus on their poorly characterized host-gene transcripts by considering their hypothetical involvement in some aspects of epigenetic regulation.

Imprinted coding RNA genes at the Prader–Willi syndrome (PWS) region

The imprinted SNURF-SNRPN domain, also called the Prader–Willi syndrome (PWS) region, maps to human chromosome 15q11q13 (mouse chromosome 7C) and extends over 3–4 Mbp in length (Nicholls and Knepper, 2001; Buiting, 2010; Chamberlain and Lalande, 2010). This is the largest imprinted gene cluster discovered so far, with two maternally expressed (UBE3A, ATP10A) and six paternally expressed (MKRN3, MAGEL2, NDN, C15orf2, SNURF-SNRPN) protein-coding genes. Imprinted expression in this domain is rather complex and under the control of an IC located upstream of the SNURF-SNRPN gene, with DNA methylation restricted to the maternal allele (Fig. 3.2). In addition to the protein-coding genes listed above, this cluster in human is also characterized by a huge C/D snoRNA array that includes SNORD107/HBII-436, SNORD64/HBII-13, SNORD108/HBII-437, SNORD116/HBII-85, SNORD115/HBII-52, SNORD109A/HBII-438A and SNORD109B/HBII-438B. Importantly, SNORD116 and SNORD115 are organized as two large clusters of highly related snoRNA gene copies, with 29 and 47 SNORD116 and SNORD115 tandemly repeated genes, respectively (Cavaille *et al.*, 2000; de los Santos *et al.*, 2000; Meguro *et al.*, 2001; Runte *et al.*, 2001). This locus also expresses other long non-coding RNA genes, such as IPW located between the SNORD116 and SNORD115 clusters (Wevrick *et al.*, 1994, 1996), as well as the poorly characterized upstream U-exons and PWRN1/PWRN2 genes transcribed upstream of the IC (Wawrzik *et al.*, 2009). Finally, the 3′ distal part of the SNORD115 cluster may also overlap with the UBE3A gene in the antisense orientation, giving rise to the so-called anti-UBE3A transcript (Rougeulle *et al.*, 1998) (see below).

The biogenesis of these snoRNAs is not entirely understood. Even though earlier studies suggested the existence of a single, huge transcription unit initiating at the IC region both in human and mouse (Runte *et al.*, 2001; Landers *et al.*, 2004), it seems more probable, as we recently proposed, that most of them are processed from at least two main alternatively spliced large non-coding transcripts (Vitali *et al.*, 2010) (see also Fig. 3.2):

1. The SNORD107/SNORD108/64/109A/116 unit widely expressed in many human tissues derives very probably from the 3′-end of the SNURF-SNRPN transcript, which might also include additional exons generated from upstream U-exons and PWRN1 transcripts (Landers *et al.*, 2004; Wawrzik *et al.*, 2009).
2. The SNORD115/SNORD109B unit, and also very probably the anti-UBE3 transcript, constitute another independent transcription unit that is mostly, if not exclusively, expressed in the brain (neuronal-specific expression).

In rodents, all snoRNA genes are mainly, if not exclusively, expressed in the brain, yet belong to two major transcripts (Vitali *et al.*, 2010). It should be noted, however, that *cis*-acting regulatory sequences and *trans*-acting factors that control the tissue-specific expression pattern of these C/D snoRNA arrays are presently unknown. In the following sections, we will discuss the potential roles of box C/D snoRNAs, particularly the SNORD116 and SNORD115 clusters that have been the focus of much attention.

The long quest for the Prader–Willi syndrome gene candidate: Is the C/D SNORD116 cluster the definitive answer?

The imprinted SNURF-SNRPN domain at 15q11q13 chromosomal region is associated with two distinct human neurogenic disorders: the Prader–Willi syndrome (PWS) and Angelman's syndrome (AS). Very briefly, PWS patients display a complex clinical picture with the most prominent features being muscular hypotonia and

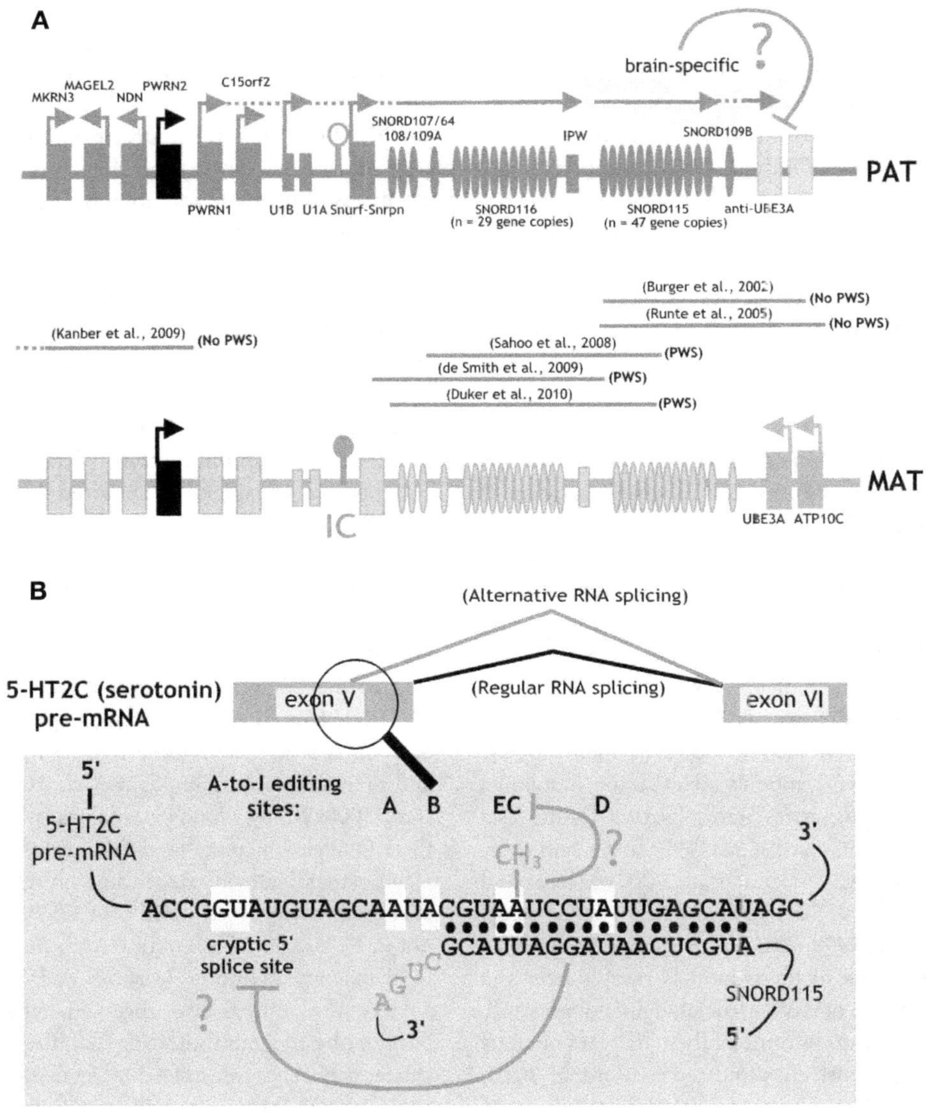

Figure 3.2 Imprinted small-RNA genes at the *Snurf-Snrpn* domain (the Prader–Willi syndrome region). (A) Schematic representation of the ~3–4-Mbp imprinted gene cluster that maps to human chromosome 15q11q13. Box C/D snoRNA and protein-coding genes are symbolized by ovals and boxes, respectively. Arrows indicate the sense of transcription at the active allele with the horizontal dotted line highlighting the notion that snoRNAs probably derive from at least two independent transcription units (see text for further details). The imprinting centre (IC) is indicated by filled and open lollipops (methylated and unmethylated status, respectively). The hypothetical role of the anti-UBE3 transcript in the silencing of the paternal UBE3A allele is also illustrated by a line and a question mark. The relative position of several informative paternally derived micro-deletions found in a few PWS and non-PWS patients is also given. Note that the cartoon is not drawn to scale. PAT, paternal chromosome; MAT, maternal chromosome. (B) Is SNORD115 involved in serotonin 2c (5-HT2C) receptor pre-mRNA processing? SNORD115, formerly called MBII-52 and HBII-52 in mouse and human, respectively, displays an 18 nt-long conserved antisense element within a portion of exon V of 5-HT2C. SNORD115 is predicted to target for 2′-O-methylation one of the five adenosines that undergoes A-to-I editing (A-E edited sites). This occurs at the so-called C-site, which faces the fifth nucleotide upstream of the D-box. This base-pairing lies 13 nucleotides upstream of a cryptic splice site. SNORD115 is thought to modulate RNA editing and/or alternative splicing of 5-HT2C pre-mRNA.

failure to thrive at birth, accompanied by hyperphagia and severe obesity in early childhood. AS patients are characterized notably by microcephaly, severe mental retardation, absence of speech and ataxia (extensively reviewed in Nicholls and Knepper, 2001; Buiting, 2010; Chamberlain and Lalande, 2010).

PWS and AS result from lack of expression of paternally and maternally expressed genes at the 15q11q13 interval, respectively. While it is accepted that AS is caused by loss of function of the UBE3A gene that encodes an E3 ubiquitin ligase, the paternally inherited gene(s) involved in PWS remain(s) uncertain, not to say controversial. One major issue that renders the formal identification of the PWS gene-candidate(s) extremely difficult is the lack of any simple, non-ambiguous genetic alteration affecting a single gene locus. Indeed, approximately 70% of PWS patients have a paternally inherited *de novo* deletion that removes the entire imprinted domain while the remaining 25–30% patients have two maternal chromosomes 15 (maternal uniparental disomy). Moreover, although many mouse knockout (KO) models deleted for individual protein-coding genes were reported [Ndn (Tsai *et al.*, 1999a; Gerard *et al.*, 1999; Muscatelli *et al.*, 2000), Magel2 (Bischof *et al.*, 2007; Mercer and Wevrick, 2009; Schaller *et al.*, 2010), Snrpn-Snurf (Yang *et al.*, 1998; Tsai *et al.*, 1999b; Bressler *et al.*, 2001)], none of these models recapitulates the full spectrum of PWS-associated phenotypes and, more importantly, none of them displays obvious obesity (extensively discussed in Buiting, 2010; Chamberlain and Lalande, 2010).

The recent identification of rare genetic alterations in a few PWS patients has nevertheless provided important insights in the aetiology of the disease. First, several balanced translocations that disrupt the SNURF-SNRPN locus, and more precisely its downstream genomic region, have been described in some PWS patients (Schulze *et al.*, 1996; Sun *et al.*, 1996; Conroy *et al.*, 1997; Wirth *et al.*, 2001; Gallagher *et al.*, 2002; Schule *et al.*, 2005). Given that the expression of protein-coding genes mapping upstream of the SNURF-SNRPN gene does not seem to be significantly affected, these clinical studies have identified a critical minimal region overlapping the box C/D snoRNA clusters. Second, a patient lacking the NDN, MKRN3 and MAGEL2 genes, but without any apparent typical PWS features, was also reported (Kanber *et al.*, 2009b), indicating that the above-mentioned upstream protein-coding genes are unlikely to contribute significantly to the disease. Third, and perhaps more tellingly, four PWS patients have been reported with a paternally inherited micro-deletion (~175–236 kb in length) overlapping the entire SNORD116 cluster as well as some neighbouring single-copy snoRNA genes, including in one case half of the SNORD115 cluster (Sahoo *et al.*, 2008; de Smith *et al.*, 2009; Duker *et al.*, 2010; M. Tauber and E. Bieth, personal communication). Studies of such patients have considerably narrowed the minimal critical region and highlight an important contribution for the SNORD116 cluster (Fig. 3.2). Two paternally inherited deletions of the SNORD115 cluster have been found in non-PWS patients, thus excluding it as a major player (Burger *et al.*, 2002; Runte *et al.*, 2005). The functional significance of SNORD116 in PWS was further confirmed by two KO mouse models generated by the Brosius and Francke groups (Skryabin *et al.*, 2007; Ding *et al.*, 2008, 2010). Upon paternal transmission of the KO allele, mice are growth-retarded and display postnatal lethality depending on the genetic background (reported in 15% of C57BL/6, but not in FVB/N, strains). In one study, in addition to deficiency in motor learning and increased anxiety, the authors also observed hyperphagia but not obesity, thus indicating that the snord116 cluster may play a role in feeding regulation (Ding *et al.*, 2008). These mice will constitute valuable models to further explore the physiological roles of the SNORD116 cluster in human.

Taken together, these studies emphasize the importance of the SNORD116 cluster in the aetiology of PWS, yet they do not exclude the contribution of additional genes within the domain. For instance, the expression status of SNORD115, which is restricted to the brain in humans, could not be assayed in patients with the micro-deletion centred around the SNORD116 cluster. Thus, we cannot formally rule out the possibility that such micro-deletions also impact (directly or indirectly) on the brain-specific expression of the downstream SNORD115 genes, possibly by removing *cis*-acting elements. As a consequence, whether the SNORD115 cluster contributes to the disease along with SNORD116 remains an open question (see also 'RNA modification, RNA splicing and RNA editing at the serotonin (5-HT2c) pre-mRNA: a "*ménage à trois*"?', below). The rationale for identifying SNORD116 genes as PWS candidate-genes is mainly justified by the fact that they represent the most conserved, functional sequences within the minimal critical region. Genetically speaking, we cannot exclude the possibility that these microdeletions remove additional still unidentified, or less studied gene loci. In this respect, we should not minimize roles for the large, spliced snoRNA host-gene transcripts that may represent functional RNA entities (see 'A role for snoRNA-host transcripts in epigenetic regulatory pathways?', below). Efforts designed to decipher the mode of action of SNORD116 (and its host gene) as well as the regulatory pathways that control differential expression of SNORD115 and SNORD116 are now essential to provide unambiguous clues regarding the complex and intricate genotype–phenotype relationships that underly PWS.

RNA modification, RNA splicing and RNA editing at the serotonin (5-HT2c) pre-mRNA: a '*ménage à trois*'?

Imprinted box C/D snoRNAs have been termed 'orphan snoRNAs' to illustrate the fact that, in contrast to the vast majority of bi-allelically expressed C/D snoRNAs encoded elsewhere in the genome, they lack obvious base-pairing potentials to ribosomal RNAs or U-spliceosomal RNAs, the two classical RNA substrates for C/D small RNAs. However, one of the PWS-encoded snoRNAs, the brain-specific SNORD115, constitutes an outstanding exception: it possesses a 18 nt-long conserved region fully complementary to a brain-specific mRNA encoding the serotonin (5-HT2C) receptor (Cavaille *et al.*, 2000). Remarkably, this antisense element is positioned upstream of the D-box: accordingly, SNORD115 is expected to function as a 2′-O-methylation RNA guide to modify the 5-HT2C mRNA, the first and only convincing mRNA target for a box C/D snoRNA described so far (Fig. 3.2).

Through this complementary sequence, SNORD115 could target a functionally important segment of 5-HT$_2$C embedded within exon V and known to undergo A-to-I RNA editing at five positions (designated as sites A-E, Fig. 3.2). Inosine is decoded as guanosine by the ribosomes. Consequently, this base modification carried out at the pre-mRNA level changes the coding potential of the 5-HT$_2$C gene, resulting in receptor isoforms with different pharmacological properties, the fully edited receptor displaying reduced constitutive activity and decreased G-coupling efficiency (Burns *et al.*, 1997; Niswender *et al.*, 1999). Remarkably, the site positioned 5 nt upstream of the D-box and predicted to be targeted for 2′-O-ribose methylation by SNORD115 is precisely the C-site whose editing was previously shown, along with that of the E-site, to play a key role in regulating serotonergic signal transduction (Burns *et al.*, 1997; Niswender *et al.*, 1999; Wang *et al.*, 2000). In addition, this exon V is also subject to alternative RNA splicing through the use of a cryptic splice site located 13 nt upstream from SNORD115 base-pairing site (see Fig. 3.2). Processing at this alternative 5′ donor splice site induces exon skipping and leads to a frame shift, resulting in the production of mRNA harbouring a premature stop codon (Canton *et al.*, 1996). This gives rise to a truncated serotonin receptor whose function, if any, remains unknown.

Several lines of evidence strengthen the view that SNORD115 has the potential to regulate 5-HT2C pre-mRNA processing, either by inhibiting RNA editing at the C-Site (Vitali *et al.*, 2005), or by promoting the use of the regular 5′ alternative splice sites (Kishore and Stamm, 2006). Although these two studies provide the first proofs of principle supporting the involvement of SNORD115 in fine-tuning 5-HT2C pre-mRNA processing, convincing data from endogenously expressed 5-HT2C gene in biologically relevant systems are still lacking: the studies cited above were performed mostly through the use of transiently transfected, genetically modified 5-HT2C minigenes, which were either artificially targeted to the nucleolus or whose splicing was optimized by mutagenesis of splice sites. Moreover, these two studies may appear contradictory since RNA editing was only inhibited by SNORD115 when 5-HT2C substrates were transcribed within the nucleolus (Vitali *et al.*, 2005), while no specific nuclear/nucleolar compartmentalization was reported for the involvement of SNORD115 in 5-HT2C alternative RNA splicing (Kishore and Stamm, 2006). The functional characterization of protein partners that specifically bind SNORD115 and of these putative truncated snoRNA forms should help clarify this complex and important issue (Kishore *et al.*, 2010; Soeno *et al.*, 2010).

Interestingly, a connection between lack of SNORD115 and 5-HT2C pre-mRNA editing efficiency was reported in a study that addressed the behaviour of a PWS mouse model where all paternally expressed genes are no longer expressed (Doe *et al.*, 2009). These SNORD115 non-expressing mice exhibit defects in specific 5-HT2C-related behaviour as well as an increase in overall levels of RNA editing across the five sites, and not specifically at the C-site as one could have predicted from the SNORD115/5-HT2C base-pairing model. Alternative 5-HT2C RNA splicing remains unchanged in these mice, in contrast to a previous report (Kishore and Stamm, 2006). Another link between altered 5-HT2C expression and PWS was evoked in a study that examined the phenotypic consequences of a mouse model in which only the fully edited isoform of 5-HT2C is expressed. Remarkably, these mice recapitulated some characteristics described in PWS and other mouse models: failure to thrive, neonatal muscular hypotonia and post-weaning hyperphagia (Morabito *et al.*, 2010).

Although these studies highlight the potential importance of 5-HT2C RNA editing (and its fine-tuning) in PWS, other authors did not report any significant and reproducible changes in the editing of 5-HT2C in human brain samples (Glatt-Deeley *et al.*, 2010), perhaps reflecting major differences in RNA editing efficiency between individuals and/or across different brain regions. In this regard, it is noteworthy that only a limited number of 5-HT2C clones were sequenced in all studies reported so far, thus limiting considerably any statistical analysis of RNA editing profiles observed in PWS-derived tissues relative to SNORD115-expressing control tissues. More exhaustive experiments involving deep sequencing technologies and a large collection of brain regions dissected at various developmental stages are now required to challenge the predicted role of SNOD115 in the control of 5-HT2C pre-mRNA editing. In addition, formal mapping of the expected 2′-O-methylation onto endogenous 5-HT2C mRNAs, which is currently technically challenging, will also undoubtedly represent another major milestone in our understanding of SNORD115 function and its mode of action.

Imprinted non-coding RNA genes at the Dlk1-Dio3 chromosomal interval

The imprinted Dlk1-Dio3 domain occupies ~1 Mbp distal chromosome 12 (12F1) in the mouse and on human 14q32 (extensively reviewed in da Rocha *et al.*, 2008). This eutherian-specific cluster contains three paternally expressed protein-coding genes: Delta-like 1 (Dlk1)/pref1, Retrotransposon-like 1 (Rtl1)/Peg11 and Deiodinase iodothyronine type 3 (Dio3). This imprinted domain also contains multiple long and short maternally expressed ncRNA genes: the Gtl2/Meg3 mRNA-like transcript (Schuster-Gossler *et al.*, 1998; Miyoshi *et al.*, 2000), a large array of box C/D snoRNAs processed from Rian (Cavaille *et al.*, 2001, 2002) and numerous miRNAs organized into two clusters within Mirg and anti-Rtl1 transcripts, respectively (Seitz *et al.*, 2003, 2004) (Fig. 3.3). The details of the biogenesis of these non-coding transcripts remain largely incomplete, and it is not totally clear whether they arise from a single large transcription unit as initially assumed (Lin *et al.*, 2003; Seitz *et al.*, 2003, 2004; Tierling *et al.*, 2006), or whether some of them might instead be generated from distinct promoters, as recently evidenced for miR-127 and miR-433 (Song and Wang, 2008a, b) or miR-453 (Caiment *et al.*, 2010).

Imprinted expression at the Dlk1-Dio3 interval is governed by a short intergenic differentially methylated region (IG-DMR) that acquired its methylation imprint in the male germline (Takada *et al.*, 2002; Lin *et al.*, 2003). The correct dosage of gene expression plays a pivotal role in embryonic development (da Rocha *et al.*, 2008). Indeed, earlier studies showed that both parental chromosomes 12, and more particularly its distal portion, are required for full-term survival (Cattanach *et al.*,

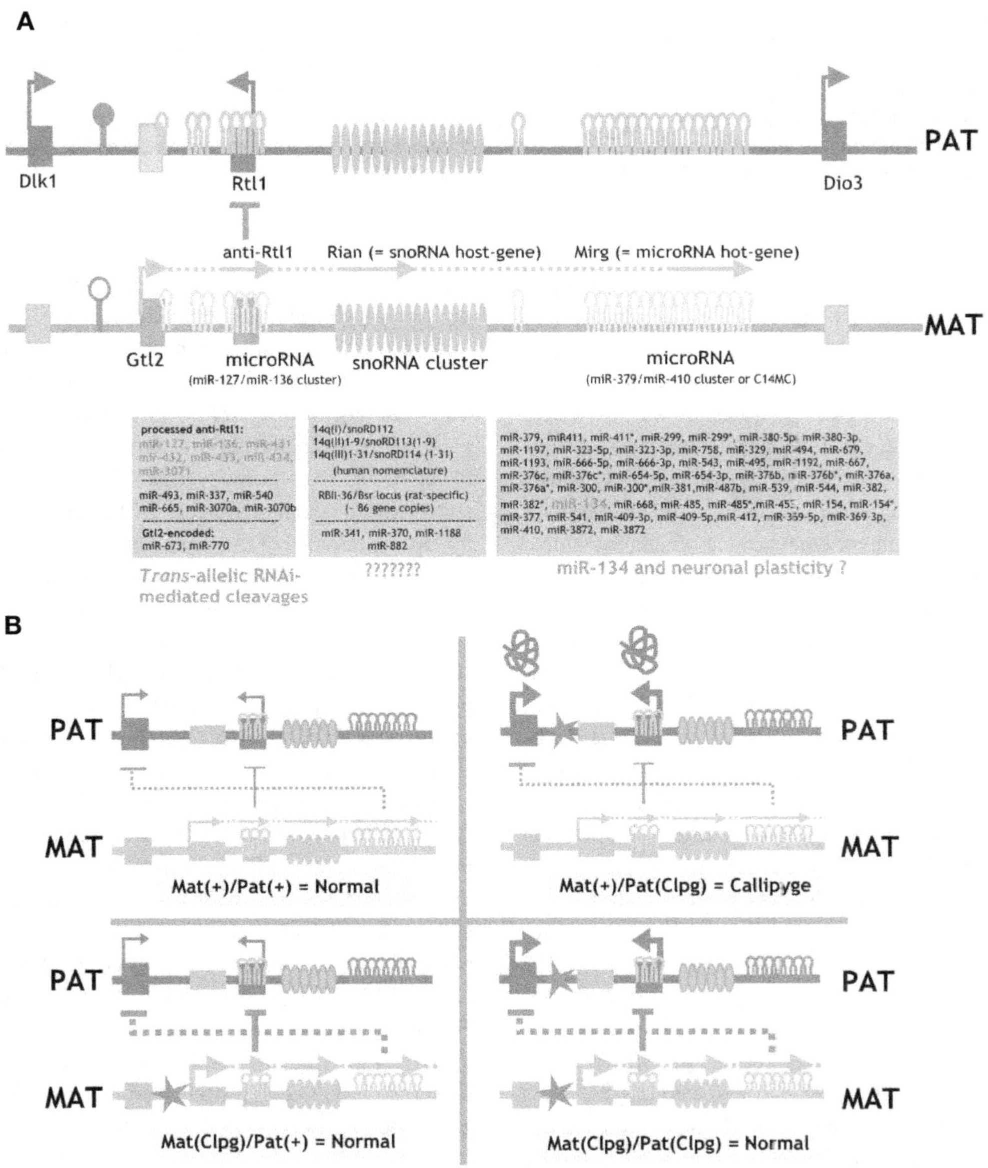

Figure 3.3 Imprinted small-RNA genes at the *Dlk1-Dio3* domain. (A) Schematic representation of the ~1-Mbp imprinted Dlk1-Dio3 gene cluster mapping to the distal end of mouse chromosome 12 (human 14q32). Protein-coding genes are symbolized by boxes while microRNAs and C/D snoRNA genes are depicted by stem loops and ovals, respectively. The horizontal dotted line illustrates the notion that *Gtl2*, anti-Rtl1, snoRNAs and miRNAs may belong to the same transcription unit; *Rian* and *Mirg* correspond to two poorly characterized ncRNAs from which some C/D snoRNAs and microRNAs are processed. Other symbols are as described in Fig. 3.2. miRNAs generated from the Dlk1-Dio3 domain are listed in the grey boxes (based on http://www.mirbase.org/). However, these miRNA sequences may not all represent functional entities. For the C/D snoRNA gene cluster, only the standard human names are given. Note that the rat snoRNA gene cluster harbours a specific array of ~100 tandemly repeated snoRNA genes, the RBII-36 locus (also referred to as the Bsr locus). (B) Polar over-dominance phenomenon at the sheep Dlk1-Dio3 domain. The Callipyge (Clpg) A-to-G point mutation, denoted by the star, up-regulates gene expression in a *cis*-acting manner

Continued overleaf

Figure 3.3 (continued) without affecting genomic imprinting. Thin and thick arrows denote normal and aberrant up-regulated expression, respectively). This model, first proposed by M. Georges's group, states that an interaction occurs *in trans* between reciprocally imprinted genes and posits that, in Clpg/Clpg individuals, increased expression of one or several maternally expressed gene(s) – which we hypothesize here to be miRNAs – antagonizes the up-regulation of the paternally expressed protein-coding genes. This results in a decreased production of the Dlk1 or Rtl1 proteins whose overexpression is believed to lead to the Callipyge phenotype (see text for further details; Georges *et al.*, 2003).

1993). In other words, embryos with two copies of maternal Chr. 12 (matUPD12) or two copies of paternal Chr.12 (patUPD12) die perinatally or during late gestation, respectively, and they display several pathological manifestations associated with placenta, cartilage, skeletal muscle and bone (Georgiades *et al.*, 2000; Tevendale *et al.*, 2006). Moreover, maternal inheritance of the IG-DMR deletion in heterozygous mice affects imprinted expression in the entire domain, with the three protein-coding genes being expressed from both alleles while non-coding RNA genes are silenced (Lin *et al.*, 2003, 2007). Such a maternal-to-paternal epigenetic switch, observed in the embryo and not in the placenta, causes perinatal lethality, providing further evidence that dosage imbalance of one or several genes encoded in the Dlk1-Dio3 domain is/are responsible for the developmental abnormalities seen in patUPD12. Studies of KO mice deleted for individual protein-coding genes showed that each of them probably contributes to the lethality phenotypes observed in UPD12 animals, as well as to the postnatal growth and metabolism homeostasis observed in surviving animals (Moon *et al.*, 2002; Hernandez *et al.*, 2006; Sekita *et al.*, 2008); reviewed in (da Rocha *et al.*, 2008). However, the biological function of imprinted non-coding RNA genes remains poorly understood, and up to now only Gtl2 and the miR-127–miR-136 cluster have been inactivated in mice through homologous recombination.

Targeted deletion at Gtl2 gene reveals complex *cis*- and *trans* effects at the Dlk1-Dio3 domain

Two recent studies reported the genetic ablation of the Gtl2 ncRNA gene (Takahashi *et al.*, 2009; Zhou *et al.*, 2010). In the first study, Takahashi *et al.* removed exons 1–5 and the Gtl2 DMR. They observed that heterozygous mutants with a deletion inherited from their mother died within 4 weeks after birth, probably because of hypoplastic pulmonary alveoli and liver necrosis. Surprisingly, upon paternal transmission, heterozygous animals were growth retarded and also died perinatally. This was largely unexpected since Gtl2 is paternally repressed. Even more surprising was the observation that homozygous Gtl2 KO mice survived and appeared normal. Although the basis for the inheritance of these phenotypes is far from clear, it is noteworthy that this deletion also affects significantly the expression of the surrounding genes: upon maternal inheritance, miRNAs and snoRNAs are down-regulated, while they are up-regulated upon paternal transmission with a concomitant decrease in Dlk1 and Rtl1 expression. In the second study, Zhou *et al.* removed exons 1–5 and the promoter region of Gtl2. They reported that heterozygous animals with maternal transmission, as well as homozygous mutants, died perinatally and displayed embryonic skeletal muscle defects, while animals carrying the paternal deletion were healthy, in contrast to the other Gtl2 KO mouse model already discussed. Furthermore, these authors also noticed dramatic misregulation of neighbouring genes upon maternal transmission, with protein-coding and ncRNA genes being up-regulated and down-regulated, respectively. Remarkably, these authors also found that, in embryos with the maternally inherited Gtl2 deletion, Dlk1 became biallelically expressed concomitantly with methylation of the maternal allele of IG-DMR. This leaves open the possibility that expression of Gtl2 (and also perhaps of the other non-coding RNA genes) may be required to maintain IG-DMR in an unmethylated active state on the maternal allele (Zhou *et al.*, 2010). Altogether, given that imprinted gene expression pattern at Dlk1-Dio3 is significantly altered in these two Gtl2 KO models, it is not easy to assign any specific function to Gtl2. Nevertheless, these studies not only reinforce the notion that gene dosage at Dlk1-Dio3 domain is critical for correct development, they also suggest the involvement of complex *cis-* and *trans*-interactions between the two parental alleles, perhaps reminiscent of the previously documented phenomenon of polar overdominance that underlies the Callipyge phenotype in sheep (see 'Polar overdominance phenomenon at the sheep Dlk1-Dio3 domain: do microRNAs shape nice buttocks?', below).

Trans-allelic RNAi-mediated gene silencing at the sense/antisense (Rtl1/miR-127-miR-136) locus regulates the feto-maternal interface in placenta

The maternally expressed miR-127–136 cluster is transcribed in the opposite orientation relative to Rtl1 gene transcription, which only occurs on the paternal chromosome (Seitz *et al.*, 2003) (see Fig. 3.3). Because they are positioned antisense to the Rtl1 gene, these miRNAs have perfect complementary binding sites within the Rtl1 mRNA and, as a consequence, they represent one of the very few examples of animal miRNAs that behave like siRNAs to trigger RNA cleavage via an RNAi-mediated pathway (Davis *et al.*, 2005). The other known example described so far involves miR-196, which silences HOXB8 mRNA (Yekta *et al.*, 2004).

The Rtl1 gene, also called paternally expressed gene 11 (Peg11), shares homology with the Sushi/Ichi-like retroelement family (Charlier *et al.*, 2001b; Lynch and Tristem, 2003; Youngson *et al.*, 2005) and provides a nice illustration of a novel cellular function acquired from transposons (i.e. exaptation) that played a major role in the evolution of placenta (Sekita *et al.*, 2008; Kaneko-Ishino and Ishino, 2010). Ishino's group generated and compared two classes of heterozygous animals that had inherited the Rtl1 KO allele, either from their father or from their mother (null for Rtl1 and miR-127–136 cluster expression, respectively). Upon paternal transmission, pups were growth-retarded and, after the F3 generation in a C57BL/6 background, died perinatally most probably because of severe placenta malformations related to defects in the maintenance of fetal capillaries that result in impaired nutritional transfer. Upon maternal inheritance of the Rtl1 KO allele, pups displayed growth retardation after weaning and, after the F6 generation, neonatal lethality predominated. Although these pups grew normally till they reached term, they also displayed placental defects in the form of 150% placentomegaly. Remarkably, and in agreement with their involvement in Rtl1 silencing, loss of expression of the miR-127–miR136 cluster upon maternal transmission of the Rtl1 (KO) allele resulted in ~3-fold overexpression of the paternal Rtl1 mRNA. Therefore, both the lack of function of Rtl1 and its overexpression (due to loss of mir-127-miR-136 function) impact on placenta development and function during late gestation (Sekita *et al.*, 2008).

The fact that maternally expressed microRNAs antagonize the function of a paternally expressed gene provides a strong argument for the conflict – or kinship – theory which states that genomic imprinting evolved from a conflict between the maternal and paternal genomes regarding the allocation of parental resources which, at least during the prenatal period, come predominantly from the pregnant mother via the placenta (Moore and Haig, 1991). Indeed, the theory predicts that paternally expressed genes (here Rtl1) act to increase the extraction of resources from the mother, while maternally expressed genes (here the mir-127–miR-136 cluster) function to limit the transfer of maternal resources (Davis *et al.*, 2005; Lewis and Redrup, 2005).

Polar overdominance phenomenon at the sheep Dlk1-Dio3 domain: do microRNAs shape nice buttocks?

The Callipyge (Clpg) phenotype in the sheep is an inherited postnatal muscular hypertrophy that results in the increase in the proportion and diameter of fast-twitch muscles fibres, predominantly evident in the hind legs of the animals, hence its name (Georges and Cockett, 1996; Georges *et al.*, 2003, 2004). The Clpg mutation maps to the imprinted Dlk1-Dio3 domain (distal chromosome 18 in sheep) and corresponds to a single base change (an A-to-G transition) within a short conserved region in the Dlk1-Gtl2 intergenic region (Freking *et al.*, 2002; Smit *et al.*, 2003). The inheritance of the Callipyge phenotype is rather unusual and deserves particular emphasis:

only heterozygous animals that inherit the Clpg mutation from their father (Clpg^{pat}) display the phenotype, while upon maternal transmission (Clpg^{mat}) or even in homozygous animals (Clpg$^{pat/mat}$), the phenotype is not expressed. This non-Mendelian mode of inheritance has been referred to as polar over-dominance (Cockett *et al.*, 1996; Georges and Cockett, 1996; see also Fig. 3.3).

The Callipyge mutation enhances imprinted gene expression *in cis*, most probably by interfering with the function of a long-range muscle-specific regulatory element since overexpression occurs in postnatal skeletal muscle (Charlier *et al.*, 2001a; Takeda *et al.*, 2006). Indeed, Clpg^{pat} animals are characterized by an increase in the expression of paternally expressed protein-coding genes encoded at Dlk1-Dio3, notably Dlk1 and Rtl1, whereas all maternally expressed non-coding RNA genes are up-regulated in Clpg^{mat} animals. In Clpg$^{pat/mat}$ heterozygotes, both the paternally and maternally expressed transcripts are aberrantly regulated. Based on these gene expression profiles, Clpg is therefore assumed to be the consequence of a gain of function of either Dlk1 or Rtl1, or both (Georges *et al.*, 2003; Bidwell *et al.*, 2004; Davis *et al.*, 2004, 2005; Kim *et al.*, 2004; Perkins *et al.*, 2006; White *et al.*, 2008; Fleming-Waddell *et al.*, 2009; Byrne *et al.*, 2010). In order to rationalize the lack of phenotype of Clpg$^{pat/mat}$ heterozygotes, which display up-regulation of Dlk1 and Rtl1 mRNA, M. Georges and colleagues proposed an appealing model involving *trans* interactions between reciprocally imprinted genes at the Dlk1-Dio3 interval: one or several maternally expressed non-coding RNA gene(s) would act as *trans*-acting repressor(s) to interfere with the function of Dlk1 and/or Rtl1 gene products (Georges *et al.*, 2003).

The regulation of Rtl1 mRNA by the antisense miR-127–miR-136 cluster comes as an evident candidate as the pattern of expression of Rtl1 and these miRNAs in Callipyge sheep fulfil the criteria for such predicted *trans*-allelic silencing (Seitz *et al.*, 2003; Davis *et al.*, 2005). This proposal is made all the more convincing by the recent identification of the RTL1 protein in Callipyge skeletal muscle, which correlates with emergence of the phenotype (Byrne *et al.*, 2010). Dlk1 may also be considered as a potential mediator of the Clpg phenotype since its ectopic expression from a myosin light-chain 3F promoter causes muscular hypertrophy in mice (Davis *et al.*, 2004). It should be noted, however, that no muscle hypertrophy was reported in another transgenic mouse model where Dlk1 expression was driven by its endogenous regulatory elements (da Rocha *et al.*, 2009). Nevertheless, another intriguing observation substantiates a potential role for Dlk1 in Clpg: in Clpg$^{pat/mat}$ animals expressing increased levels of Dlk1 mRNA but not the Clpg phenotype, the DLK1 protein is not detected. It was thus rapidly seen that up-regulation of miRNAs in Clpg$^{pat/mat}$ heterozygotes might counteract post-transcriptionally the phenotypic consequences of up-regulation of Dlk1 mRNA (Davis *et al.*, 2004; Seitz *et al.*, 2004). In agreement with this notion, a recent exhaustive analysis of the sheep microRNA repertoire expressed in muscle coupled to bio-informatics revealed suggestive, but not experimentally validated, evidence that some miRNAs encoded within the Dlk1-Dio3 domain could account for the putative translational down-regulation of Dlk1 observed in Clpg$^{pat/mat}$ animals (Caiment *et al.*, 2010). While decisive data are still lacking and although we cannot formally rule out a role for Gtl2, box C/D snoRNAs or any other still unidentified genes in the domain, miRNAs from Dlk1-Dio3 constitute the most obvious candidates for mediating the direct or indirect *trans-allelic* inhibition effects that are predicted to occur in Clpg$^{pat/mat}$ muscles, either by acting on Rtl1 or Dlk1 mRNAs – or both.

Do miR-134 and the miR-379–miR-410 cluster play a role in higher brain functions?

Changes in the strength of connections between neurons, involving dendritogenesis and spine development, are widely recognized as the primary mechanism that underlies learning and memory formation and storage (Flavell and Greenberg, 2008). In a first series of studies performed by Schratt and colleagues in primary cultures of dissociated neurons, a member of the miR-379–miR-410 cluster, miR-134, was shown to regulate negatively the size of dendritic spines (Schratt *et al.*, 2006; Tai and Schuman, 2006). miR-134 is particularly enriched in the synapto-dendritic compartment where it is believed to repress locally the translation of Limk1 mRNA, which encodes a protein kinase involved in the control of actin filament dynamics in dendrites. Remarkably, treating neurons with brain-derived neurotrophic factor (BDNF) inhibits miR-134-mediated translational repression. It was hypothesized that, upon synaptic stimulation, BDNF is released and, through still poorly understood mechanisms that might involve the mTOR signalling pathways, triggers spine growth through the conversion of 'dormant Limk1 mRNA' to 'active Limk1 mRNA' at the spines (Schratt *et al.*, 2006). Of note is the fact that neuronal activity, as judged by BDNF treatment or membrane depolarization by low concentrations of KCl, also increases the expression of the miR-379–miR-410 cluster, including miR-134. This transcriptional regulation is mediated by Myocyte Enhancing factor 2 (MEF2) which binds upstream of the cluster (Fiore *et al.*, 2009). Interestingly, expression of at least three microRNAs – miR-134, miR-329 and miR-381 – is necessary for activity-dependent dendritic outgrowth of hippocampal neurons *in vitro* and one of the identified miR-134 mRNA targets encodes for the translational repressor Pumillo 2. Therefore, BDNF increases miR-134 expression required for dendritic arborization, but also inhibits its function within dendrites to promote spine growth, thus revealing sophisticated and complex fine-tuning of neuronal activities. Confirming a significant role for miR-134 in neuronal regulation, a recent study proposed a model whereby the SIRT1 histone deacetylase modulates synaptic plasticity by silencing the miR-134 gene through a complex containing the YY1 transcription factor. This results in the increased expression of another miR-134 mRNA target, the cAMP response-binding protein (CREB), which plays a pivotal role in the formation of long-lasting memory (Gao *et al.*, 2010). Finally, miR-134 was also recently found to target, among others, transcripts of the BMP antagonist Chrdl-1 gene and of the microtubule modulator Dcx gene to regulate embryonic neuronal maturation, neuron migration and cortical development (Gaughwin *et al.*, 2011). The imprinted mir-379–miR-410 cluster (also referred to as C14MC in human), with miR-134 as a leader, thus now emerges as a key player in cerebral development and function in eutherians. Defects in neuronal circuitry are involved in a growing list of cerebral deficiencies. As a case in point, the expression of miR-134 and other members of the miR-379–miR-410 cluster is altered in a mouse model of schizophrenia exhibiting defective dendritic size and complexity (Stark *et al.*, 2008).

A role for imprinted microRNA genes in the biology of ES and iPS cells?

miRNAs have been shown to play varied roles in the maintenance of self-renewal and pluripotency of mouse embryonic stem (ES) cells (reviewed in Mallanna and Rizzino, 2010; Martinez and Gregory, 2010). miR-134, whose expression is up-regulated upon retinoic acid treatment, promotes the differentiation of mouse ES cells to the ectodermal lineage *in vitro*, most probably by fine-tuning the expression of Sox2, LHRH1 and Nanog mRNAs (Tay *et al.*, 2008a,b).

In addition, recent work has also highlighted a potential contribution of microRNAs during the reprogramming of somatic cells into induced pluripotent stem (iPS) cells. This process requires the ectopic expression of a selected set of transcription factors, notably Oct4, sox2, c-myc and Klf4, the so-called 'Yamanaka cocktail' (Takahashi and Yamanaka, 2006). Although relatively simple in appearance, transcription factor-based reprogramming is nevertheless relatively inefficient and involves an intricate superposition of genetic, epigenetic and stochastic events (Hanna *et al.*, 2009). Considerable efforts worldwide bear on better understanding the molecular and cellular differences between iPS and ES cells, since most generated iPS cells appear trapped in partially reprogrammed states (Mikkelsen *et al.*, 2008). In this context, it is of considerable interest that transcriptome analysis of genetically matched mouse ES and iPS cells showed that the maternally expressed Gtl2, box C/D snoRNAs/Rian and the microRNA/Mirg genes are silenced in most, but not all, iPS cells, because of aberrant DNA methylation patterns that occur *in vitro* during the reprogramming process

(Liu *et al.*, 2010; Stadtfeld *et al.*, 2010). More importantly, iPS clones that retain their ability to faithfully express the ncRNA genes cited above generate viable 'all-iPS cells' mice after injection into tetraploid blastocysts, and thus display full developmental potential. Overall, this study puts forward the notion that normal gene expression at the Dlk1-Dio3 domain, and particularly that of the maternally expressed ncRNA genes, is a prerequisite to obtain fully pluripotent iPS clones. Through unbiased deep sequencing approaches, another independent study also reported that correct expression of miRNA genes encoded at the Dlk1-Dio3 domain correlates with the pluripotency levels of ES and iPS cells. The authors further hypothesized that some of these miRNAs might regulate pluripotency by targeting the Polycomb repressive complex 2 (PRC2) (Liu *et al.*, 2010). Although the biological meaning of these observations still remains unclear, the correct expression of ncRNA genes at Dlk1-Dio3, and perhaps that of its miRNAs, can be considered as a key signature for the isolation of fully reprogrammed iPS cells (Liu *et al.*, 2010; Stadtfeld *et al.*, 2010).

Imprinted microRNA genes at the C19MC and Sfmbt2 loci

More recently, two additional large imprinted microRNA clusters have been discovered: the C19MC and the Sfmbt2 clusters (also named C2MC). C19MC is a 100 kbp-long, primate-specific miRNA gene cluster positioned on human chromosome19q13.41, upstream of the stem-cell specific miR-371–miR-373 cluster. It contains 46 tandemly repeated microRNA genes of related sequences that are flanked by antisense-oriented Alu elements (Alus) (Bentwich *et al.*, 2005). A subsequent in-depth analysis showed that many C19MC miRNAs are intron-encoded and processed by the DGCR8-Drosha (Microprocessor) complex from a previously unidentified, non-protein-coding Pol II transcript (C19MC-HG) which is mainly, if not exclusively, expressed in the placenta from only the paternally derived allele (Bortolin-Cavaille *et al.*, 2009; Noguer-Dance *et al.*, 2010).

The rodent-specific C2MC microRNA gene cluster in the proximal region of mouse chromosome 2 represents the largest miRNA cluster identified to date in the mouse (Calabrese *et al.*, 2007; Landgraf *et al.*, 2007). Its genes are grouped in intron 10 of the Polycomb group (PcG) Sfmbt2 gene [Sex comb on midleg (drosophila)-like with four MBT domains 2] preferentially expressed from the paternal allele in early embryos and in later-stage extraembryonic tissues (Kuzmin *et al.*, 2008). Thus, although not definitively proven, the intron-encoded C2MC cluster is also expected to be maternally repressed as well (Fig. 3.4).

Due to their recent discoveries and their poor conservation through evolution, the functional characterization of these miRNA clusters remains scarce. However, some interesting clues as to their function are beginning to be reported. Apart from being expressed mostly in placenta, C19MC has been found to be frequently amplified in certain aggressive brain tumours (Li *et al.*, 2009) and is translocated in a subgroup of thyroid adenomas (Rippe *et al.*, 2010). Moreover, one of the C19MC-encoded microRNA, miR-520c, was shown to act as a tumour-promoting microRNA when ectopically expressed in a human breast-tumour cell line (Huang *et al.*, 2008). These results suggest that some C19MC-encoded microRNAs might play important roles in cancer. C19MC miRNAs are also suspected to regulate important cellular functions in human ES cells since their expression is rapidly down-regulated upon cell differentiation (Bar *et al.*, 2008; Laurent *et al.*, 2008; Tzur *et al.*, 2008; Ren *et al.*, 2009; Stadler *et al.*, 2010). Remarkably, several C19MC- and C2MC-encoded microRNAs (the miR-519–miR-520 and miR-467 families, respectively) share a common 'AAGUGC' seed sequence also found in other ES-specific microRNAs residing in the miR-290–295 and miR-302b–367 clusters. Moreover, the same seed sequence is also harboured within the large miR-430 family whose functions are crucial for clearance of maternal transcripts during early embryogenesis in zebrafish (Giraldez *et al.*, 2005; Laurent *et al.*, 2008; Wang *et al.*, 2008). Because the seed region of microRNAs plays a critical role in mRNA targeting and silencing, these

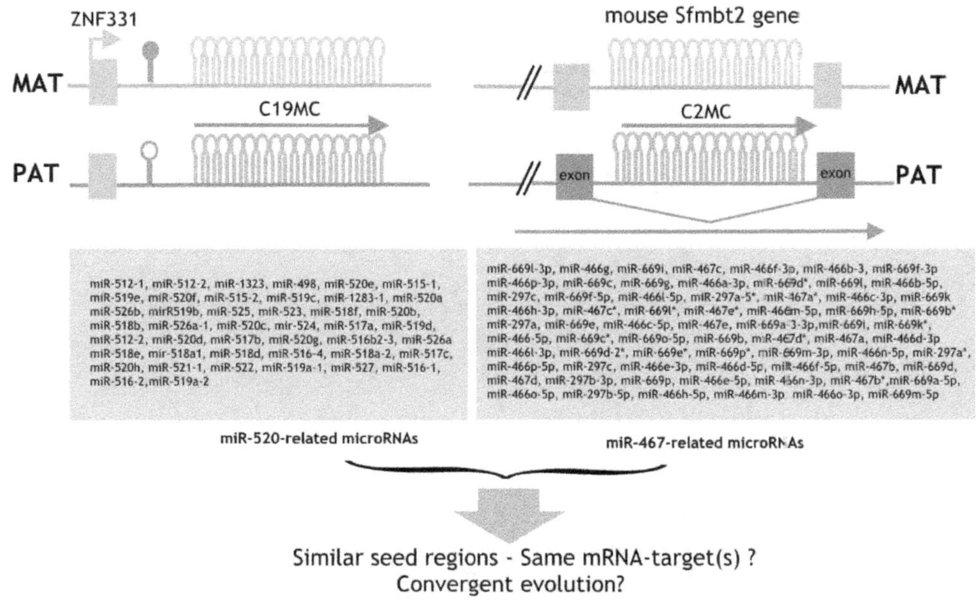

Figure 3.4 C9MC and C2MC: two poorly conserved imprinted microRNA gene clusters. Schematic representation of the 100-kb-long C19MC (primate-specific, left) and C2MC (rodent-specific, right) embedded within intron 10 of the paternally expressed Sfbmt2 gene. Because C19MC is located downstream of the maternally expressed ZNF331, it may belong to a larger imprinted gene cluster. miRNAs generated at both loci are listed in the grey boxes (based on http://www.mirbase.org/). Whether all these microRNA sequences represent functional entities remains an open question, particularly for several C2MC-encoded microRNAs that display low-complexity sequences. Interestingly, some C19MC- and C2MC-encoded microRNAs share a similar seed sequence, leaving open the possibility that they may target the same mRNA target(s) or subsets thereof.

observations raise the possibility that some C2MC- and C19MC-encoded microRNAs may exert similar functions in human and mouse, respectively, sharing the same mRNA targets – or families thereof – that may be related to placenta function and/or development (Fig. 3.4). Their similarly imprinted expression in the placenta reinforces this appealing hypothesis.

Imprinted small-RNA host-genes: only snoRNA vehicles or functional RNA entities?

Many miRNA and box C/D snoRNA genes in the imprinted Dlk1-Dio3 and Snurf-snrpn domains are embedded within introns of larger non-coding RNA genes extending over large genomic distances up to several hundreds of kbp. These transcription units therefore also produce a complex population of long, unspliced and alternatively spliced mRNA-like transcripts (Cavaille *et al.*, 2000, 2001, 2002; de los Santos *et al.*, 2000; Meguro *et al.*, 2001; Runte *et al.*, 2001; Landers *et al.*, 2004; Royo *et al.*, 2007; Vitali *et al.*, 2010; Fig. 3.5). Small RNAs are not thought to be transcribed independently, but derive instead from post-transcriptional maturation: box C/D small RNAs result from exonucleolytic trimming of linearized introns while miRNAs are processed through the canonical microRNA pathway involving two sequential endonucleolytic cleavages mediated by Drosha and Dicer complexes. Hence, these small-RNA host gene transcripts, relatively poorly conserved between mammalian species, have been mostly considered as small-RNA vehicles without any obvious functions.

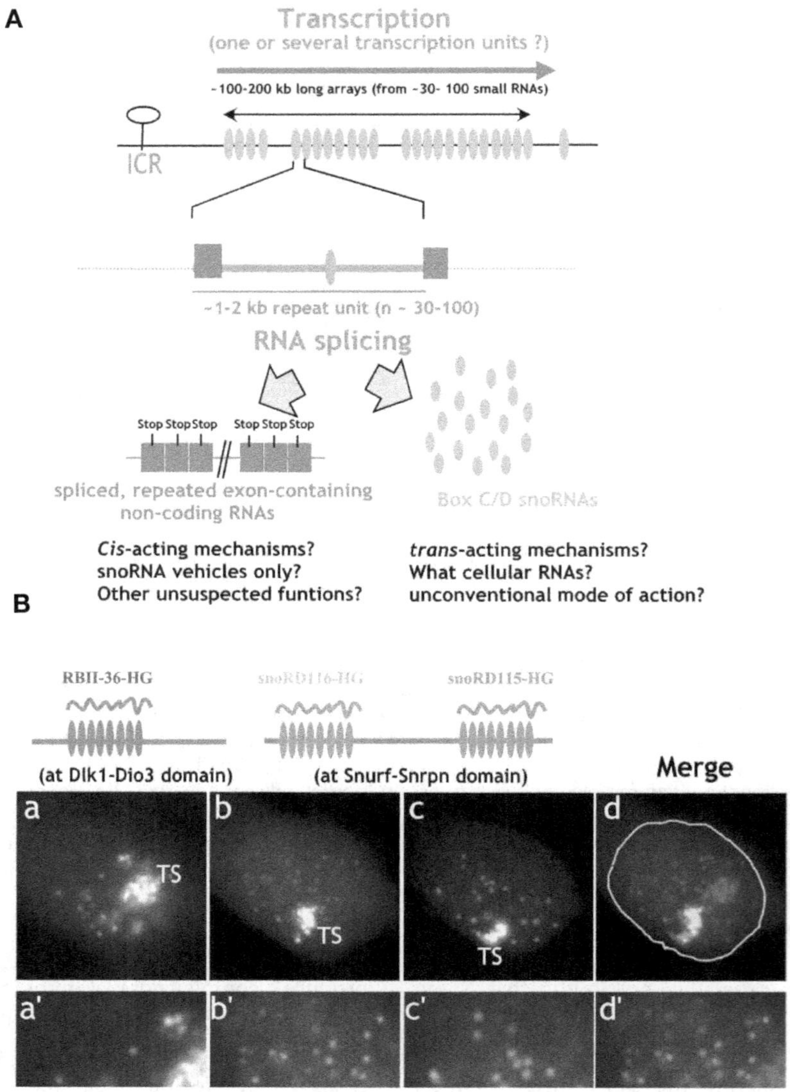

Figure 3.5 SnoRNA host-genes generate spliced, nuclear retained-transcripts. (A) A model for biogenesis of tandemly repeated C/D snoRNA genes. Most, if not all, imprinted C/D snoRNA genes are localized within introns of larger non-coding RNA genes. In the rodent *Snurf-Snrpn* domain (snord115 and snord116) and rat Dlk1-Dio3 domain (RBII-36), each snoRNA gene copy (ovals) is embedded within highly related 1 to 2 kbp-long repeat units comprising one short non-coding exon (grey boxes) and flanking intronic sequences (grey line). As a result, in addition to the fully processed snoRNAs, such transcription units also generate a complex population of alternatively spliced, repeated exon-containing transcripts whose function, if any, is unknown. (B) Spliced snoRNA host-gene transcripts are nuclear-localized RNA species. Rat primary neurons were hybridized to oligo-probes matching the spliced exon–exon junctions of the RBII-36 host-gene transcript (panel a), the snord116 host gene (panel b) and the snord 115 host gene (panel c). These spliced RNA species mostly localize at the transcription sites as large, structurally constrained FISH signals (TS) in addition to numerous dot-like signals dispersed in the entire nucleoplasm (enlarged panels a′–c′). Reproduced with permission of *Journal of Cell Science* (Vitali, P., Royo, H., Marty, V., Bortolin-Cavaillé, M. L., and Cavaillé, J. (2010). Long nuclear-retained non-coding RNAs and allele-specific higher-order chromatin organization at imprinted snoRNA gene arrays. J. Cell. Sci. *123*, 70–83).

Long nuclear-retained non-coding RNAs are generated from highly decondensed arrays of box C/D snoRNA genes

The use of highly sensitive RNA FISH approaches performed in primary rodent cells (fibroblasts or neurons) revealed a peculiar intra-nuclear fate for these spliced, repeated exon-containing transcripts, which deserves further consideration. Freshly made spliced RNA species generated from the snoRNA gene arrays accumulate as elongated, large signals around their transcription sites, while fully released RNA species are visualized as punctate and stable nuclear RNA foci dispersed throughout the nucleoplasm, without any preferential co-localization with known nuclear subcompartments (Royo *et al.*, 2007; Vitali *et al.*, 2010). Hence, despite being faithfully spliced, these snoRNA host-transcripts are not efficiently transported into the cytoplasm and may represent a novel class of nuclear-retained transcripts. Remarkably, spliced RNA species visualized at transcription sites appear to be prevented from freely diffusing. As illustrated in Fig. 3.5, this is particularly impressive at the SNORD115 and SNORD116 gene clusters, located ~50 kbp apart, but whose spliced RNA signals do not overlap. Although such a counterintuitive observation is difficult to explain, related structurally constrained RNA species in the vicinity of their transcription sites are also observed at C19MC, suggesting that this phenomenon may be a hallmark of repeated imprinted small-RNA gene arrays (Noguer-Dance *et al.*, 2010).

Finally, our and LaSalle's groups have reported another outstanding cell-imaging observation: a remarkable allele-specific chromatin decondensation phenomenon can be observed in interphase nuclei, where the transcriptionally active allele is preferentially visualized as extended DNA FISH signals. In contrast, the silent but genetically identical allele in the same nucleus is revealed, as expected, mostly as a single pinpoint DNA FISH signal (Leung *et al.*, 2009; Vitali *et al.*, 2010). This selective chromatin decondensation process is developmentally regulated and, surprisingly, appears to correlate with the size of the nucleolus during neuronal maturation (Leung *et al.*, 2009). While LaSalle's group found that transcription from the IC is required for establishing chromatin decondensation at the Snurf-snrpn arrays, substantial decondensed DNA FISH signals were also described in snoRNA-non-expressing cells, and to some extent in mitotic cells (Vitali *et al.*, 2010). The significance of these apparent discrepancies remains unknown and may result from the differential sensitivity of the DNA FISH protocols used.

A role for snoRNA-host transcripts in epigenetic regulatory pathways?

The contribution of large imprinted non-coding RNAs in the regulation of genomic imprinting is now relatively well documented, as illustrated by the Airn and Kcnq1ot1/Lit1 ncRNAs that function at the Igf2R and Kcnq1 chromosomal domains, respectively (Koerner *et al.*, 2009; Kanduri, 2011). At both loci, transcription of these two macroRNA genes initiates at the IC region from the unmethylated, paternally inherited chromosomes, and these two ncRNA genes display reciprocal imprinted expression relative to the flanking protein-coding genes that are only expressed from the maternal allele. Truncation experiments in mice demonstrated that these ncRNAs act bidirectionally as *cis*-acting silencers over long distances to negatively control the expression of surrounding protein-coding genes in a developmentally and lineage-specific manner (Sleutels *et al.*, 2002; Mancini-Dinardo *et al.*, 2006; Green *et al.*, 2007; Pandey *et al.*, 2008; Santoro and Barlow, 2011). Although their detailed molecular modes of action remain to be elucidated, one emerging notion is that Airn and Kcnq1ot1 are mostly nuclear and accumulate locally through interactions with chromatin, allowing recruitment of the histone-modifying activities (G9a, PRC2 complex) or DNA methyl-transferases (Dnmt1) that ultimately trigger epigenetic silencing (Nagano *et al.*, 2008; Pandey *et al.*, 2008; Terranova *et al.*, 2008; Nagano and Fraser, 2009; Mohammad *et al.*, 2010; Kanduri, 2011).

The cell imaging observations discussed above for the Snurf-Snrpn and Dlk1-Dio3 domains highlight the higher-order organization of snoRNA-expressing chromatin that is accompanied by highly compartmentalized nuclear RNAs (Fig. 3.5), in this case the presence of large amounts of RNA in close proximity to chromatin. Taken together, with the *caveat* that convincing functional data are still lacking, these observations raise the intriguing possibility that expression of small-RNA host transcripts, and also more broadly the entire non-coding array from which they are generated, may play a role in establishing or maintaining silencing of flanking protein-coding genes. Interestingly, and as discussed previously, loss of expression of the maternally expressed Gtl2 ncRNA genes and downstream small RNA-containing transcripts is associated with defects in the expression of the paternally expressed Dlk1 gene as well as in the methylation status at the IG-DMR (Zhou *et al.*, 2010). Moreover, genome-wide analysis of PRC2-associated ncRNA in ES cells has recently uncovered a hypothetical role for Gtl2 in targeting the PRC2 silencing complex to the Dlk1 gene (Zhao *et al.*, 2010). However, one must note that Gtl2 and Dlk1 are generally not co-expressed in the same tissues (da Rocha *et al.*, 2007). These findings suggest a function for maternally expressed genes in imprinting control.

Other arguments also favour a role for long, paternally expressed ncRNAs at the Snurf-Snrpn domain in the silencing of the maternally expressed Ube3a gene, whose imprinted expression is limited to brain. First, UBE3A is associated with a paternally expressed antisense transcript, anti-UBE3A (Rougeulle *et al.*, 1998). Second, lack of paternal expression of anti-Ube3a results in an increase in the paternal expression of Ube3a (Chamberlain and Brannan, 2001), while ectopic expression of anti-Ube3a from the maternally inherited allele is associated with reduced expression of Ube3a (Johnstone *et al.*, 2006). Third, the Ube3a gene is only imprinted in neurons where the Snurf-Snrpn-anti-Ube3A transcription unit is highly expressed; in contrast, in glial cells that express neither snoRNA-containing ncRNAs nor anti-Ube3a, Ube3a is biallelically expressed (Yamasaki *et al.*, 2003; Kishino, 2006). Fourth, neuronal up-regulation of transcription of the SNORD115 and anti-UBE3A transcripts parallels silencing of the UBE3A gene in human iPS-derived neurons (Chamberlain *et al.*, 2010). Taken together, these correlative lines of evidence support a model whereby the Ube3a gene is paternally repressed in a lineage-specific manner, perhaps due to the neuronal-specific expression of the paternally expressed Snurf-Snrpn-anti-Ube3A transcription unit, and more specifically its more distal part that includes the snoRD115 host gene and anti-Ube3A. Because the Ube3A gene is not associated with any allele-specific DNA methylation marks, how such a hypothetical regulation may occur is still unclear. Although we do not exclude the possibility that high levels of transcription across the paternal snord115-anti-Ube3A region impacts – directly or indirectly – on the expression of the downstream Ube3A allele, we would like to speculate that highly organized RNA species accumulating at the maternal allele may create a 'repressive subnuclear compartment' where the paternal Ube3a allele is preferentially localized and silenced. A related model, supported by experimental data, was already proposed for the Xist ncRNA in X-chromosome inactivation (Chaumeil *et al.*, 2006). In addition, such a hypothesis is in line with the emerging concept that some long nuclear-localized RNAs, and more precisely the factors that transcribe them, have an essential architectural role in the formation of nuclear bodies, as recently demonstrated for the Neat1 (MEN ε/β) ncRNA and nuclear paraspeckles (Mao *et al.*, 2011; Shevtsov and Dundr, 2011, and references therein).

Repeated small non-coding RNA genes and genomic imprinting: the chicken or the egg?

Imprinted small ncRNA genes share two important characteristics. First, they constitute a recent innovation during the course of mammalian evolution, being either eutherian specific (the Snurf-Snrpn and Dlk1-Dio3 domains), restricted to a group of species in the mammalian lineage (C19MC and C2MC are primate and rodent specific, respectively) or even species-specific (RBII-36 is found only in the rat Dlk1-Dio3 domain; Cavaille *et al.*, 2001). Second, they tend to be grouped in large clusters that host numerous related genes (from ~30 to more than 100 gene copies), probably originating from segmental duplication from a single or a few ancestral copies. Genomic imprinting is also a mammal-specific phenomenon. More precisely, the vast majority of imprinted genes have been described in eutherian species, as well as to some extent in marsupials, but never in monotremes (Renfree *et al.*, 2009). Surprisingly, until shown otherwise, no large repeated small-RNA gene clusters escape genomic imprinting, thus pointing to a potential co-evolution between the genesis of these repeated units and the emergence of genomic imprinting.

Progress made over the last year in sequencing several metazoan genomes has been most precious to help rebuild the evolutionary steps that created modern imprinted loci. In the case of the Snurf-Snrpn domain, remarkable work led by Graves and colleagues suggests that acquisition of genomic imprinting occurred in eutherian species when the current Snurf-Snrpn domain was assembled via chromosomal translocations followed by insertions of snoRNA clusters and several retroposed genes (Rapkins *et al.*, 2006; Hore *et al.*, 2007). The timing of imprinting acquisition at the Dlk1-Dio3 locus also corresponds to the therian-eutherian split. An in-depth analysis carried out by Ferguson-Smith and colleagues showed that the Gtl2 and Rtl1 genes were present before the advent of modern eutherian species and suggests that establishment of genomic imprinting in eutherians was contemporary with the insertion of both snoRNA and microRNA gene clusters, as well as with a decrease in SINE content (Hore *et al.*, 2007; Edwards *et al.*, 2008).

A comparably thorough analysis has not yet been conducted for the recently discovered C19MC and C2MC loci. That C19MC is present only in primates (Zhang *et al.*, 2008) indicates that genomic imprinting at this locus evolved very recently. We demonstrated that C19MC is maternally repressed in human placenta (Noguer-Dance *et al.*, 2010), and it will be of interest to analyse its imprinted status in other primate species. Preliminary data indicate that C19MC is monoallelically expressed in placenta tissues from the Rhesus monkey *M. mulatta* (Labialle *et al.*, unpublished). Of equal interest is the Sfmbt2 gene. Data originating from our lab and from Varmuza's group suggest that Sfmbt2 is imprinted only in rat and mouse (Kuzmin *et al.*, 2008; Wang *et al.*, 2011). Moreover, gene synteny is stable in this region across vertebrate genomes, apart from the insertion of the C2MC locus in a rodent ancestor, propelling the latter as the major event correlating with establishment of local genomic imprinting.

What makes a genomic region a target for genomic imprinting? A reasonable possibility is that the local genetic landscape favours the establishment of genomic imprinting. Such a view was suggested early on when, based on the role of DNA methylation in silencing transposed elements and in controlling genomic imprinting, D. Barlow proposed that both processes might actually be functionally linked. This is the 'host defence hypothesis' which theorized that genomic imprinting derives from an epigenetic mechanism dedicated to silence repeated DNA (Barlow, 1993; Yoder *et al.*, 1997). Strong support has been provided since that time, notably by the identification of numerous imprinted retrotransposon-derived genes (Wood and Oakey, 2006; Suzuki *et al.*, 2007; Wood *et al.*, 2007; Kanber *et al.*, 2009a) and the demonstration of the key role of the *de novo* DNA methyltransferase Dnmt3L in both silencing of transposable elements in the male germline and establishment of imprinting marks in female germlines (Bourc'his *et al.*, 2001; Bourc'his and Bestor, 2004). As a eutherian-specific gene, Dnmt3L thus provides a functional and evolutionary link between transposable-element silencing and genomic imprinting in mammals (Bourc'his and Bestor, 2006; Bourc'his and Proudhon, 2008).

As discussed above, there is an overall relationship between acquisition of large clusters of small ncRNA genes and establishment of genomic imprinting at the corresponding sites of amplification. It is therefore tempting to speculate that acquisition of large clusters of repeated small ncRNA genes, probably perceived as foreign DNA, might have been targeted by the host defence machinery in eutherian genomes (Labialle and Cavaillé, 2011). However, these correlations do not constitute proof *per se* and the possibility of a reverse scenario where imprinted loci are prone to insert and amplify repeated small-RNA genes could also be envisaged. Further clarification of these intricate sets of potential connections between genomic imprinting and repeated small-RNA genes will require strong input, notably by investigating more 'neglected and exotic' mammalian genomes.

Conclusions

A great proportion of C/D box snoRNA and microRNA genes is regulated by genomic imprinting in eutherians (Royo and Cavaille, 2008; Noguer-Dance *et al.*, 2010). This intriguing observation suggests an evolutionary selection for targeting genomic imprinting to these small ncRNA genes, most probably in relation with their unconventional physical clustering in large repeated arrays that might be part of a signal which promotes epigenetic silencing (Labialle and Cavaillé, 2011).

These events have occurred several times during metazoan evolution, suggesting that they represent recurrent selective advantages. Together with providing shared regulatory elements *in cis*, this peculiar genomic organization may support enhanced global expression levels required for ncRNAs to perform their functions, such as related small RNAs targeting the same cellular mRNAs or families thereof. Moreover, duplications may also promote novel functions if one of the duplicates has the opportunity to accumulate gain-of-function mutations, as has been widely described for protein-coding genes. Indeed, several microRNA gain and loss events as well as microRNA divergences can be inferred by comparing human and mouse C14MC/miR-379–miR410 (Kircher *et al.*, 2008), suggesting that the evolution of these clusters is an ongoing process. Finally, for undetermined reasons – perhaps the requirement for a basic level of redundancy – the size of large clusters might reflect their biological importance, as illustrated by the miR-430 family of microRNAs in zebrafish (Giraldez *et al.*, 2006).

Efforts to understand the striking relationship between small-RNA gene arrays and genomic imprinting will be a key for future progress in the comprehension of the biology of small RNAs and the significance of genomic imprinting in mammalian evolution, as well as for decoding the genetic ground that organizes epigenetic constraints in mammalian germlines. This constitutes an exciting but technically and conceptually challenging task. Difficulties arise from the lack of confident RNA target predictions for these poorly evolutionary conserved small RNAs. In addition, the degenerate repetitive structure of the small ncRNA clusters leaves one only guessing at a complex and intermingled redundancy and divergence in their individual gene function. Researchers wishing to use knockout approaches will have to deal with these limitations and decide: (i) to invalidate one (or a few) gene(s) in the array with the *caveat* of leaving some level of redundancy, thus decreasing the power of their analysis; (ii) to delete the whole cluster but possibly cumulate different biological effects together with reducing the ability to define the role of individual small-RNA genes. Moreover, classical gene-targeting strategies will also impact on the metabolism of small-RNA host-gene transcripts, adding another layer of complexity to our understating of the role(s) of each molecule.

The coordinate and imprinted regulation of numerous small regulatory non-coding RNA genes gives a new twist to the parent-of-origin control exerted on mammalian genomes. Given the functional importance of genomic imprinting in development and behaviour (Davies *et al.*, 2007, 2008), it is not overreaching to consider that imprinted small RNAs might play decisive functions, perhaps in the control of resource acquisition between the mother and her offspring as well as in higher brain function. Understanding important aspects of mammalian genome regulation, developmental processes, physiopathology and evolution will therefore depend in part on the efforts that will be invested in these questions. In a companion paper that commented on the first characterization of imprinted small-RNA genes in the Snurf-Snrpn domain (Cavaille *et al.*, 2000), W. Filipowicz wrote that it was 'Time for RNomics' to emphasize our need to better characterize the full repertoire of small RNAs encoded within mammalian genomes (Filipowicz, 2000). More than one decade later, we must accept that our knowledge of the physiology of imprinted small-RNA genes still remains, unfortunately so, quite limited, and we will simply conclude by saying that it is 'Time for functions!'

Acknowledgements

We thank E. Käs for careful reading and editing of the manuscript and G. Canal for helpful comments. We are indebted to Dr Varmuza for sharing with us unpublished observations. This work was supported by grants from EMBO (EMBO Young Investigator Programme), from ARC (Association pour la Recherche sur le Cancer) and from ANR (Agence Nationale de la Recherche, projet ImpMir).

References

Aravin, A.A., and Bourc'his, D. (2008). Small RNA guides for *de novo* DNA methylation in mammalian germ cells. Genes Dev. *22*, 970–975.

Arnaud, P. (2010). Genomic imprinting in germ cells: imprints are under control. Reproduction *140*, 411–423.

Bachellerie, J.P., Cavaille, J., and Huttenhofer, A. (2002). The expanding snoRNA world. Biochimie *84*, 775–790.

Bar, M., Wyman, S.K., Fritz, B.R., Qi, J., Garg, K.S., Parkin, R.K., Kroh, E.M., Bendoraite, A., Mitchell, P.S., Nelson, A.M., *et al.* (2008). MicroRNA discovery and profiling in human embryonic stem cells by deep sequencing of small RNA libraries. Stem Cells *26*, 2496–2505.

Barlow, D.P. (1993). Methylation and imprinting: from host defense to gene regulation? Science *260*, 309–310.

Bartolomei, M.S. (2009). Genomic imprinting: employing and avoiding epigenetic processes. Genes Dev. *23*, 2124–2133.

Bentwich, I., Avniel, A., Karov, Y., Aharonov, R., Gilad, S., Barad, O., Barzilai, A., Einat, P., Einav, U., Meiri, E., *et al.* (2005). Identification of hundreds of conserved and nonconserved human microRNAs. Nat. Genet. *37*, 766–770.

Bidwell, C.A., Kramer, L.N., Perkins, A.C., Hadfield, T.S., Moody, D.E., and Cockett, N.E. (2004). Expression of PEG11 and PEG11AS transcripts in normal and callipyge sheep. BMC Biol. *2*, 17.

Birney, E., Stamatoyannopoulos, J.A., Dutta, A., Guigo, R., Gingeras, T.R., Margulies, E.H., Weng, Z., Snyder, M., Dermitzakis, E.T., Thurman, R.E., *et al.* (2007). Identification and analysis of functional elements in 1% of the human genome by the ENCODE pilot project. Nature *447*, 799–816.

Bischof, J.M., Stewart, C.L., and Wevrick, R. (2007). Inactivation of the mouse Magel2 gene results in growth abnormalities similar to Prader–Willi syndrome. Hum. Mol. Genet. *16*, 2713–2719.

Bortolin-Cavaille, M.L., Dance, M., Weber, M., and Cavaille, J. (2009). C19MC microRNAs are processed from introns of large Pol-II, non-protein-coding transcripts. Nucleic Acids Res. *37*, 3464–3473.

Bourc'his, D., and Bestor, T.H. (2004). Meiotic catastrophe and retrotransposon reactivation in male germ cells lacking Dnmt3L. Nature *431*, 96–99.

Bourc'his, D., and Bestor, T.H. (2006). Origins of extreme sexual dimorphism in genomic imprinting. Cytogenet. Genome Res. *113*, 36–40.

Bourc'his, D., and Proudhon, C. (2008). Sexual dimorphism in parental imprint ontogeny and contribution to embryonic development. Mol. Cell. Endocrinol. *282*, 87–94.

Bourc'his, D., Xu, G.L., Lin, C.S., Bollman, B., and Bestor, T.H. (2001). Dnmt3L and the establishment of maternal genomic imprints. Science *294*, 2536–2539.

Brannan, C.I., Dees, E.C., Ingram, R.S., and Tilghman, S.M. (1990). The product of the H19 gene may function as an RNA. Mol. Cell. Biol. *10*, 28–36.

Bressler, J., Tsai, T.F., Wu, M.Y., Tsai, S.F., Ramirez, M.A., Armstrong, D., and Beaudet, A.L. (2001). The SNRPN promoter is not required for genomic imprinting of the Prader–Willi/Angelman domain in mice. Nat. Genet. *28*, 232–240.

Buiting, K. (2010). Prader–Willi syndrome and Angelman syndrome. Am. J. Med. Genet. C. Semin. Med. Genet. *154C*, 365–376.

Burger, J., Horn, D., Tonnies, H., Neitzel, H., and Reis, A. (2002). Familial interstitial 570 kbp deletion of the UBE3A gene region causing Angelman syndrome but not Prader–Willi syndrome. Am. J. Med. Genet. *111*, 233–237.

Burns, C.M., Chu, H., Rueter, S.M., Hutchinson, L.K., Canton, H., Sanders-Bush, E., and Emeson, R.B. (1997). Regulation of serotonin-2C receptor G-protein coupling by RNA editing. Nature *387*, 303–308.

Bushati, N., and Cohen, S.M. (2007). microRNA functions. Annu. Rev. Cell. Dev. Biol. *23*, 175–205.

Byrne, K., Colgrave, M.L., Vuocolo, T., Pearson, R., Bidwell, C.A., Cockett, N.E., Lynn, D.J., Fleming-Waddell, J.N., and Tellam, R.L. (2010). The imprinted retrotransposon-like gene PEG11 (RTL1) is expressed as a full-length protein in skeletal muscle from Callipyge sheep. PLoS One *5*, e8638.

Caiment, F., Charlier, C., Hadfield, T., Cockett, N., Georges, M., and Baurain, D. (2010). Assessing the effect of the CLPG mutation on the microRNA catalog of skeletal muscle using high-throughput sequencing. Genome Res. *20*, 1651–1662.

Calabrese, J.M., Seila, A.C., Yeo, G.W., and Sharp, P.A. (2007). RNA sequence analysis defines Dicer's role in mouse embryonic stem cells. Proc. Natl. Acad. Sci. U.S.A. *104*, 18097–18102.

Canton, H., Emeson, R.B., Barker, E.L., Backstrom, J.R., Lu, J.T., Chang, M.S., and Sanders-Bush, E. (1996). Identification, molecular cloning, and distribution of a short variant of the 5-hydroxytryptamine2C receptor produced by alternative splicing. Mol. Pharmacol. *50*, 799–807.

Carninci, P., Kasukawa, T., Katayama, S., Gough, J., Frith, M.C., Maeda, N., Oyama, R., Ravasi, T., Lenhard, B., Wells, C., *et al.* (2005). The transcriptional landscape of the mammalian genome. Science *309*, 1559–1563.

Cattanach, B.M., Burtenshaw, M.D., Rasberry, C., and Evans, E.P. (1993). Large deletions and other gross forms of chromosome imbalance compatible with viability and fertility in the mouse. Nat. Genet. *3*, 56–61.

Cavaille, J., Buiting, K., Kiefmann, M., Lalande, M., Brannan, C.I., Horsthemke, B., Bachellerie, J.P., Brosius, J., and Huttenhofer, A. (2000). Identification of brain-specific and imprinted small nucleolar RNA genes exhibiting an unusual genomic organization. Proc. Natl. Acad. Sci. U.S.A. *97*, 14311–14316.

Cavaille, J., Vitali, P., Basyuk, E., Huttenhofer, A., and Bachellerie, J.P. (2001). A novel brain-specific box C/D small nucleolar RNA processed from tandemly repeated introns of a noncoding RNA gene in rats. J. Biol. Chem. *276*, 26374–26383.

Cavaille, J., Seitz, H., Paulsen, M., Ferguson-Smith, A.C., and Bachellerie, J.P. (2002). Identification of tandemly repeated C/D snoRNA genes at the imprinted human 14q32 domain reminiscent of those at the Prader–Willi/Angelman syndrome region. Hum. Mol. Genet. *11*, 1527–1538.

Chamberlain, S.J., and Brannan, C.I. (2001). The Prader–Willi syndrome imprinting center activates the paternally expressed murine Ube3a antisense transcript but represses paternal Ube3a. Genomics *73*, 316–322.

Chamberlain, S.J., and Lalande, M. (2010). Neurodevelopmental disorders involving genomic imprinting at human chromosome 15q11-q13. Neurobiol. Dis. *39*, 13–20.

Chamberlain, S.J., Chen, P.F., Ng, K.Y., Bourgois-Rocha, F., Lemtiri-Chlieh, F., Levine, E.S., and Lalande, M. (2010). Induced pluripotent stem cell models of the genomic imprinting disorders Angelman and Prader–Willi syndromes. Proc. Natl. Acad. Sci. U.S.A. *107*, 17668–17673.

Charlier, C., Segers, K., Karim, L., Shay, T., Gyapay, G., Cockett, N., and Georges, M. (2001a). The callipyge mutation enhances the expression of coregulated imprinted genes *in* cis without affecting their imprinting status. Nat. Genet. *27*, 367–369.

Charlier, C., Segers, K., Wagenaar, D., Karim, L., Berghmans, S., Jaillon, O., Shay, T., Weissenbach, J., Cockett, N., Gyapay, G., *et al.* (2001b). Human-ovine comparative sequencing of a 250-kb imprinted domain encompassing the callipyge (clpg) locus and identification of six imprinted transcripts: DLK*1*, DAT, GTL2, PEG*11*, antiPEG*11*, and MEG8. Genome Res. *11*, 850–862.

Chaumeil, J., Le Baccon, P., Wutz, A., and Heard, E. (2006). A novel role for Xist RNA in the formation of a repressive nuclear compartment into which genes are recruited when silenced. Genes Dev. *20*, 2223–2237.

Cockett, N.E., Jackson, S.P., Shay, T.L., Farnir, F., Berghmans, S., Snowder, G.D., Nielsen, D.M., and Georges, M. (1996). Polar overdominance at the ovine callipyge locus. Science *273*, 236–238.

Conroy, J.M., Grebe, T.A., Becker, L.A., Tsuchiya, K., Nicholls, R.D., Buiting, K., Horsthemke, B., Cassidy, S.B., and Schwartz, S. (1997). Balanced translocation *46*,XY,t(2;15) (q37.2;q11.2) associated with atypical Prader–Willi syndrome. Am. J. Hum. Genet. *61*, 388–394.

Cowley, M., and Oakey, R.J. (2010). Retrotransposition and genomic imprinting. Brief Funct. Genomics *9*, 340–346.

Davis, E., Jensen, C.H., Schroder, H.D., Farnir, F., Shay-Hadfield, T., Kliem, A., Cockett, N., Georges, M., and Charlier, C. (2004). Ectopic expression of DLK1 protein in skeletal muscle of padumnal heterozygotes causes the callipyge phenotype. Curr. Biol. *14*, 1858–1862.

Davis, E., Caiment, F., Tordoir, X., Cavaille, J., Ferguson-Smith, A., Cockett, N., Georges, M., and Charlier, C. (2005). RNAi-mediated allelic trans-interaction at the imprinted Rtl1/Peg11 locus. Curr. Biol. *15*, 743–749.

Davies, W., Isles, A.R., Humby, T., and Wilkinson, L.S. (2007). What are imprinted genes doing in the brain? Epigenetics 2, 201–206.

Davies, W., Lynn, P.M., Relkovic, D., and Wilkinson, L.S. (2008). Imprinted genes and neuroendocrine function. Front Neuroendocrinol. *29*, 413–427.

Ding, F., Li, H.H., Zhang, S., Solomon, N.M., Camper, S.A., Cohen, P., and Francke, U. (2008). SnoRNA Snord116 (Pwcr1/MBII-85) deletion causes growth deficiency and hyperphagia in mice. PLoS One *3*, e1709.

Ding, F., Li, H.H., Li, J., Myers, R.M., and Francke, U. (2010). Neonatal maternal deprivation response and developmental changes in gene expression revealed by hypothalamic gene expression profiling in mice. PLoS One *5*, e9402.

Dinger, M.E., Amaral, P.P., Mercer, T.R., and Mattick, J.S. (2009). Pervasive transcription of the eukaryotic genome: functional indices and conceptual implications. Brief Funct. Genomic Proteomic *8*, 407–423.

Doe, C.M., Relkovic, D., Garfield, A.S., Dalley, J.W., Theobald, D.E., Humby, T., Wilkinson, L.S., and Isles, A.R. (2009). Loss of the imprinted snoRNA mbii-52 leads to increased 5htr2c pre-RNA editing and altered 5HT2CR-mediated behaviour. Hum. Mol. Genet. *18*, 2140–2148.

Duker, A.L., Ballif, B.C., Bawle, E.V., Person, R.E., Mahadevan, S., Alliman, S., Thompson, R., Traylor, R., Bejjani, B.A., Shaffer, L.G., *et al.* (2010). Paternally inherited microdeletion at 15q11.2 confirms a significant role for the SNORD116 C/D box snoRNA cluster in Prader–Willi syndrome. Eur. J. Hum. Genet. *18*, 1196–1201.

Edwards, C.A., Mungall, A.J., Matthews, L., Ryder, E., Gray, D.J., Pask, A.J., Shaw, G., Graves, J.A., Rogers, J., Dunham, I., *et al.* (2008). The evolution of the DLK1-DIO3 imprinted domain in mammals. PLoS Biol. *6*, e135.

Filipowicz, W. (2000). Imprinted expression of small nucleolar RNAs in brain: time for RNomics. Proc. Natl. Acad. Sci. U.S.A. *97*, 14035–14037.

Filipowicz, W., Bhattacharyya, S.N., and Sonenberg, N. (2008). Mechanisms of post-transcriptional regulation by microRNAs: are the answers in sight? Nat. Rev. Genet. *9*, 102–114.

Fiore, R., Khudayberdiev, S., Christensen, M., Siegel, G., Flavell, S.W., Kim, T.K., Greenberg, M.E., and

Schratt, G. (2009). Mef2-mediated transcription of the miR379–410 cluster regulates activity-dependent dendritogenesis by fine-tuning Pumilio2 protein levels. Embo J. *28*, 697–710.

Flavell, S.W., and Greenberg, M.E. (2008). Signaling mechanisms linking neuronal activity to gene expression and plasticity of the nervous system. Annu. Rev. Neurosci. *31*, 563–590.

Fleming-Waddell, J.N., Olbricht, G.R., Taxis, T.M., White, J.D., Vuocolo, T., Craig, B.A., Tellam, R.L., Neary, M.K., Cockett, N.E., and Bidwell, C.A. (2009). Effect of DLK1 and RTL1 but not MEG3 or MEG8 on muscle gene expression in Callipyge lambs. PLoS One *4*, e7399.

Freking, B.A., Murphy, S.K., Wylie, A.A., Rhodes, S.J., Keele, J.W., Leymaster, K.A., Jirtle, R.L., and Smith, T.P. (2002). Identification of the single base change causing the callipyge muscle hypertrophy phenotype, the only known example of polar overdominance in mammals. Genome Res. *12*, 1496–1506.

Gallagher, R.C., Pils, B., Albalwi, M., and Francke, U. (2002). Evidence for the role of PWCR1/HBII-85 C/D box small nucleolar RNAs in Prader–Willi syndrome. Am. J. Hum. Genet. *71*, 669–678.

Gao, J., Wang, W.Y., Mao, Y.W., Graff, J., Guan, J.S., Pan, L., Mak, G., Kim, D., Su, S.C., and Tsai, L.H. (2010). A novel pathway regulates memory and plasticity via SIRT1 and miR-134. Nature *466*, 1105–1109.

Garzon, R., Marcucci, G., and Croce, C.M. (2010). Targeting microRNAs in cancer: rationale, strategies and challenges. Nat. Rev. Drug Discov. *9*, 775–789.

Gaughwin, P., Ciesla, M., Yang, H., Lim, B., and Brundin, P. (2011). Stage-specific modulation of cortical neuronal development by Mmu-miR-134. Cereb Cortex. Epub ahead of print.

Georges, M., and Cockett, N. (1996). The ovine callipyge locus: a paradigm illustrating the importance of non-Mendelian genetics in livestock. Reprod. Nutr. Dev. *36*, 651–657.

Georges, M., Charlier, C., and Cockett, N. (2003). The callipyge locus: evidence for the trans interaction of reciprocally imprinted genes. Trends Genet. *19*, 248–252.

Georges, M., Charlier, C., Smit, M., Davis, E., Shay, T., Tordoir, X., Takeda, H., Caiment, F., and Cockett, N. (2004). Toward molecular understanding of polar overdominance at the ovine callipyge locus. Cold Spring Harb. Symp. Quant. Biol. *69*, 477–483.

Georgiades, P., Watkins, M., Surani, M.A., and Ferguson-Smith, A.C. (2000). Parental origin-specific developmental defects in mice with uniparental disomy for chromosome 12. Development *127*, 4719–4728.

Gerard, M., Hernandez, L., Wevrick, R., and Stewart, C.L. (1999). Disruption of the mouse necdin gene results in early post-natal lethality. Nat. Genet. *23*, 199–202.

Giraldez, A.J., Cinalli, R.M., Glasner, M.E., Enright, A.J., Thomson, J.M., Baskerville, S., Hammond, S.M., Bartel, D.P., and Schier, A.F. (2005). MicroRNAs regulate brain morphogenesis in zebrafish. Science *308*, 833–838.

Giraldez, A.J., Mishima, Y., Rihel, J., Grocock, R.J., Van Dongen, S., Inoue, K., Enright, A.J., and Schier, A.F. (2006). Zebrafish MiR-430 promotes deadenylation and clearance of maternal mRNAs. Science *312*, 75–79.

Glatt-Deeley, H., Bancescu, D.L., and Lalande, M. (2010). Prader–Willi syndrome, Snord*115*, and Htr2c editing. Neurogenetics *11*, 143–144.

Green, K., Lewis, A., Dawson, C., Dean, W., Reinhart, B., Chaillet, J.R., and Reik, W. (2007). A developmental window of opportunity for imprinted gene silencing mediated by DNA methylation and the Kcnq1ot1 noncoding RNA. Mamm. Genome *18*, 32–42.

Hanna, J., Saha, K., Pando, B., van Zon, J., Lengner, C.J., Creyghton, M.P., van Oudenaarden, A., and Jaenisch, R. (2009). Direct cell reprogramming is a stochastic process amenable to acceleration. Nature *462*, 595–601.

Henras, A.K., Dez, C., and Henry, Y. (2004). RNA structure and function in C/D and H/ACA s(no) RNPs. Curr. Opin. Struct. Biol. *14*, 335–343.

Hernandez, A., Martinez, M.E., Fiering, S., Galton, V.A., and St Germain, D. (2006). Type 3 deiodinase is critical for the maturation and function of the thyroid axis. J. Clin. Invest. *116*, 476–484.

Hore, T.A., Rapkins, R.W., and Graves, J.A. (2007). Construction and evolution of imprinted loci in mammals. Trends Genet. *23*, 440–448.

Huang, Q., Gumireddy, K., Schrier, M., le Sage, C., Nagel, R., Nair, S., Egan, D.A., Li, A., Huang, G., Klein-Szanto, A.J., *et al.* (2008). The microRNAs miR-373 and miR-520c promote tumour invasion and metastasis. Nat. Cell. Biol. *10*, 202–210.

Jacquier, A. (2009). The complex eukaryotic transcriptome: unexpected pervasive transcription and novel small RNAs. Nat. Rev. Genet. *10*, 833–844.

Johnstone, K.A., DuBose, A.J., Futtner, C.R., Elmore, M.D., Brannan, C.I., and Resnick, J.L. (2006). A human imprinting centre demonstrates conserved acquisition but diverged maintenance of imprinting in a mouse model for Angelman syndrome imprinting defects. Hum. Mol. Genet. *15*, 393–404.

Kanber, D., Berulava, T., Ammerpohl, O., Mitter, D., Richter, J., Siebert, R., Horsthemke, B., Lohmann, D., and Buiting, K. (2009a). The human retinoblastoma gene is imprinted. PLoS Genet. *5*, e1000790.

Kanber, D., Giltay, J., Wieczorek, D., Zogel, C., Hochstenbach, R., Caliebe, A., Kuechler, A., Horsthemke, B., and Buiting, K. (2009b). A paternal deletion of MKRN3, MAGEL2 and NDN does not result in Prader–Willi syndrome. Eur. J. Hum. Genet. *17*, 582–590.

Kanduri, C. (2011). Kcnq1ot*1*, A chromatin regulatory RNA. Semin. Cell. Dev. Biol. Epub ahead of print.

Kaneko-Ishino, T., and Ishino, F. (2010). Retrotransposon silencing by DNA methylation contributed to the evolution of placentation and genomic imprinting in mammals. Dev. Growth Differ. *52*, 533–543.

Kim, K.S., Kim, J.J., Dekkers, J.C., and Rothschild, M.F. (2004). Polar overdominant inheritance of a DLK1 polymorphism is associated with growth and fatness in pigs. Mamm. Genome *15*, 552–559.

Kircher, M., Bock, C., and Paulsen, M. (2008). Structural conservation versus functional divergence of maternally expressed microRNAs in the Dlk1/Gtl2 imprinting region. BMC Genomics *9*, 346.

Kishino, T. (2006). Imprinting in neurons. Cytogenet Genome Res. *113*, 209–214.

Kishore, S., and Stamm, S. (2006). The snoRNA HBII-52 regulates alternative splicing of the serotonin receptor 2C. Science *311*, 230–232.

Kishore, S., Khanna, A., Zhang, Z., Hui, J., Balwierz, P.J., Stefan, M., Beach, C., Nicholls, R.D., Zavolan, M., and Stamm, S. (2010). The snoRNA MBII-52 (SNORD 115) is processed into smaller RNAs and regulates alternative splicing. Hum. Mol. Genet. *19*, 1153–1164.

Kiss, T. (2002). Small nucleolar RNAs: an abundant group of noncoding RNAs with diverse cellular functions. Cell *109*, 145–148.

Koerner, M.V., Pauler, F.M., Huang, R., and Barlow, D.P. (2009). The function of non-coding RNAs in genomic imprinting. Development *136*, 1771–1783.

Kuzmin, A., Han, Z., Golding, M.C., Mann, M.R., Latham, K.E., and Varmuza, S. (2008). The PcG gene Sfmbt2 is paternally expressed in extraembryonic tissues. Gene Expr. Patterns *8*, 107–116.

Labaille, S. and Cavaillé, J. (2011). Do repeated arrays of regulatory small-RNA genes elicit genomic imprinting?: Cocurrent emergence of large clusters of small non-coding RNAs and genomic imprinting at four evolutionarily distinct eutherian chromosomal loci. Bioessays. Epub ahead of print.

Landers, M., Bancescu, D.L., Le Meur, E., Rougeulle, C., Glatt-Deeley, H., Brannan, C., Muscatelli, F., and Lalande, M. (2004). Regulation of the large (approximately 1000 kb) imprinted murine Ube3a antisense transcript by alternative exons upstream of Snurf/Snrpn. Nucleic Acids Res. *32*, 3480–3492.

Landgraf, P., Rusu, M., Sheridan, R., Sewer, A., Iovino, N., Aravin, A., Pfeffer, S., Rice, A., Kamphorst, A.O., Landthaler, M., *et al.* (2007). A mammalian microRNA expression atlas based on small RNA library sequencing. Cell *129*, 1401–1414.

Laurent, L.C., Chen, J., Ulitsky, I., Mueller, F.J., Lu, C., Shamir, R., Fan, J.B., and Loring, J.F. (2008). Comprehensive microRNA profiling reveals a unique human embryonic stem cell signature dominated by a single seed sequence. Stem Cells *26*, 1506–1516.

Lee, J.T. (2010). The X as model for RNA's niche in epigenomic regulation. Cold Spring Harb. Perspect. Biol. *2*, a003749.

Leung, K.N., Vallero, R.O., DuBose, A.J., Resnick, J.L., and LaSalle, J.M. (2009). Imprinting regulates mammalian snoRNA-encoding chromatin decondensation and neuronal nucleolar size. Hum. Mol. Genet. *18*, 4227–4238.

Lewis, A., and Redrup, L. (2005). Genetic imprinting: conflict at the Callipyge locus. Curr. Biol. *15*, R291–294.

Li, M., Lee, K.F., Lu, Y., Clarke, I., Shih, D., Eberhart, C., Collins, V.P., Van Meter, T., Picard, D., Zhou, L., *et al.* (2009). Frequent amplification of a chr19q13.41 microRNA polycistron in aggressive primitive neuroectodermal brain tumors. Cancer Cell *16*, 533–546.

Lin, S.P., Youngson, N., Takada, S., Seitz, H., Reik, W., Paulsen, M., Cavaille, J., and Ferguson-Smith, A.C. (2003). Asymmetric regulation of imprinting on the maternal and paternal chromosomes at the Dlk1-Gtl2 imprinted cluster on mouse chromosome 12. Nat. Genet. *35*, 97–102.

Lin, S.P., Coan, P., da Rocha, S.T., Seitz, H., Cavaille, J., Teng, P.W., Takada, S., and Ferguson-Smith, A.C. (2007). Differential regulation of imprinting in the murine embryo and placenta by the Dlk1-Dio3 imprinting control region. Development *134*, 417–426.

Liu, L., Luo, G.Z., Yang, W., Zhao, X., Zheng, Q., Lv, Z., Li, W., Wu, H.J., Wang, L., Wang, X.J., *et al.* (2010). Activation of the imprinted Dlk1-Dio3 region correlates with pluripotency levels of mouse stem cells. J. Biol. Chem. *285*, 19483–19490.

Lynch, C., and Tristem, M. (2003). A co-opted gypsy-type LTR-retrotransposon is conserved in the genomes of humans, sheep, mice, and rats. Curr. Biol. *13*, 1518–1523.

Mallanna, S.K., and Rizzino, A. (2010). Emerging roles of microRNAs in the control of embryonic stem cells and the generation of induced pluripotent stem cells. Dev. Biol. *344*, 16–25.

Mancini-Dinardo, D., Steele, S.J., Levorse, J.M., Ingram, R.S., and Tilghman, S.M. (2006). Elongation of the Kcnq1ot1 transcript is required for genomic imprinting of neighboring genes. Genes Dev. *20*, 1268–1282.

Mao, Y.S., Sunwoo, H., Zhang, B., and Spector, D.L. (2011). Direct visualization of the co-transcriptional assembly of a nuclear body by noncoding RNAs. Nat. Cell. Biol. *13*, 95–101.

Martinez, N.J., and Gregory, R.I. (2010). MicroRNA gene regulatory pathways in the establishment and maintenance of ESC identity. Cell Stem Cell *7*, 31–35.

Mattick, J.S. (2009). The genetic signatures of noncoding RNAs. PLoS Genet. *5*, e1000459.

Meguro, M., Mitsuya, K., Nomura, N., Kohda, M., Kashiwagi, A., Nishigaki, R., Yoshioka, H., Nakao, M., Oishi, M., and Oshimura, M. (2001). Large-scale evaluation of imprinting status in the Prader–Willi syndrome region: an imprinted direct repeat cluster resembling small nucleolar RNA genes. Hum. Mol. Genet. *10*, 383–394.

Mercer, R.E., and Wevrick, R. (2009). Loss of magel2, a candidate gene for features of Prader–Willi syndrome, impairs reproductive function in mice. PLoS One *4*, e4291.

Mercer, T.R., Dinger, M.E., and Mattick, J.S. (2009). Long non-coding RNAs: insights into functions. Nat. Rev. Genet. *10*, 155–159.

Mikkelsen, T.S., Hanna, J., Zhang, X., Ku, M., Wernig, M., Schorderet, P., Bernstein, B.E., Jaenisch, R., Lander, E.S., and Meissner, A. (2008). Dissecting direct reprogramming through integrative genomic analysis. Nature *454*, 49–55.

Miyoshi, N., Wagatsuma, H., Wakana, S., Shiroishi, T., Nomura, M., Aisaka, K., Kohda, T., Surani, M.A., Kaneko-Ishino, T., and Ishino, F. (2000). Identification of an imprinted gene, Meg3/Gtl2 and its human homologue MEG3, first mapped on mouse distal chromosome 12 and human chromosome 14q. Genes Cells *5*, 211–220.

Moazed, D. (2009). Small RNAs in transcriptional gene silencing and genome defence. Nature *457*, 413–420.

Mohammad, F., Mondal, T., Guseva, N., Pandey, G.K., and Kanduri, C. (2010). Kcnq1ot1 noncoding RNA mediates transcriptional gene silencing by interacting with Dnmt1. Development *137*, 2493–2499.

Moon, Y.S., Smas, C.M., Lee, K., Villena, J.A., Kim, K.H., Yun, E.J., and Sul, H.S. (2002). Mice lacking paternally expressed Pref-1/Dlk1 display growth retardation and accelerated adiposity. Mol. Cell Biol. *22*, 5585–5592.

Moore, T., and Haig, D. (1991). Genomic imprinting in mammalian development: a parental tug-of-war. Trends Genet. *7*, 45–49.

Morabito, M.V., Abbas, A.I., Hood, J.L., Kesterson, R.A., Jacobs, M.M., Kump, D.S., Hachey, D.L., Roth, B.L., and Emeson, R.B. (2010). Mice with altered serotonin 2C receptor RNA editing display characteristics of Prader–Willi syndrome. Neurobiol Dis. *39*, 169–180.

Morris, K.V. (2011). The emerging role of RNA in the regulation of gene transcription in human cells. Semin. Cell. Dev. Biol. Epub ahead of print.

Muscatelli, F., Abrous, D.N., Massacrier, A., Boccaccio, I., Le Moal, M., Cau, P., and Cremer, H. (2000). Disruption of the mouse Necdin gene results in hypothalamic and behavioral alterations reminiscent of the human Prader–Willi syndrome. Hum. Mol. Genet. *9*, 3101–3110.

Nagano, T., and Fraser, P. (2009). Emerging similarities in epigenetic gene silencing by long noncoding RNAs. Mamm. Genome *20*, 557–562.

Nagano, T., Mitchell, J.A., Sanz, L.A., Pauler, F.M., Ferguson-Smith, A.C., Feil, R., and Fraser, P. (2008). The Air noncoding RNA epigenetically silences transcription by targeting G9a to chromatin. Science *322*, 1717–1720.

Nicholls, R.D., and Knepper, J.L. (2001). Genome organization, function, and imprinting in Prader–Willi and Angelman syndromes. Annu. Rev. Genomics Hum. Genet. *2*, 153–175.

Niswender, C.M., Copeland, S.C., Herrick-Davis, K., Emeson, R.B., and Sanders-Bush, E. (1999). RNA editing of the human serotonin 5-hydroxytryptamine 2C receptor silences constitutive activity. J. Biol. Chem. *274*, 9472–9478.

Noguer-Dance, M., Abu-Amero, S., Al-Khtib, M., Lefevre, A., Coullin, P., Moore, G.E., and Cavaille, J. (2010). The primate-specific microRNA gene cluster (C19MC) is imprinted in the placenta. Hum. Mol. Genet. *19*, 3566–3582.

Pandey, R.R., Mondal, T., Mohammad, F., Enroth, S., Redrup, L., Komorowski, J., Nagano, T., Mancini-Dinardo, D., and Kanduri, C. (2008). Kcnq1ot1 antisense noncoding RNA mediates lineage-specific transcriptional silencing through chromatin-level regulation. Mol. Cell *32*, 232–246.

Perkins, A.C., Kramer, L.N., Spurlock, D.M., Hadfield, T.S., Cockett, N.E., and Bidwell, C.A. (2006). Postnatal changes in the expression of genes located in the callipyge region in sheep skeletal muscle. Anim. Genet. *37*, 535–542.

Ponting, C.P., Oliver, P.L., and Reik, W. (2009). Evolution and functions of long noncoding RNAs. Cell *136*, 629–641.

Prasanth, K.V., and Spector, D.L. (2007). Eukaryotic regulatory RNAs: an answer to the 'genome complexity' conundrum. Genes Dev. *21*, 11–42.

Rapkins, R.W., Hore, T., Smithwick, M., Ager, E., Pask, A.J., Renfree, M.B., Kohn, M., Hameister, H., Nicholls, R.D., Deakin, J.E., *et al.* (2006). Recent assembly of an imprinted domain from non-imprinted components. PLoS Genet. *2*, e182.

Reichow, S.L., Hamma, T., Ferre-D'Amare, A.R., and Varani, G. (2007). The structure and function of small nucleolar ribonucleoproteins. Nucleic Acids Res. *35*, 1452–1464.

Reik, W., and Walter, J. (2001). Genomic imprinting: parental influence on the genome. Nat. Rev. Genet. *2*, 21–32.

Ren, J., Jin, P., Wang, E., Marincola, F.M., and Stroncek, D.F. (2009). MicroRNA and gene expression patterns in the differentiation of human embryonic stem cells. J. Transl. Med. *7*, 20.

Renfree, M.B., Hore, T.A., Shaw, G., Graves, J.A., and Pask, A.J. (2009). Evolution of genomic imprinting: insights from marsupials and monotremes. Annu. Rev. Genomics Hum. Genet. *10*, 241–262.

Rippe, V., Dittberner, L., Lorenz, V.N., Drieschner, N., Nimzyk, R., Sendt, W., Junker, K., Belge, G., and Bullerdiek, J. (2010). The two stem cell microRNA gene clusters C19MC and miR-371–3 are activated by specific chromosomal rearrangements in a subgroup of thyroid adenomas. PLoS One *5*, e9485.

da Rocha, S.T., Tevendale, M., Knowles, E., Takada, S., Watkins, M., and Ferguson-Smith, A.C. (2007). Restricted co-expression of Dlk1 and the reciprocally imprinted non-coding RNA, Gtl2, implications for cis-acting control. Dev. Biol. *306*, 810–823.

da Rocha, S.T., Edwards, C.A., Ito, M., Ogata, T., and Ferguson-Smith, A.C. (2008). Genomic imprinting at the mammalian Dlk1-Dio3 domain. Trends Genet. *24*, 306–316.

da Rocha, S.T., Charalambous, M., Lin, S.P., Gutteridge, I., Ito, Y., Gray, D., Dean, W., and Ferguson-Smith, A.C. (2009). Gene dosage effects of the imprinted delta-like homologue 1 (dlk1/pref1) in development: implications for the evolution of imprinting. PLoS Genet. *5*, e1000392.

Rougeulle, C., Cardoso, C., Fontes, M., Colleaux, L., and Lalande, M. (1998). An imprinted antisense RNA overlaps UBE3A and a second maternally expressed transcript. Nat. Genet. *19*, 15–16.

Royo, H., and Cavaille, J. (2008). Non-coding RNAs in imprinted gene clusters. Biol. Cell *100*, 149–166.

Royo, H., Bortolin, M.L., Seitz, H., and Cavaille, J. (2006). Small non-coding RNAs and genomic imprinting. Cytogenet. Genome Res. *113*, 99–108.

Royo, H., Basyuk, E., Marty, V., Marques, M., Bertrand, E., and Cavaille, J. (2007). Bsr, a nuclear-retained RNA with monoallelic expression. Mol. Biol. Cell *18*, 2817–2827.

Runte, M., Huttenhofer, A., Gross, S., Kiefmann, M., Horsthemke, B., and Buiting, K. (2001). The IC-SNURF-SNRPN transcript serves as a host for multiple small nucleolar RNA species and as an antisense RNA for UBE3A. Hum. Mol. Genet. *10*, 2687–2700.

Runte, M., Varon, R., Horn, D., Horsthemke, B., and Buiting, K. (2005). Exclusion of the C/D box snoRNA gene cluster HBII-52 from a major role in Prader–Willi syndrome. Hum. Genet. *116*, 228–230.

Sahoo, T., del Gaudio, D., German, J.R., Shinawi, M., Peters, S.U., Person, R.E., Garnica, A., Cheung, S.W., and Beaudet, A.L. (2008). Prader–Willi phenotype caused by paternal deficiency for the HBII-85 C/D box small nucleolar RNA cluster. Nat. Genet. *40*, 719–721.

Santoro, F., and Barlow, D.P. (2011). Developmental control of imprinted expression by macro non-coding RNAs. Semin. Cell Dev. Biol. Epub ahead of print.

de los Santos, T., Schweizer, J., Rees, C.A., and Francke, U. (2000). Small evolutionarily conserved RNA, resembling C/D box small nucleolar RNA, is transcribed from PWCR*1*, a novel imprinted gene in the Prader–Willi deletion region, which Is highly expressed in brain. Am. J. Hum. Genet. *67*, 1067–1082.

Schaller, F., Watrin, F., Sturny, R., Massacrier, A., Szepetowski, P., and Muscatelli, F. (2010). A single postnatal injection of oxytocin rescues the lethal feeding behaviour in mouse newborns deficient for the imprinted Magel2 gene. Hum. Mol. Genet. *19*, 4895–4905.

Schratt, G.M., Tuebing, F., Nigh, E.A., Kane, C.G., Sabatini, M.E., Kiebler, M., and Greenberg, M.E. (2006). A brain-specific microRNA regulates dendritic spine development. Nature *439*, 283–289.

Schule, B., Albalwi, M., Northrop, E., Francis, D.I., Rowell, M., Slater, H.R., Gardner, R.J., and Francke, U. (2005). Molecular breakpoint cloning and gene expression studies of a novel translocation t(4;15) (q27;q11.2) associated with Prader–Willi syndrome. BMC Med. Genet. *6*, 18.

Schulze, A., Hansen, C., Skakkebaek, N.E., Brondum-Nielsen, K., Ledbeter, D.H., and Tommerup, N. (1996). Exclusion of SNRPN as a major determinant of Prader–Willi syndrome by a translocation breakpoint. Nat. Genet. *12*, 452–454.

Schuster-Gossler, K., Bilinski, P., Sado, T., Ferguson-Smith, A., and Gossler, A. (1998). The mouse Gtl2 gene is differentially expressed during embryonic development, encodes multiple alternatively spliced transcripts, and may act as an RNA. Dev. Dyn. *212*, 214–228.

Seitz, H., Youngson, N., Lin, S.P., Dalbert, S., Paulsen, M., Bachellerie, J.P., Ferguson-Smith, A.C., and Cavaille, J. (2003). Imprinted microRNA genes transcribed antisense to a reciprocally imprinted retrotransposon-like gene. Nat. Genet. *34*, 261–262.

Seitz, H., Royo, H., Bortolin, M.L., Lin, S.P., Ferguson-Smith, A.C., and Cavaille, J. (2004). A large imprinted microRNA gene cluster at the mouse Dlk1-Gtl2 domain. Genome Res. *14*, 1741–1748.

Sekita, Y., Wagatsuma, H., Nakamura, K., Ono, R., Kagami, M., Wakisaka, N., Hino, T., Suzuki-Migishima, R., Kohda, T., Ogura, A., *et al.* (2008). Role of retrotransposon-derived imprinted gene, Rtl*1*, in the feto–maternal interface of mouse placenta. Nat. Genet. *40*, 243–248.

Shevtsov, S.P., and Dundr, M. (2011). Nucleation of nuclear bodies by RNA. Nat. Cell. Biol. *13*, 167–173.

Skryabin, B.V., Gubar, L.V., Seeger, B., Pfeiffer, J., Handel, S., Robeck, T., Karpova, E., Rozhdestvensky, T.S., and Brosius, J. (2007). Deletion of the MBII-85 snoRNA gene cluster in mice results in postnatal growth retardation. PLoS Genet. *3*, e235.

Sleutels, F., Zwart, R., and Barlow, D.P. (2002). The non-coding Air RNA is required for silencing autosomal imprinted genes. Nature *415*, 810–813.

Smit, M., Segers, K., Carrascosa, L.G., Shay, T., Baraldi, F., Gyapay, G., Snowder, G., Georges, M., Cockett, N., and Charlier, C. (2003). Mosaicism of Solid Gold supports the causality of a noncoding A-to-G transition in the determinism of the callipyge phenotype. Genetics *163*, 453–456.

de Smith, A.J., Purmann, C., Walters, R.G., Ellis, R.J., Holder, S.E., Van Haelst, M.M., Brady, A.F., Fairbrother, U.L., Dattani, M., Keogh, J.M., *et al.* (2009). A deletion of the HBII-85 class of small nucleolar RNAs (snoRNAs) is associated with hyperphagia, obesity and hypogonadism. Hum. Mol. Genet. *18*, 3257–3265.

Soeno, Y., Taya, Y., Stasyk, T., Huber, L.A., Aoba, T., and Huttenhofer, A. (2010). Identification of novel ribonucleo–protein complexes from the brain-specific snoRNA MBII-52. RNA *16*, 1293–1300.

Song, G., and Wang, L. (2008a). MiR-433 and miR-127 arise from independent overlapping primary transcripts encoded by the miR-433–127 locus. PLoS One *3*, e3574.

Song, G., and Wang, L. (2008b). Transcriptional mechanism for the paired miR-433 and miR-127 genes by nuclear receptors SHP and ERRgamma. Nucleic Acids Res. *36*, 5727–5735.

Stadler, B.M., Ivanovska, I., Mehta, K., Song, S., Nelson, A., Tan, Y., Mathieu, J., Darby, G.C., Blau, C.A., Ware, C., *et al.* (2010). Characterization of microRNAs involved in embryonic stem cell states. Stem Cells Dev. *19*, 935–950.

Stadtfeld, M., Apostolou, E., Akutsu, H., Fukuda, A., Follett, P., Natesan, S., Kono, T., Shioda, T., and Hochedlinger, K. (2010). Aberrant silencing of imprinted genes on chromosome 12qF1 in mouse induced pluripotent stem cells. Nature *465*, 175–181.

Stark, K.L., Xu, B., Bagchi, A., Lai, W.S., Liu, H., Hsu, R., Wan, X., Pavlidis, P., Mills, A.A., Karayiorgou, M., *et al.* (2008). Altered brain microRNA biogenesis contributes to phenotypic deficits in a 22q11-deletion mouse model. Nat. Genet. *40*, 751–760.

Sun, Y., Nicholls, R.D., Butler, M.G., Saitoh, S., Hainline, B.E., and Palmer, C.G. (1996). Breakage in the SNRPN locus in a balanced *46,XY,t(15;19)* Prader–Willi syndrome patient. Hum. Mol. Genet. *5*, 517–524.

Suzuki, S., Ono, R., Narita, T., Pask, A.J., Shaw, G., Wang, C., Kohda, T., Alsop, A.E., Marshall Graves, J.A., Kohara, Y., *et al.* (2007). Retrotransposon silencing by DNA methylation can drive mammalian genomic imprinting. PLoS Genet. *3*, e55.

Tai, H.C., and Schuman, E.M. (2006). MicroRNA: microRNAs reach out into dendrites. Curr. Biol. *16*, R121–123.

Takada, S., Paulsen, M., Tevendale, M., Tsai, C.E., Kelsey, G., Cattanach, B.M., and Ferguson-Smith, A.C. (2002). Epigenetic analysis of the Dlk1-Gtl2 imprinted domain on mouse chromosome *12*, implications for imprinting control from comparison with Igf2-H19. Hum. Mol. Genet. *11*, 77–86.

Takahashi, K., and Yamanaka, S. (2006). Induction of pluripotent stem cells from mouse embryonic and adult fibroblast cultures by defined factors. Cell *126*, 663–676.

Takahashi, N., Okamoto, A., Kobayashi, R., Shirai, M., Obata, Y., Ogawa, H., Sotomaru, Y., and Kono, T. (2009). Deletion of Gtl2, imprinted non-coding RNA, with its differentially methylated region induces lethal parent-origin-dependent defects in mice. Hum. Mol. Genet. *18*, 1879–1888.

Takeda, H., Caiment, F., Smit, M., Hiard, S., Tordoir, X., Cockett, N., Georges, M., and Charlier, C. (2006). The callipyge mutation enhances bidirectional long-range DLK1-GTL2 intergenic transcription in cis. Proc. Natl. Acad. Sci. U.S.A. *103*, 8119–8124.

Tay, Y., Zhang, J., Thomson, A.M., Lim, B., and Rigoutsos, I. (2008a). MicroRNAs to Nanog, Oct4 and Sox2 coding regions modulate embryonic stem cell differentiation. Nature 455, 1124–1128.

Tay, Y.M., Tam, W.L., Ang, Y.S., Gaughwin, P.M., Yang, H., Wang, W., Liu, R., George, J., Ng, H.H., Perera, R.J., *et al.* (2008b). MicroRNA-134 modulates the differentiation of mouse embryonic stem cells, where it causes post-transcriptional attenuation of Nanog and LRH1. Stem Cells *26*, 17–29.

Terranova, R., Yokobayashi, S., Stadler, M.B., Otte, A.P., van Lohuizen, M., Orkin, S.H., and Peters, A.H. (2008). Polycomb group proteins Ezh2 and Rnf2 direct genomic contraction and imprinted repression in early mouse embryos. Dev. Cell *15*, 668–679.

Tevendale, M., Watkins, M., Rasberry, C., Cattanach, B., and Ferguson-Smith, A.C. (2006). Analysis of mouse conceptuses with uniparental duplication/deficiency for distal chromosome *12*, comparison with chromosome 12 uniparental disomy and implications for genomic imprinting. Cytogenet Genome Res. *113*, 215–222.

Tierling, S., Dalbert, S., Schoppenhorst, S., Tsai, C.E., Oliger, S., Ferguson-Smith, A.C., Paulsen, M., and Walter, J. (2006). High-resolution map and imprinting analysis of the Gtl2-Dnchc1 domain on mouse chromosome 12. Genomics *87*, 225–235.

Tsai, T.F., Armstrong, D., and Beaudet, A.L. (1999a). Necdin-deficient mice do not show lethality or the obesity and infertility of Prader–Willi syndrome. Nat. Genet. *22*, 15–16.

Tsai, T.F., Jiang, Y.H., Bressler, J., Armstrong, D., and Beaudet, A.L. (1999b). Paternal deletion from Snrpn to Ube3a in the mouse causes hypotonia, growth retardation and partial lethality and provides evidence for a gene contributing to Prader–Willi syndrome. Hum. Mol. Genet. *8*, 1357–1364.

Turner, A.M., and Morris, K.V. (2010). Controlling transcription with noncoding RNAs in mammalian cells. Biotechniques *48*, ix–xvi.

Tzur, G., Levy, A., Meiri, E., Barad, O., Spector, Y., Bentwich, Z., Mizrahi, L., Katzenellenbogen, M., Ben-Shushan, E., Reubinoff, B.E., *et al.* (2008). MicroRNA expression patterns and function in endodermal differentiation of human embryonic stem cells. PLoS One 3, e3726.

Verona, R.I., Mann, M.R., and Bartolomei, M.S. (2003). Genomic imprinting: intricacies of epigenetic regulation in clusters. Annu. Rev. Cell. Dev. Biol. *19*, 237–259.

Vitali, P., Basyuk, E., Le Meur, E., Bertrand, E., Muscatelli, F., Cavaille, J., and Huttenhofer, A. (2005). ADAR2-mediated editing of RNA substrates in the nucleolus is inhibited by C/D small nucleolar RNAs. J. Cell. Biol. *169*, 745–753.

Vitali, P., Royo, H., Marty, V., Bortolin-Cavaille, M.L., and Cavaille, J. (2010). Long nuclear-retained non-coding RNAs and allele-specific higher-order chromatin organization at imprinted snoRNA gene arrays. J. Cell. Sci. *123*, 70–83.

Wang, Q., O'Brien, P.J., Chen, C.X., Cho, D.S., Murray, J.M., and Nishikura, K. (2000). Altered G protein-coupling functions of RNA editing isoform and splicing variant serotonin2C receptors. J. Neurochem. *74*, 1290–1300.

Wang, Q., Chow, J., Hong, J., Ferguson Smith, A., Moreno, C., Seaby, P., Vrana, P., Miri, K., Tak, J., Chung, E., *et al.* (2011). Recent acquisition of imprinting at the rodent Sfmbt2 locus correlates with insertion of a large block of miRNAs. BMC Genomics *12*, 204.

Wang, X., Song, X., Glass, C.K., and Rosenfeld, M.G. (2010). The Long Arm of Long Noncoding RNAs: Roles as Sensors Regulating Gene Transcriptional Programs. Cold Spring Harb. Perspect. Biol.

Wang, Y., Baskerville, S., Shenoy, A., Babiarz, J.E., Baehner, L., and Blelloch, R. (2008). Embryonic stem cell-specific microRNAs regulate the G1-S transition and promote rapid proliferation. Nat. Genet. *40*, 1478–1483.

Wawrzik, M., Spiess, A.N., Herrmann, R., Buiting, K., and Horsthemke, B. (2009). Expression of SNURF-SNRPN upstream transcripts and epigenetic regulatory genes during human spermatogenesis. Eur. J. Hum. Genet. *17*, 1463–1470.

Wevrick, R., Kerns, J.A., and Francke, U. (1994). Identification of a novel paternally expressed gene in the Prader–Willi syndrome region. Hum. Mol. Genet. 3, 1877–1882.

Wevrick, R., Kerns, J.A., and Francke, U. (1996). The IPW gene is imprinted and is not expressed in the Prader–Willi syndrome. Acta Genet. Med. Gemellol. (Roma) *45*, 191–197.

White, J.D., Vuocolo, T., McDonagh, M., Grounds, M.D., Harper, G.S., Cockett, N.E., and Tellam, R. (2008). Analysis of the callipyge phenotype through skeletal muscle development; association of Dlk1 with muscle precursor cells. Differentiation *76*, 283–298.

Wirth, J., Back, E., Huttenhofer, A., Nothwang, H.G., Lich, C., Gross, S., Menzel, C., Schinzel, A., Kioschis, P., Tommerup, N., *et al.* (2001). A translocation breakpoint cluster disrupts the newly defined 3′ end of the SNURF-SNRPN transcription unit on chromosome 15. Hum. Mol. Genet. *10*, 201–210.

Wood, A.J., and Oakey, R.J. (2006). Genomic imprinting in mammals: emerging themes and established theories. PLoS Genet. *2*, e147.

Wood, A.J., Roberts, R.G., Monk, D., Moore, G.E., Schulz, R., and Oakey, R.J. (2007). A screen for retrotransposed imprinted genes reveals an association between X chromosome homology and maternal germ-line methylation. PLoS Genet. *3*, e20.

Yamasaki, K., Joh, K., Ohta, T., Masuzaki, H., Ishimaru, T., Mukai, T., Niikawa, N., Ogawa, M., Wagstaff, J., and Kishino, T. (2003). Neurons but not glial cells show reciprocal imprinting of sense and antisense transcripts of Ube3a. Hum. Mol. Genet. *12*, 837–847.

Yang, T., Adamson, T.E., Resnick, J.L., Leff, S., Wevrick, R., Francke, U., Jenkins, N.A., Copeland, N.G., and Brannan, C.I. (1998). A mouse model for Prader–Willi syndrome imprinting-centre mutations. Nat. Genet. *19*, 25–31.

Yekta, S., Shih, I.H., and Bartel, D.P. (2004). MicroRNA-directed cleavage of HOXB8 mRNA. Science *304*, 594–596.

Yoder, J.A., Walsh, C.P., and Bestor, T.H. (1997). Cytosine methylation and the ecology of intragenomic parasites. Trends Genet. *13*, 335–340.

Youngson, N.A., Kocialkowski, S., Peel, N., and Ferguson-Smith, A.C. (2005). A small family of sushi-class retrotransposon-derived genes in mammals and their relation to genomic imprinting. J. Mol. Evol. *61*, 481–490.

Zhang, R., Wang, Y.Q., and Su, B. (2008). Molecular evolution of a primate-specific microRNA family. Mol. Biol. Evol. *25*, 1493–1502.

Zhao, J., Ohsumi, T.K., Kung, J.T., Ogawa, Y., Grau, D.J., Sarma, K., Song, J.J., Kingston, R.E., Borowsky, M., and Lee, J.T. (2010). Genome-wide identification of polycomb-associated RNAs by RIP-seq. Mol. Cell *40*, 939–953.

Zhou, Y., Cheunsuchon, P., Nakayama, Y., Lawlor, M.W., Zhong, Y., Rice, K.A., Zhang, L., Zhang, X., Gordon, F.E., Lidov, H.G., *et al.* (2010). Activation of paternally expressed genes and perinatal death caused by deletion of the Gtl2 gene. Development *137*, 2643–2652.

Part II

Non-coding RNAs: Gene Regulation and Epigenetics

The X-chromosome Archetype for Non-coding RNA Regulation of the Epigenome

4

Daniel H. Kim and Jeannie T. Lee

Abstract

X-chromosome inactivation provides a model for discovering the still emerging functions of non-coding RNAs, which exhibit diverse roles in regulating the epigenome. Combining recent insights into X-chromosome non-coding RNAs with the advent of next-generation sequencing technologies promises many new discoveries in interrogating the functions of non-coding RNAs at the genomic level. Global RNA–protein interaction information can now be queried by utilizing RNA immunoprecipitation followed by high-throughput sequencing (RIP-Seq), which has recently uncovered the diversity of the non-coding Polycomb transcriptome. In this chapter, we discuss non-coding RNAs of the X-inactivation centre that have provided unique insights into RNA molecules as molecular switches, guides and tethers to the epigenome, and key participants in diverse regulatory pathways, including RNA interference and Polycomb silencing.

Introduction

While the vast majority of the mammalian genome is probably transcribed (70–90%), only 1–2% is considered protein coding, revealing a complexity that is only beginning to unfold (Carninci *et al.*, 2005; Kapranov *et al.*, 2007; Mercer *et al.*, 2009). Noncoding RNAs (ncRNAs) play important roles in regulating the epigenome, and recent findings suggest essential roles for ncRNAs as sequence- and allele-specific guides for DNA and histone modifications. In the mammalian germline, small ncRNAs known as Piwi-interacting RNAs (piRNAs) act in RNA-directed DNA methylation pathways to silencing transposable elements (Aravin *et al.*, 2007). Similarly, during X-chromosome inactivation (XCI), the Tsix ncRNA associates with the *de novo* DNA methyltransferase Dnmt3a to facilitate DNA methylation of the non-coding gene *Xist*, which coats the entire inactive X (Xi) chromosome *in cis* (Sun *et al.*, 2006; Payer and Lee, 2008). At the human *HOX-C* locus, a ncRNA known as HOTAIR associates with Polycomb repressive complex 2 (PRC2) and mediates epigenetic regulation of the *HOX-D* region in *trans* (Rinn *et al.*, 2007). Furthermore, the Xist, Tsix, and RepA ncRNAs bind PRC2 and act *in cis* during the initiation phase of XCI in mouse ES cells (Zhao *et al.*, 2008). Using XCI as a roadmap this chapter will highlight the diversity of the ncRNA repertoire in regulating the epigenome.

Non-coding RNAs of the X-inactivation centre

The X-inactivation centre (*Xic*) features several prominent long ncRNAs that regulate XCI (Fig. 4.1) (Payer and Lee, 2008). Xist is a 17-kb ncRNA that spreads along the X-chromosome and coats the entire Xi *in cis* (Borsani *et al.*, 1991; Brown *et al.*, 1991, 1992; Brockdorff *et al.*, 1992; Clemson *et al.*, 1996; Penny *et al.*, 1996; Marahrens *et al.*, 1997). The Xist-coated domain is characterized by the presence of histone H3 lysine 27 trimethylation (H3K27me3), which is catalysed by PRC2 (Plath *et al.*, 2003; Silva *et al.*, 2003; Kohlmaier *et al.*, 2004; Zhao *et al.*, 2008). Given the direct

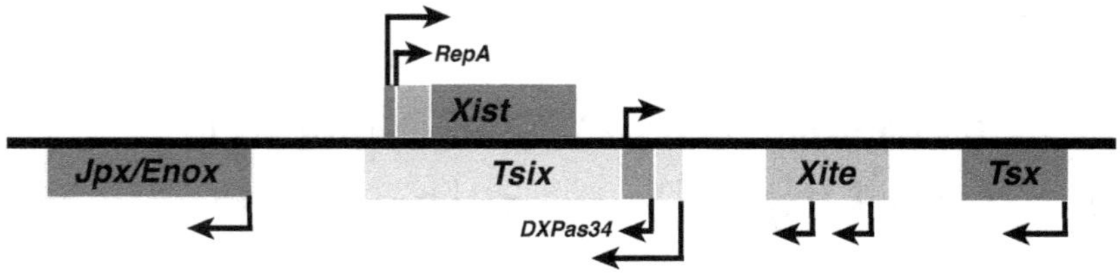

Figure 4.1 Non-coding RNAs of the X-inactivation centre. Long non-coding RNAs are key regulators of X-chromosome inactivation. Furthermore, the *Xist* and *Tsix* non-coding genes also contain shorter transcription units within their respective loci.

interaction between Xist and PRC2, Xist may serve to recruit PRC2 specifically to the Xi during the process of XCI (Zhao *et al.*, 2008). *Xist* is regulated by both positive and negative factors, namely other long ncRNAs of the *Xic*. The 40-kb Tsix ncRNA is antisense to *Xist* and acts to repress *Xist* expression on the active X (Xa) chromosome (Lee *et al.*, 1999; Lee and Lu, 1999; Sado *et al.*, 2001).

Non-coding RNAs bind Polycomb

Xist ncRNA initiates the formation of facultative heterochromatin on the Xi, but its mechanisms of action have remained largely unknown until recently. A role for Xist RNA in recruiting chromatin modifiers to the Xi, however, has long been speculated. Given the involvement of the heterochromatin mark H3K27me3 in silencing the Xi, one candidate interacting factor has been PRC2 (Plath *et al.*, 2003; Silva *et al.*, 2003; Kohlmaier *et al.*, 2004). Indeed, recent work has shown that the PRC2 subunit, Ezh2, associates with Xist RNA in mouse ES cells (Zhao *et al.*, 2008). These components of PRC2 associate with both single- and double-stranded regions of Xist RNA, as shown by sensitivity of RIP experiments to RNase I and RNase V1. Xist RNA binds PRC2 through a unique motif known as 'Repeat A', located at the 5′ end of the RNA. In the mouse, Repeat A is a 28-nt element repeated 7.5 times in tandem. The element has been proposed to form various secondary structures, including a double stem–loop and more complex structures (Wutz *et al.*, 2002; Maenner *et al.*, 2010).

It was recently discovered that a second transcript crosses the Repeat A motif. A short 1.6 kb transcript called 'RepA' originates just upstream of Repeat A and is transcribed in the same orientation as Xist. It is thought that Repeat A co-transcriptionally recruits PRC2 to the 5′ end of *Xist* and that, paradoxically, trimethylation of H3-K27 by PRC2 facilitates the transcriptional activation of *Xist* (Fig. 4.2) (Zhao *et al.*, 2008). Full-length Xist RNA also binds PRC2 and spreads the repressive chromatin complex throughout the X-chromosome as the RNA accumulates along it *in cis*. Interestingly, the antisense sequence to RepA RNA also binds to PRC2 both *in vitro*, as shown by RNA gel shifts, and *in vivo*, as shown by strand-specific RIP. Antisense Tsix RNA may regulate the interaction between PRC2 and RepA/Xist RNA by titrating PRC2 away from RepA RNA and creating a potential block to the progression of XCI. With respect to XCI, knockdowns of RepA RNA and Ezh2 phenocopy each other, both blocking the progression of XCI, as evidenced by failure to up-regulate Xist RNA and by the inability of embryoid bodies to differentiate properly. The observation that PRC2 utilizes an RNA cofactor as a sequence-specific guide provides a new paradigm for epigenomic regulation by ncRNAs.

Non-coding RNA discovery through next-generation sequencing

Many ncRNAs have been implicated in epigenomic regulation through the application of next-generation sequencing technologies. In particular, high-throughput RNA sequencing (RNA-Seq) has enabled transcriptome profiling at unprecedented depths, revealing the complexity of the mammalian transcriptome (Wang *et al.*, 2009). Next-generation sequencing

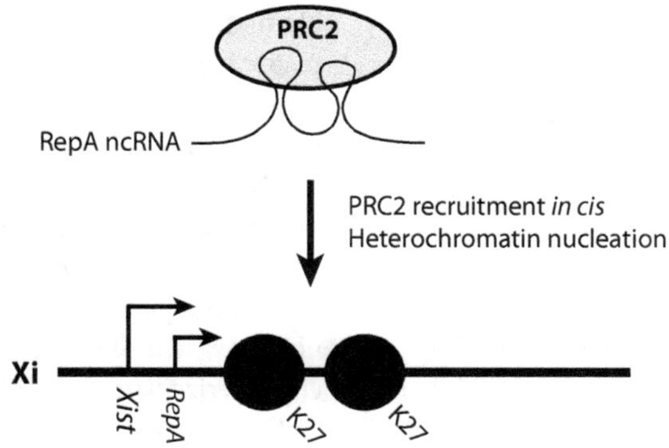

Figure 4.2 Polycomb recruitment via RepA non-coding RNA. The RepA non-coding RNA binds to PRC2 prior to the initiation of X-chromosome inactivation and recruits PRC2 to the future inactive X-chromosome. PRC2 then methylates H3K27 and nucleates the formation of heterochromatin.

methods provide some advantages over traditional, hybridization-based approaches (e.g. microarrays). For example, RNA-Seq provides the ability to map and quantify transcriptomes, including novel species of ncRNA (Trapnell *et al.*, 2010). To generate RNA-Seq libraries, cDNAs of total or partitioned RNA (poly(A)+, size-selected, RNA-immunoprecipitation, etc.) with attached adaptors are sequenced from one or both ends using high-throughput sequencing technology. The use of RNA-Seq is particularly attractive for discovering novel ncRNAs from unannotated regions of the genome. RNA-Seq approaches are not prone to the kind of background seen in hybridization-based approaches and are sensitive enough to detect transcripts spanning five orders of magnitude, allowing for low-abundance ncRNAs to be discovered. RNA-Seq also provides single-base resolution and a digital readout of transcript expression levels genome-wide.

Non-coding Polycomb transcriptome

RNA-seq technology can be combined with IP to identify transcriptomes associated with protein complexes. In one recent application, native RNA immunoprecipitation (RIP) was combined with RNA-Seq (RIP-Seq), creating a powerful technique to uncover novel, Polycomb-associated ncRNAs genome-wide (Fig. 4.3) (Zhao *et al.*, 2010). In this method, 200–1200 nt nuclear RNAs that associate with Ezh2 are immunoprecipitated from mES cells, and libraries are generated with cDNAs having strand-specific adaptors for sequencing on the Illumina platform. Mapping Polycomb-associated ncRNA reads yield alignments on every chromosome in the mouse genome, indicative of a broad-spectrum sampling of the PRC2 transcriptome. Transcripts originating from the X-chromosome are well represented, with positive controls, Tsix, RepA, and Xist ncRNAs, being present multiple times within PRC2 RIP-Seq libraries. In mES cells, at least 9000 distinct transcripts bind with PRC2, revealing a highly complex and abundant population of long ncRNAs that may serve to direct PRC2 to its target loci throughout the genome. The PRC2 transcriptome may represent 5%–25% of all mouse transcripts, depending on the total number of transcripts within the mouse genome. Thus, a sizeable fraction of the transcriptional output of mES cells may be devoted to the regulation of epigenetic pathways involving Polycomb.

Approximately one-fifth of all bivalent loci (Bernstein *et al.*, 2006) in mES cells are associated with a PRC2-interacting ncRNA. A similar fraction of Suz12 binding sites also contain ncRNAs in the PRC2 transcriptome (Boyer *et al.*, 2006; Ku *et al.*, 2008). Some large intergenic ncRNAs (lincRNAs) (Guttman *et al.*, 2009) have also been associated with Polycomb

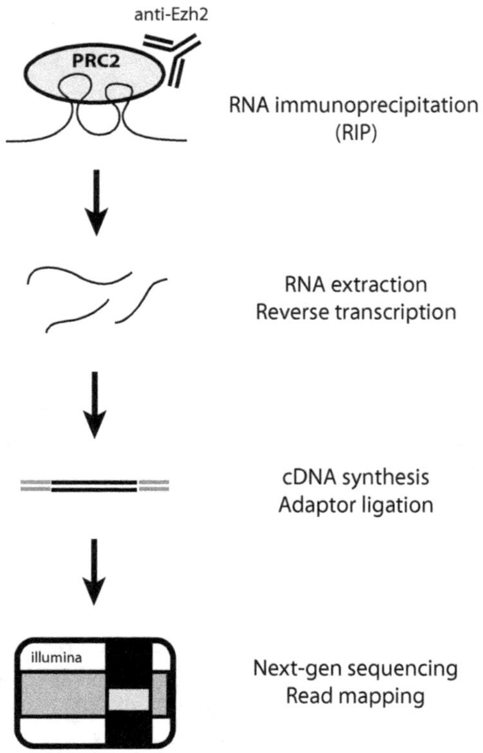

Figure 4.3 RIP-Seq: RNA immunoprecipitation and RNA-Seq. RNA immunoprecipitation of PRC2 and its associated RNA is performed with an Ezh2-specific antibody, upon which the RNA is extracted and reverse transcribed for cDNA synthesis, adaptor ligation, next-generation sequencing, and bioinformatic analysis.

proteins via a RIP-chip approach (Khalil *et al.*, 2009). Only ~2% of the PRC2 transcriptome of Zhao *et al.* (2010) intersect with lincRNAs and few if any intersect with the Polycomb-associated transcripts identified by Khalil *et al.* (2009), suggesting that different cell types might employ different sets of Polycomb RNAs. Interestingly, ~40% of known oncogenes and tumour-suppressor genes are targeted by a member of the PRC2 transcriptome, which is notable, given the role of Polycomb misregulation in various cancers (Mills). These cancer-related ncRNAs may serve as novel therapeutic or diagnostic targets for cancer treatment and screening.

Non-coding RNA as molecular switches: the case of Tsix and Jpx

Importantly, ncRNAs serve to regulate the expression not only of coding genes but also of other ncRNAs. The *Xic* is enriched for ncRNA genes, and in particular, the Xist ncRNA is regulated by another long ncRNA known as Jpx (Tian *et al.*). *Xist* is both positively and negatively regulated by ncRNAs, with Tsix ncRNA serving as the well-characterized negative controller of *Xist*. *Tsix* is further regulated by the upstream *Xite*, which acts as a positive regulator of *Tsix* expression (Ogawa and Lee, 2003; Stavropoulos *et al.*, 2005), and indirectly as a repressor of *Xist*. *Xist* is positively regulated by *Jpx*, which lies ~10 kb upstream of *Xist*. In female mES cells, *Jpx* expression increases during ES cell differentiation and is expressed from both alleles. *Jpx* is thus not subject to XCI.

Jpx serves as an essential activator of *Xist* and is critical for the progression of XCI in female cells. Deletion of one *Jpx* allele dramatically reduces the expression of *Jpx* during female mES cell differentiation by 80–90%, suggesting that expression from the intact *Jpx* allele is also compromised and that *Jpx* expression may be subject to cross-talk regulation between the two alleles. Differentiating embryoid bodies (Silva *et al.*) with deficiencies in *Jpx* expression exhibit poor cellular outgrowth and fail to properly up-regulate *Xist*, as evidenced by the absence of Xist clouds after 8–12 days of differentiation in culture, when the majority of cells exhibit Xist clouds. Additionally, *Jpx* mutant cells are characterized by massive cell death during differentiation. Expressing *Jpx* from an autosomal transgene restores *Xist* up-regulation and rescues the cell death phenotype, indicating that Jpx RNA acts *in trans*, unlike other RNAs expressed from the X-inactivation centre. Knockdown using RNA interference against the Jpx RNA recapitulates the heterozygous *Jpx* deletion phenotype of massive cell death and loss of proper *Xist* up-regulation. Taken together, these data demonstrate that Jpx acts as a functional RNA molecule rather than as a traditional *cis*-regulatory enhancer. Consistent with a role in activating *Xist* and XCI, deleting *Jpx* in male mES cells yields no noticeable phenotypes, with Xist and Tsix RNA levels remaining unperturbed. *Xist* regulation is thus a complex process controlled by both positive and negative RNA factors, Tsix and Jpx appear to independently control *Xist* expression as parallel non-coding RNA switches, where Tsix functions to silence *Xist* on the Xa, while Jpx induces *Xist* expression on the Xi.

Non-coding RNA-directed DNA methylation

A parallel mechanism of *Xist* regulation occurs through DNA methylation at the *Xist* promoter via *Tsix* (Sado *et al.*, 2005; Sun *et al.*, 2006). In male ES cells, the *Xist* promoter is fully methylated in the undifferentiated state, even in *Tsix*-deficient cells. However, in female ES cells, the *Xist* promoter is partially methylated in the undifferentiated state, with the promoter region becoming fully methylated on the Xa upon ES cell differentiation. In *Tsix*$^{-}$/$^{-}$ female ES cells, the *Xist* promoter is significantly under-methylated, indicating that *Tsix* plays a role in methylating *Xist*. Interestingly, Tsix RNA associates with the *de novo* DNA methyltransferase, Dnmt3a, in both male and female ES cells, as determined by RNA-chromatin immunoprecipitation (RNA-ChIP) (Sun *et al.*, 2006). Tsix RNA may therefore either direct Dnmt3a to the *Xist* promoter or promote Dnmt3a's catalytic activity at the *Xist* promoter. This type of mechanism may be a form of RNA-directed DNA methylation, especially given the potential for RNAi regulation (see next section). However, it should be noted that Dnmt3a is not absolutely necessary to silence *Xist*, since Dnmt3a-deficiency does not result in *Xist* transcriptional activation (Sado *et al.*, 2004). DNA methylation is probably just one of multiple mechanisms involved in silencing *Xist*.

Non-coding RNAs and RNA interference

The action of Tsix RNA at the *Xist* promoter is reminiscent of small RNA-directed DNA methylation via RNA interference (RNAi)-related pathways (Chan *et al.*, 2005). In both male and female ES cells, a small fraction of Tsix and Xist ncRNAs form long double-stranded RNAs (dsRNAs) (Ogawa *et al.*, 2008). As male and female ES cell differentiation proceeds, the amounts of Xist/Tsix dsRNAs decrease in both cell types. Interestingly, small RNAs of ~25–42 nt in length (referred to as xiRNAs for their *Xic* location) are increasingly found at the *Xist*/*Tsix* locus in both orientations as male and female ES cells differentiate. xiRNAs are present at the *Xist* promoter, and given the role of Tsix RNA in *Xist* promoter DNA methylation, xiRNAs processed from Tsix RNA may potentially contribute to RNA-directed DNA methylation, reminiscent of similarly sized piRNAs in the mammalian germline (Aravin *et al.*, 2007).

Further evidence that xiRNAs are derived from Tsix RNA comes from ES cells where *Tsix* or *Xite*, which regulates *Tsix*, have been deleted. Absence of *Tsix* and *Xite* both correspond to the dramatic loss of antisense-stranded xiRNAs, as well as sense-stranded xiRNAs. Given the presence of xiRNAs in male ES cells, this suggests that xiRNAs are produced from the Xa. Supporting this idea, Xist/Tsix dsRNAs are stabilized and detectable when they originate from the Xi, indicating that only Xa-originating Xist/Tsix dsRNAs may be processed into xiRNAs. Given the role of the RNAi enzyme Dicer in processing dsRNAs into small RNAs, Dicer may be involved in generating xiRNAs from Xist/Tsix duplexes on the Xa. Female ES cells where Dicer is deleted exhibit loss of xiRNAs, as well as precocious expression of Xist RNA by ~5–10-fold prior to the initiation of XCI (Kanellopoulou *et al.*, 2009; Nesterova *et al.*, 2008; Ogawa *et al.*, 2008). Similarly, male ES cells lacking Dicer show *Xist* up-regulation during early differentiation. Several possible mechanisms may explain the role of Dicer in *Xist* regulation: (1) Dicer may process Xist/Tsix dsRNAs into xiRNAs, which would reduce the levels of Xist RNA, (2) xiRNAs generated by Dicer may be required to transcriptionally silence *Xist* expression through RNA-directed DNA methylation at the *Xist* promoter, (3) Dicer may affect the levels of pluripotency factors, namely Oct4, Sox2, and Nanog, which regulate *Xist* expression and the kinetics of XCI (Navarro *et al.*, 2008, 2010; Donohoe *et al.*, 2009) and/or (4) Dicer may regulate the levels of DNA methyltransferases required for *Xist* promoter methylation (Nesterova *et al.*, 2008).

Supporting the role of Dicer and RNAi in XCI, loss of Dicer affects the ability of Xist RNA to induce heterochromatin formation on the Xi, even when *Xist* levels are elevated. In *Dicer-Tsix* double mutant female ES cells, *Xist* is dramatically up-regulated, and during ES cell differentiation, *Xist* levels are similar to wild-type levels and Xist RNA localizes to the Xi. However, 30–40% of cells that exhibit Xist clouds are unable to induce H3K27me3 and heterochromatin formation on the Xi. Dicer may thus be required for proper silencing of Xi chromatin. The role of Dicer in XCI is further supported by the presence of LINE-1-derived small interfering RNAs (siRNAs) on the X-chromosome during female ES cell differentiation (Chow *et al.*). Taken together, these data provide strong support for the intersection between the XCI and RNAi pathways.

Conclusions

Investigating the mechanisms of X-chromosome silencing have continued to yield interesting and novel discoveries concerning the roles of ncRNAs in epigenome regulation. In particular, long ncRNAs seem uniquely suited to the phenomena of XCI and allele-specific regulation, given that they remain bound to the locus from which they are transcribed. The connection between *Xic* ncRNAs and RNAi-related pathways may also extend to other regions of the genome and may provide additional insights into ncRNA regulation of the epigenome. In as much as many long ncRNAs in mammalian genomes remain mysterious, X-chromosome ncRNAs will serve as archetype for investigating their functions.

Acknowledgements

We thank all laboratory members for valuable discussions. D.H.K. is supported by a Damon Runyon Cancer Research Foundation Fellowship (DRG-#2027-09) and the Beckman Fellows Program at Caltech. J.T.L is an Investigator of the Howard Hughes Medical Institute.

References

Aravin, A.A., Hannon, G.J., and Brennecke, J. (2007). The Piwi-piRNA pathway provides an adaptive defense in the transposon arms race. Science *318*, 761–764.

Bernstein, B.E., Mikkelsen, T.S., Xie, X., Kamal, M., Huebert, D.J., Cuff, J., Fry, B., Meissner, A., Wernig, M., Plath, K., *et al.* (2006). A bivalent chromatin structure marks key developmental genes in embryonic stem cells. Cell *125*, 315–326.

Borsani, G., Tonlorenzi, R., Simmler, M.C., Dandolo, L., Arnaud, D., Capra, V., Grompe, M., Pizzuti, A., Muzny, D., Lawrence, C., *et al.* (1991). Characterization of a murine gene expressed from the inactive X chromosome. Nature *351*, 325–329.

Boyer, L.A., Plath, K., Zeitlinger, J., Brambrink, T., Medeiros, L.A., Lee, T.I., Levine, S.S., Wernig, M., Tajonar, A., Ray, M.K., *et al.* (2006). Polycomb complexes repress developmental regulators in murine embryonic stem cells. Nature *441*, 349–353.

Brockdorff, N., Ashworth, A., Kay, G.F., McCabe, V.M., Norris, D.P., Cooper, P.J., Swift, S., and Rastan, S. (1992). The product of the mouse Xist gene is a 15 kb inactive X-specific transcript containing no conserved ORF and located in the nucleus. Cell *71*, 515–526.

Brown, C.J., Ballabio, A., Rupert, J.L., Lafreniere, R.G., Grompe, M., Tonlorenzi, R., and Willard, H.F. (1991). A gene from the region of the human X inactivation centre is expressed exclusively from the inactive X chromosome. Nature *349*, 38–44.

Brown, C.J., Hendrich, B.D., Rupert, J.L., Lafreniere, R.G., Xing, Y., Lawrence, J., and Willard, H.F. (1992). The human XIST gene: analysis of a 17 kb inactive X-specific RNA that contains conserved repeats and is highly localized within the nucleus. Cell *71*, 527–542.

Carninci, P., Kasukawa, T., Katayama, S., Gough, J., Frith, M.C., Maeda, N., Oyama, R., Ravasi, T., Lenhard, B., Wells, C., *et al.* (2005). The transcriptional landscape of the mammalian genome. Science *309*, 1559–1563.

Chan, S.W., Henderson, I.R., and Jacobsen, S.E. (2005). Gardening the genome: DNA methylation in *Arabidopsis thaliana*. Nat. Rev. Genet. *6*, 351–360.

Chow, J.C., Ciaudo, C., Fazzari, M.J., Mise, N., Servant, N., Glass, J.L., Attreed, M., Avner, P., Wutz, A., Barillot, E., *et al.* LINE-1 activity in facultative heterochromatin formation during X chromosome inactivation. Cell *141*, 956–969.

Clemson, C.M., McNeil, J.A., Willard, H.F., and Lawrence, J.B. (1996). XIST RNA paints the inactive X chromosome at interphase: evidence for a novel RNA involved in nuclear/chromosome structure. J. Cell. Biol. *132*, 259–275.

Donohoe, M.E., Silva, S.S., Pinter, S.F., Xu, N., and Lee, J.T. (2009). The pluripotency factor Oct4 interacts with Ctcf and also controls X-chromosome pairing and counting. Nature *460*, 128–132.

Guttman, M., Amit, I., Garber, M., French, C., Lin, M.F., Feldser, D., Huarte, M., Zuk, O., Carey, B.W., Cassady, J.P., *et al.* (2009). Chromatin signature reveals over a thousand highly conserved large non-coding RNAs in mammals. Nature *458*, 223–227.

Kanellopoulou, C., Muljo, S.A., Dimitrov, S.D., Chen, X., Colin, C., Plath, K., and Livingston, D.M. (2009). X chromosome inactivation in the absence of Dicer. Proc. Natl. Acad. Sci. U.S.A. *106*, 1122–1127.

Kapranov, P., Cheng, J., Dike, S., Nix, D.A., Duttagupta, R., Willingham, A.T., Stadler, P.F., Hertel, J., Hackermuller, J., Hofacker, I.L., *et al.* (2007). RNA maps reveal new RNA classes and a possible function for pervasive transcription. Science *316*, 1484–1488.

Khalil, A.M., Guttman, M., Huarte, M., Garber, M., Raj, A., Rivea Morales, D., Thomas, K., Presser, A., Bernstein, B.E., van Oudenaarden, A., *et al.* (2009). Many human large intergenic noncoding RNAs associate with chromatin-modifying complexes and affect gene expression. Proc. Natl. Acad. Sci. U.S.A. *106*, 11667–11672.

Kohlmaier, A., Savarese, F., Lachner, M., Martens, J., Jenuwein, T., and Wutz, A. (2004). A chromosomal memory triggered by Xist regulates histone methylation in X inactivation. PLoS Biol. *2*, E171.

Ku, M., Koche, R.P., Rheinbay, E., Mendenhall, E.M., Endoh, M., Mikkelsen, T.S., Presser, A., Nusbaum, C., Xie, X., Chi, A.S., *et al.* (2008). Genomewide analysis of PRC1 and PRC2 occupancy identifies two classes of bivalent domains. PLoS Genet. *4*, e1000242.

Lee, J.T., and Lu, N. (1999). Targeted mutagenesis of Tsix leads to nonrandom X inactivation. Cell *99*, 47–57.

Lee, J.T., Davidow, L.S., and Warshawsky, D. (1999). Tsix, a gene antisense to Xist at the X-inactivation centre. Nat. Genet. *21*, 400–404.

Maenner, S., Blaud, M., Fouillen, L., Savoye, A., Marchand, V., Dubois, A., Sanglier-Cianferani, S., Van Dorsselaer, A., Clerc, P., Avner, P., *et al.* (2010). 2-D structure of the A region of Xist RNA and its implication for PRC2 association. PLoS Biol. *8*, e1000276.

Marahrens, Y., Panning, B., Dausman, J., Strauss, W., and Jaenisch, R. (1997). Xist-deficient mice are defective in dosage compensation but not spermatogenesis. Genes Dev. *11*, 156–166.

Mercer, T.R., Dinger, M.E., and Mattick, J.S. (2009). Long non-coding RNAs: insights into functions. Nat. Rev. Genet. *10*, 155–159.

Mills, A.A. (2010). Throwing the cancer switch: reciprocal roles of polycomb and trithorax proteins. Nat. Rev. Cancer *10*, 669–682.

Navarro, P., Chambers, I., Karwacki-Neisius, V., Chureau, C., Morey, C., Rougeulle, C., and Avner, P. (2008). Molecular coupling of Xist regulation and pluripotency. Science *321*, 1693–1695.

Navarro, P., Oldfield, A., Legoupi, J., Festuccia, N., Dubois, A., Attia, M., Schoorlemmer, J., Rougeulle, C., Chambers, I., and Avner, P. (2010). Molecular coupling of Tsix regulation and pluripotency. Nature *468*, 457–460.

Nesterova, T.B., Popova, B.C., Cobb, B.S., Norton, S., Senner, C.E., Tang, Y.A., Spruce, T., Rodriguez, T.A., Sado, T., Merkenschlager, M., *et al.* (2008). Dicer regulates Xist promoter methylation in ES cells indirectly through transcriptional control of Dnmt3a. Epigen. Chromatin. *1*, 2.

Ogawa, Y., and Lee, J.T. (2003). Xite, X-inactivation intergenic transcription elements that regulate the probability of choice. Mol. Cell *11*, 731–743.

Ogawa, Y., Sun, B.K., and Lee, J.T. (2008). Intersection of the RNA interference and X-inactivation pathways. Science *320*, 1336–1341.

Payer, B., and Lee, J.T. (2008). X chromosome dosage compensation: how mammals keep the balance. Annu. Rev. Genet. *42*, 733–772.

Penny, G.D., Kay, G.F., Sheardown, S.A., Rastan, S., and Brockdorff, N. (1996). Requirement for Xist in X chromosome inactivation. Nature *379*, 131–137.

Plath, K., Fang, J., Mlynarczyk-Evans, S.K., Cao, R., Worringer, K.A., Wang, H., de la Cruz, C.C., Otte, A.P., Panning, B., and Zhang, Y. (2003). Role of histone H3 lysine 27 methylation in X inactivation. Science *300*, 131–135.

Rinn, J.L., Kertesz, M., Wang, J.K., Squazzo, S.L., Xu, X., Brugmann, S.A., Goodrough, L.H., Helms, J.A., Farnham, P.J., Segal, E., *et al.* (2007). Functional demarcation of active and silent chromatin domains in human HOX loci by noncoding RNAs. Cell *129*, 1311–1323.

Sado, T., Wang, Z., Sasaki, H., and Li, E. (2001). Regulation of imprinted X-chromosome inactivation in mice by Tsix. Development *128*, 1275–1286.

Sado, T., Okano, M., Li, E., and Sasaki, H. (2004). *De novo* DNA methylation is dispensable for the initiation and propagation of X chromosome inactivation. Development *131*, 975–982.

Sado, T., Hoki, Y., and Sasaki, H. (2005). Tsix silences Xist through modification of chromatin structure. Dev. Cell *9*, 159–165.

Silva, J., Mak, W., Zvetkova, I., Appanah, R., Nesterova, T.B., Webster, Z., Peters, A.H., Jenuwein, T., Otte, A.P., and Brockdorff, N. (2003). Establishment of histone h3 methylation on the inactive X chromosome requires transient recruitment of Eed-Enx1 polycomb group complexes. Dev. Cell *4*, 481–495.

Stavropoulos, N., Rowntree, R.K., and Lee, J.T. (2005). Identification of developmentally specific enhancers for Tsix in the regulation of X chromosome inactivation. Mol. Cell. Biol. *25*, 2757–2769.

Sun, B.K., Deaton, A.M., and Lee, J.T. (2006). A transient heterochromatic state in Xist preempts X inactivation choice without RNA stabilization. Mol. Cell *21*, 617–628.

Tian, D., Sun, S., and Lee, J.T. (2010). The long noncoding RNA, Jpx, is a molecular switch for X chromosome inactivation. Cell *143*, 390–403.

Trapnell, C., Williams, B.A., Pertea, G., Mortazavi, A., Kwan, G., van Baren, M.J., Salzberg, S.L., Wold, B.J., and Pachter, L. (2010). Transcript assembly and quantification by RNA-Seq reveals unannotated transcripts and isoform switching during cell differentiation. Nat. Biotechnol. *28*, 511–515.

Wang, Z., Gerstein, M., and Snyder, M. (2009). RNA-Seq: a revolutionary tool for transcriptomics. Nat. Rev. Genet. *10*, 57–63.

Wutz, A., Rasmussen, T.P., and Jaenisch, R. (2002). Chromosomal silencing and localization are mediated by different domains of Xist RNA. Nat. Genet. *30*, 167–174.

Zhao, J., Ohsumi, T.K., Kung, J.T., Ogawa, Y., Grau, D.J., Sarma, K., Song, J.J., Kingston, R.E., Borowsky, M., and Lee, J.T. (2010). Genome-wide identification of polycomb-associated RNAs by RIP-seq. Mol. Cell *40*, 939–953.

Zhao, J., Sun, B.K., Erwin, J.A., Song, J.J., and Lee, J.T. (2008). Polycomb proteins targeted by a short repeat RNA to the mouse X chromosome. Science *322*, 750–756.

5 Roles of Non-coding RNAs in the Control of the Coupling between Transcription and Alternative Splicing

Mariano Alló and Alberto R. Kornblihtt

Since the discovery of splicing in 1977, alternative splicing was seen for more than two decades as an interesting mechanism to generate protein diversity but with limited genome-wide influence because it was thought to affect only 20% of mammalian genes. The sequencing of the human and other mammals' genomes in the early 2000s together with more recent high throughput analyses of splicing isoforms generated a renewed interest in alternative splicing. We know now that alternative splicing affects more than 90% of human genes (Wang *et al.*, 2008), that normal and pathological cell differentiation not only depends on differential gene transcription but also on alternative splicing patterns, that mutations that either create or abolish alternative splicing regulatory sequences, named splicing enhancers and silencers, are a widespread source of human disease (Caceres and Kornblihtt, 2002) and that alternative splicing factors can be misregulated in cancer (Srebrow and Kornblihtt, 2006).

The relationship between splicing and non-coding RNAs has emerged recently in the middle of an avalanche of papers showing how chromatin context could affect splicing choices. The convergence of these previously unrelated areas (non-coding RNAs, chromatin and splicing) has presented a novel and intriguing scenario that will be covered in this chapter, starting with a brief overview of transcription and splicing introducing the understanding of how these processes could be modulated by external factors. Then we will focus on our work using non-coding small RNAs (ncsRNAs) to regulate alternative splicing in human cells. Finally, we will discuss the evidence supporting the potential activity of endogenous non-coding RNAs as modulators of alternative splicing.

Transcription and splicing

Transcription involves a series of biological reactions carrying the information from DNA to a mature mRNA molecule through different processing steps. Most mRNA encoding eukaryotic genes are transcribed by RNA polymerase II (Pol II), whereas RNA polymerase I transcribes ribosomal RNAs (28S, 18S and 5.8S) and RNA polymerase III transcribes transfer RNA together with ribosomal 5S.

Formerly it was believed that transcription and the mRNA processing steps were pathways compartmentalized both temporally and spatially within the nucleus. The processing steps include:

1. capping: a modified GTP nucleotide is added to nascent mRNA to form a 7 methyl-guanosine cap;
2. splicing: the removal of intron and ligation of exons;
3. polyadenylation: The nascent mRNA is cut in the 3′ end with the subsequent addition of the poly(A) tail.

However, this view has changed dramatically over the last decade by evidence that all processing steps occur co-transcriptionally and are coupled to transcription. It has been proposed that co-transcriptionality gives an structure for step by step quality control (Orphanides and Reinberg,

2002). In this actual view, transcription of protein coding genes starts with the formation of a PIC (pre-initiation complex) over the promoter region including the recruitment of Pol II and other general transcription factors. But Pol II is not ready to start transcription yet, it needs a phosphorylation step carried out by CdK7 (large subunit of general transcription factor TFIIH) which phosphorylates the carboxy-terminal domain (CTD) of the large subunit of RNA polymerase II (Akoulitchev *et al.*, 1995). After phosphorylation of serines 5 of the CTD and the concomitant conformational change, Pol II moves away from the promoter transcribing the first 20–40 nucleotides before remaining paused. Most factors taking place in the transcription initiation process stay assembled to form a scaffold of basal factors (Yudkovsky *et al.*, 2000) favouring higher rate of initiation in subsequent rounds of transcription (Jiang and Gralla, 1993).

RNA polymerase II tends to release from template DNA (abortive initiation) or stay paused until the action of transcription factors such as P-TEFb (Price, 2000), formed by the cyclin dependent kinase (CdK9) and other cyclins (T1, T2 or K). CdK9 phosphorylates serines 2 of the CTD promoting the release of the paused state and the beginning of the elongation phase of transcription (Mancebo *et al.*, 1997; Yang *et al.*, 1997; Peng *et al.*, 1998).

RNA polymerase II elongation

The rate of transcription of Pol II during elongation is affected by different factors including template chromatin structure and the action of positive and negative auxiliary regulators. Nevertheless, these factors not only affect the elongation rate but also the closely related process of processivity. While elongation rate is defined as the mean of nucleotides transcribed/time interval, processivity involves the capacity of Pol II for transcribe genes without releasing the template (mean of nucleotides added by initiation event). Albeit, both parameters are deeply interconnected and can be affected either in a coordinated or differentiated manner by diverse kinds of factors (Mason and Struhl, 2005). Usually both are used as a common feature of transcription quality.

RNA polymerase II is considered as a highly processive enzyme, capable of transcribing giant genes as large as million bases long. However, Pol II is an enzyme prone to transcriptional pauses and stalls (Sims *et al.*, 2004). Those pauses could be rescued or reverted by different positive elongation factors such as TFIIF, elongin or ELL (Bengal *et al.*, 1991; Bradsher *et al.*, 1993; Shilatifard *et al.*, 1996; Bourgeois *et al.*, 2002; Morillon *et al.*, 2003; Rondon *et al.*, 2003). But also, negative elongation factors exist. NELF is very well known complex of five proteins that promotes Pol II pausing (Yamaguchi *et al.*, 1999).

During transcription, chromatin, but more specifically nucleosomes are considered a barrier for Pol II moving forward. We will discuss this issue in detail later.

Splicing

Typical animal and plant genes consist of sequences of exons separated by introns which are excised from the pre-mRNA during the process known as splicing (Black, 2003). Exon-intron boundaries are defined by donor (5′ site) and acceptor (3′ site) sequences and the direct interaction between these sequences and basal components of splicing machinery. It has been shown that the relationship between exon and flanking introns length play an important role during exon recognition (Marais *et al.*, 2005). More than 300 proteins are sequentially assembled together with the ribonucleoprotein complexes (snRNPs) U1, U2, U4, U5 and U6 to form the complex known as 'the splicesome' (Jurica and Moore, 2003).

Two models have been proposed for splicing site recognition (Berget, 1995): (i) intron definition, when the recognition is due to direct identification of intronic 5′ and 3′ splice sites as a splicing unit. (ii) exon definition, when the recognition begins through exon boundaries.

Intron definition is used when intron sizes are in the range of 200–250 nucleotides long. As introns become longer this kind of recognition turns inefficient and exon definition operates preferentially (Hertel, 2008). In any case, the splicing process will be done by the dynamic mega-complex called spliceosome (Staley and Guthrie, 1998; Black, 2003; House and Lynch, 2008).

As we have already mentioned, for many years transcription and mRNA processing were considered as independent processes taking place in the cellular nucleus. Through biochemical, cytological and functional evidences, it has been demonstrated that the three main mRNA processing reactions (capping, splicing and polyadenylation) are strongly coupled to RNA polymerase II transcription (Bentley, 1999; Hirose and Manley, 2000; Proudfoot, 2000; Proudfoot *et al.*, 2002). The central role of this coupling is played by the unusual carboxy-terminal domain of Pol II (CTD), which is composed in mammals by 52 heptads repeats of Tyr-Ser-Pro-Thr-Ser-Pro-Ser (Y-S-P-T-S-P-S), 45 repeats in flies and 26 in yeasts (Palancade and Bensaude, 2003). During transcription initiation the CTD is hyper-phosporylated in serine 5 (Ser-5) by TFIIH, producing the stalling of Pol II. At the same time capping factors associates to the CTD and add the 7-methylguanosine cap to the nascent transcript. Then, pTEFb is recruited and phosphorylates Ser-2 of CTD favouring the entrance in elongation phase of transcription (Orphanides and Reinberg, 2002).

Splicing can occur either co-transcriptionally or post-transcriptionally, and in some particular cases, in both ways. A good example is represented by the balbiani rings genes in *Chironomus tentans*, where first introns are removed co-transcriptionally, while last ones are spliced post-transcriptionally (Neugebauer, 2002).

There are several examples showing co-transcriptional splicing *in vivo*, from the observation of intronic loops in nacsent mRNA with electronic microscopy, (Beyer and Osheim, 1988) to the co-immunoprecipitation of splicing proteins associated to chromatin of active genes (Kotovic *et al.*, 2003; Gornemann *et al.*, 2005; Lacadie and Rosbash, 2005; Listerman *et al.*, 2006). It has been also shown in-vivo that Pol II is able to recruit splicing factors to transcriptional foci (Misteli and Spector, 1999).

Curiously, one of the first examples reported in humans of co-transcriptionality was discovered in one of the biggest (2400 bp) human genes: distrofin (Tennyson *et al.*, 1995). In this sense, co-transcriptional splicing represents an explanatory framework to understand how is the recognition of small exons achieved in a sea of giant introns (Goldstrohm *et al.*, 2001).

Alternative splicing and its regulation

Traditionally, alternative splicing has been assumed to be regulated by the action of splicing enhancers and silencers. These are short sequences (10 nt long), positioned either in exons or introns stimulating (enhancers) or inhibiting (silencers) the use of splice sites through the specific binding of regulatory proteins such as SR proteins (serine/arginine rich proteins) or heterogeneous nuclear RNPs (Martinez-Contreras *et al.*, 2007; Lin *et al.*, 2008). Additionally, some silencers, instead of recruiting regulatory proteins, act by altering a particular pre-mRNA secondary structure, impeding the recognition of a neighbouring splicing enhancer by SR proteins (Buratti and Baralle, 2004). Since alternative splicing affects so many essential processes its regulation should be highly coordinated. It is worth nothing that this regulation involves thousands of synchronic steps in an extremely complex regulatory network. As part of this network a regulatory splicing code has been proposed by Blencowe and Frey, as a collection of RNA features that can be used for prediction of isoform abundances in a tissue specific manner (Barash *et al.*, 2010).

But alternative splicing regulation not only depends on the recognition of RNA features. As a consequence of the previously described coupling between transcription and splicing (Neugebauer, 2002; Proudfoot *et al.*, 2002; Kornblihtt *et al.*, 2004) variations that affect transcription can also lead to changes in mRNA processing efficiency, particularly influencing the outcome of splicing events.

Others and we have studied the mechanisms by which transcription affects alternative splicing, essentially after the surprising finding that promoter structure and occupation control alternative splicing (Cramer *et al.*, 1997, 1999; Auboeuf *et al.*, 2002). Two different but not exclusive models have been proposed (Kornblihtt, 2007):

1 The *recruitment model*, by which different factors associated with the transcription

machinery, in particular Pol II CTD, regulate directly or indirectly splicing choices.

It has been reported that both co-activators recruited to promoters and splicing factors associated to transcriptional machinery are able to regulate alternative splicing (Lai *et al.*, 1999; Monsalve *et al.*, 2000; Guillouf *et al.*, 2006). The CTD dependent effect of the SR protein SRp20 on alternative splicing of fibronectin gene exon 33 (FN-E33), is an illustrative example of the recruitment model (de la Mata and Kornblihtt, 2006).

More recently, it has been shown that the adaptor protein MRG15 binds to chromatin locations with lysine 36 trimethylated (H3K36me3) preferentially situated closer to alternative splicing event and recruits PTB (negative splicing regulator) favouring exon skipping (Luco *et al.*, 2010).

2 The *kinetic coupling model*, where the rate of Pol II elongation influences the outcome of the alternative splicing process by providing different windows of opportunity in the usage of weak and strong splice sites depending on the timing of their emergence (Kornblihtt, 2005). Early indirect evidence for a kinetic coupling (Eperon *et al.*, 1988; Roberts *et al.*, 1998) was stregthened by later findings that alternative splicing is regulated when Pol II elongation is modulated by the use of transcription factors with differential activities on elongation (Kadener *et al.*, 2001, 2002; Nogues *et al.*, 2002) or, even most directly, using slow Pol II mutants (de la Mata *et al.*, 2003; Howe *et al.*, 2003).

Although there is accumulated evidence supporting both models (Kornblihtt, 2007) we will focus here on the kinetic coupling model. The mechanistic framework underlying this coupling is based in the 'first come, first served' hypothesis (Fig. 5.1). Skipping of many cassette exons, such as fibronectin exon 33 (also known as EDI or EDA) occurs because the acceptor site (3′ splice site) of the upstream intron is suboptimal compared with the acceptor site of the downstream intron. Recent experiments of our group have shown that slow elongation would favour recruitment of splicing factors to the pre-mRNA that would favour the recognition of exon EDI and the commitment to subsequent inclusion (de la Mata *et al.*, 2010). Once commitment is achieved, the order of intron removal becomes irrelevant to guarantee inclusion. According to our interpretation, 'first served' would not be equivalent to 'first excised' but to 'first committed' (Fig. 5.1C).

The stronger support for Pol II elongation in the regulation of alternative splicing has come from evidence showing that alternative splicing is affected by inhibition of Pol II elongation resulting from Pol II hyperphosphorylation triggered by the DNA damage signalling that follows irradiation of cells with UV light (Muñoz *et al.*, 2009).

On the other hand, a role for chromatin affecting Pol II elongation and alternative splicing has also become clear. One of the first evidences suggesting this possible role for chromatin was provided in 1991 (Adami and Babiss, 1991) when it was found that two copies of the same Adenovirus genome in the same nucleus gave rise to different alternatively spliced RNAs each. Staggered infections of two discernible viral genomes resulted in the simultaneous presence of yet-unreplicated and replicated genomes in the same nucleus, each of them eliciting two different mRNA splicing variants of the E1a gene. Since the only difference resided in a property of the template DNAs, the authors speculated that the molecular bases for the change in splicing should be in the chromatin organization acquired upon replication. If viral replication promoted a more compact chromatin structure that, in turn, would reduce Pol II elongation rates, more time would be given to assemble splicing complexes at the upstream E1a 5′ splice site, favouring its use compared with the downstream 5′ splice site, and giving rise to a shorter E1a mRNA. In the same way, transiently transfected minigenes are known to assemble in a physiological chromatin context (Cereghini and Yaniv, 1984; Tong *et al.*, 2006), whose compaction increases upon replication. Our group has shown that replication favours a more compact chromatinization of alternative splicing reporters and caused lower elongation rates (Kadener *et al.*, 2001) which, in turn, increase the inclusion of a cassette exon into mature mRNA. This idea was reinforced by finding that trichostatin A (TSA), a potent inhibitor

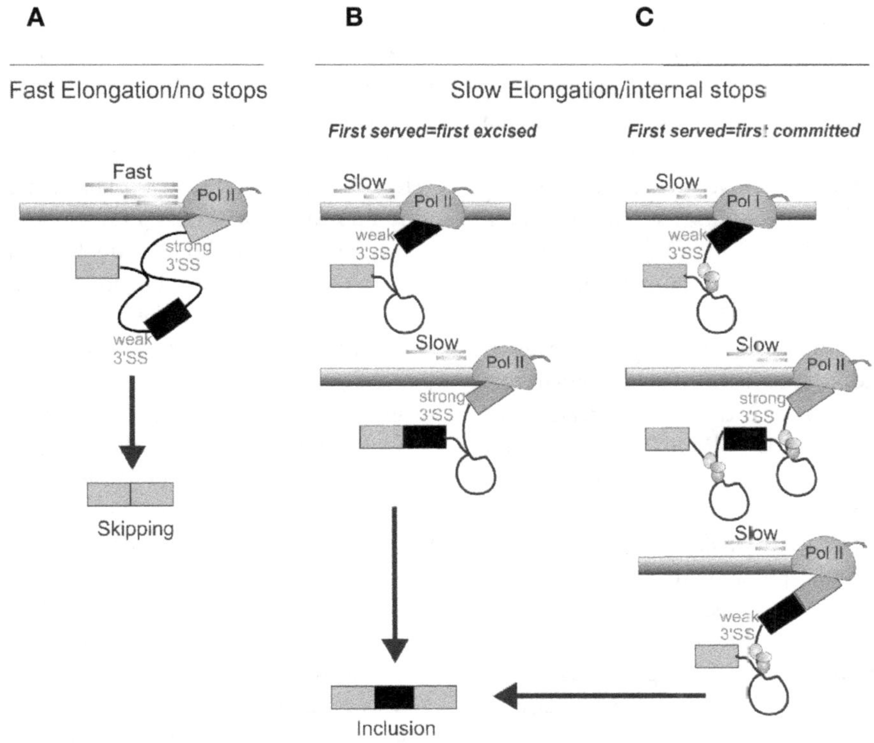

Figure 5.1 Two different models for the 'first come, first served' mechanism of alternative splicing. (A) Fast elongation promotes the use of the stronger 3′ splice site located in the downstream intron. (B) Slow elongation causes preferential excision of the upstream intron (first served = first excised). (C) Slow elongation causes commitment to inclusion of the alternative cassette exon via recruitment of splicing factors.

of histone deacetylation, favoured skipping of alternative exons whose inclusion is stimulated by slow elongation (Nogues *et al.*, 2002).

Finally, the first strong evidence that chromatin changes could affect alternative splicing of an endogenous gene was provided by Batsché *et al.* (2006) who studied the mechanism by which the chromatin remodelling factor SWI/SNF regulates alternative splicing of the CD44 gene. SWI/SNF promotes inclusion into mRNA of the block of ten consecutive alternative exons located in the centre of the gene. The Brahma (Brm) subunit of SWI/SNF interacts with Pol II, the spliceosomal snRNPs U1 and U5 and the RNA-binding protein Sam 68. The latter binds to splicing regulatory elements present in the CD44 variable exons and were known to stimulate their inclusion. The macromolecular complex accumulates on the CD44 gene, peaking at the intragenic alternative spliced region and causing Pol II stalling, which favours inclusion of the variable exons, in agreement with the kinetic coupling model.

Non-coding RNAs enter the field

Probably, the first link between non-coding RNAs and splicing has emerged, unexpectedly, from the microRNA (miRNA) field. The group of Javier Cáceres, in Edinburgh, has found that the heterogeneous nuclear ribonucleoprotein A1 (hnRNP A1), a very well-known splicing regulator, binds to pre-miR-18a facilitating Drosha processing (Guil and Caceres, 2007; Michlewski *et al.*, 2008). But additionally, Caceres' group has also shown that the same protein is capable of generating the opposite effect in another context, i.e. pre-microRNA processing inhibition (Michlewski and Caceres, 2010).

Other groups have shown that miRNAs can affect alternative splicing. miR-133 is expressed in muscular tissue exclusively, and during myoblasts differentiation it represses the expression of nPTB, another well-known splicing regulator, controlling the alternative splicing of a set of muscle specific exons (Boutz *et al.*, 2007). In addition, another miRNA (miR-124) plays a similar role during nervous system development repressing PTBP1 (PTB/hnRNPA1), a neuronal specific modulator of alternative splicing (Makeyev *et al.*, 2007). More recently it has been shown that miR-23a/b has a similar effect during heart development, inhibiting CUGBP and ETR-3-like factor (CELF). Both proteins together regulate half of alternative splicing events modulated during heart development (Kalsotra *et al.*, 2010).

But, until this point there were no reports showing non-coding RNAs able to alter directly splicing decisions. In 2004 Kevin Morris opened the door for this finding. He has shown that non-coding small RNAs directed against promoter regions are able to affect chromatin state at the genomic target DNA, inhibiting transcription initiation (Morris *et al.*, 2004). Then, this finding was supported by a growing number of reports showing that exogenously applied ncsRNAs targeting promoters could trigger transcriptional gene silencing (TGS) in human cells through heterochromatin formation. The process involves recruitment of chromatin modifying enzymes resulting in specific histone 3 methylations usually associated to facultative heterochromatin (H3K9me2 and H3K27me3), DNA methylation and histone deacetylation (Morris *et al.*, 2004; Kim *et al.*, 2006) and needs of the Argonaute proteins AGO1 and AGO2 (Janowski *et al.*, 2006; Kim *et al.*, 2006, 2008).

Taking into account our previous knowledge about the links between chromatin, Pol II elongation rate and alternative splicing, we hypothesized that, if instead of targeting promoters, ncsRNAs were directed to sequences in the body of genes, mapping close to alternative exons, they could generate an heterochromatin context and prevent efficient pol II elongation, consequently affecting splicing choices. Accordingly, we have recently reported that exogenous ncsRNAs targeting intronic or exonic sequences downstream of the alternative EDI exon of the endogenous fibronectin gene stimulate FN-E33 inclusion into the mature mRNA (Allo *et al.*, 2009).

When we transfected human cell lines (Hep3B) with ncsRNAs (21–25 mer) mapping to intron downstream of E33 (I33) we observed an enrichment of EDI+/EDI– ratio, but only when the ncsRNA was designed to load preferentially the antisense strand in the silencing complex (I33as). We've used different strategies to ensure that the guide strand was loaded in sense (I33s) or antisense orientation respectively in each case.

Using ncsRNAs targeting the exon 34 we were able to measure both RNA interference pathways acting simultaneously, i.e. knock-down of fibronectin gene by the classical post-transcriptional gene silencing (PTGS) pathway in the cytoplasm and in the remaining mRNA we observed the increment of EDI+/EDI- ratio as a consequence of the TGS process in the nucleus.

This effect depends on AGO1, the Argonaute protein necessary to guide small RNAs to their genomic targets leading to chromatin modifications characteristic of TGS (Kim *et al.*, 2006). Likewise, the effect is abolished by treatments that promote chromatin relaxation, inhibit key silencing proteins, or increase Pol II elongation such as 5-aza-2 deoxycytidine (5azadC: DNA methyl transferase inhibitor), Trichostatin A (TSA: a histone deacetylase inhibitor) and BIX (BIX-01294 [2-(hexahydro-4-methyl-1H-1,4-diazepin-1-yl)–6,7-dimethoxy-N-[1-(phenylmethyl)-4-piperidinyl]4-quinazolinamine]]: histone methyl-transferase inhibitor). Moreover, the effect of ncsRNAs is also abrogated by depletion of the heterochromatin protein HP1-alpha.

Accordingly, and as we hypothesizes previously, ncsRNAs transfection also increases H3K9me2 and H3K27me3 marks at the endogenous targeted intragenic region with no changes observed at the promoter.

But... was the increment of both facultative heterochromatin marks followed by an elongation disruption? Using two different systems with reporters for alternative splicing we observed lower elongation rates when we used ncsRNAs targeting E33 downstream intron than when we use a non-related ncsRNA control.

The effect of ncsRNAs needs the typical RNA:RNA hybridization of TGS pathway between the ncsRNA guide strand loaded in the silencing complex and the nascent transcript.

To assess the cell specificity of small RNAs' effect we transfected HeLa cells with the same ncsRNAs. As in Hep 3B cells, I33as induced and increase in E33 exon inclusion that was also gene-specific. Remarkably, in these cells we observed the unexpected effect of I33s stimulating E33 inclusion. Other groups have shown that the antisense strand of siRNAs directs histone modifications and TGS induction (Weinberg *et al.*, 2006). In this context, the effect of I33s could be explained if an antisense FN transcript, able to hybridize with it, existed in HeLa cells. Indeed, we discovered the presence of a long-non-coding RNA in antisense orientation to FN gene in HeLa cells, consisting of three exons and overlapping three-quarters of FN1 including E33 alternative region. We speculate that the existence of this long non-coding RNA could provide the template for RNA::RNA hybridization with the ncsRNA whose guide strand was designed to be loaded preferentially in sense orientation.

We have named this novel mechanism 'TGS-AS' for transcriptional gene silencing regulation of alternative splicing.

Our working model for TGS-AS is (Fig. 5.2):

1. A gene with an exon cassette is transcribed by a Pol II (orange line) at high elongation rates. The splicing machinery does not have enough time to recognize the weak splice site of the alternative exon favouring its skipping.
2. ncsRNAs targeting the downstream intron are transfected.
3. Guide strand is loaded in the silencing complex containing AGO1. Once reach the nucleus this complex starts to scan nascent transcripts searching for a complementary sequence to the guide strands.
4. When it finds the sequences, it would produce the recruitment of chromatin factors, methyl-transferases and HP1 alpha.
5. A heterochromatin structure is generated.
6. This heterochromatin island provokes and steric impediment for Pol II elongation reducing its rate. This gives more time to splicing machinery for the recognition of the weak splice site, now, favouring its inclusion in the mature mRNA.

Endogenous non-coding RNAs and splicing

The physiological relevance of TGS has been highlighted by recent findings showing that endogenous ncsRNAs and microRNAs can regulate gene expression in mammalian cells through histone modifications and DNA methylation (Aravin and Bourc'his, 2008; Gonzalez *et al.*, 2008; Zhang *et al.*, 2008a,b). Moreover, a recently published article (Khraiwesh *et al.*) shows that endogenous microRNAs can trigger TGS in the moss *Physcomitrella patens*. Can these or other small RNAs act as the raw material for an endogenous TGS-AS pathway?

Trying to answer this question we speculated that if some of the above scenarios were modulating alternative splicing endogenously via TGS, the knockdown of AGO1 would affect them. To test this, we used an automated high-throughput RT-PCR platform to screen 96 cancer-related alternative splicing events (Klinck *et al.*, 2008), in cells transfected with siAGO1. Interestingly, isoform patterns were significantly affected by AGO1 knockdown in 35% and 38% of the alternative splicing events in Hep3B and HeLa cells respectively, suggesting a widespread nature for TGS-AS. But, at the same time, these results are the first evidence showing that AGO1, a key protein of RNA interference pathways, is also important for alternative splicing regulation.

In the same direction, we performed ChIP-seq using antibodies against AGO1 protein and histone H3 covalent modifications (H3K9me2, H3K27me3, H3K36me3 and H3). Preliminary data allowed the identification of 24,000 target regions for AGO1, 23,000 for H3K9me2 and 51,000 for H3K27me3 in the genome of the human mammary cell line MCF7. Approximately 50% of the targets for each mark map within genes and, among these, AGO1 targets are enriched in exons, preferentially in genes with low expression levels. About 17% of the H3K9me2 targets overlap with H3K27me3 targets when analysed

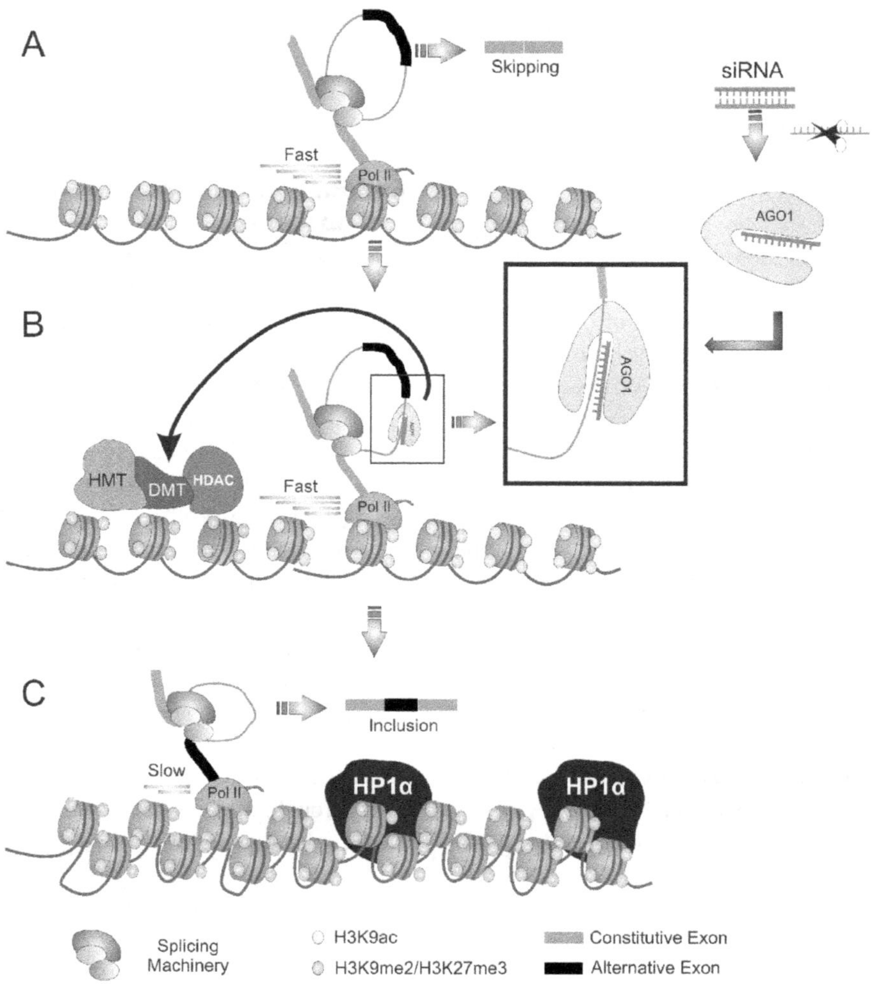

Figure 5.2 Model for TGS-AS. Transfection with ncsRNAs targeting the intron mapping downstream of the alternative exon (black) promotes dimethylation and trimethylation of H3K9 and H3K27 respectively (nucleosomes in grey) triggered by the ncsRNA guide strand entering a silencing complex containing AGO1. HP1α is recruited and the resulting condensed chromatin structure generates roadblocks to Pol II elongation causing higher inclusion of the alternative exon according to the kinetic coupling model.

genome-wide. However the overlapping increases to 55% when only AGO1 target regions are considered. These results indicate that AGO1 is found in the nucleus and located at specific regions of the genome. Whether this association is mediated by RNA or is causative of TGS remains to be determined.

Another potential raw material for TGS-AS is the sense–antisense transcription couples. It has been demonstrated that regions with bidirectional and overlapping transcription are often epigenetically silenced (Tufarelli *et al.*, 2003; Yu *et al.*, 2008). In a related pathway, repetitive sequences and other 'genome parasites' are usually silenced by being embedded in heterochromatin regions. When this occurs in the vicinity of a promoter, it results in the regulation of host gene transcriptional initiation. Our results open the question whether these sense–antisense couples would regulate splicing if they occur intragenically and closer to an alternative splicing event.

From small to long non-coding RNAs as elicitors of TGS-AS

The central point in the search of endogenous examples of TGS-AS lies in the certainty that any non-coding RNA capable of affecting intragenic chromatin modifications would also be able to affect transcription quality and RNA processing. These concepts are reinforced by the following evidence:

1. Intragenic DNA methylation causes local decreases in histone acetylation and chromatin accessibility, resulting in a decline in Pol II elongation, without affecting transcription levels (Lorincz *et al.*, 2004).
2. When transfected plasmids that are templates for alternative splicing transcripts are prompted to replicate, a condition that favours chromatinization and in consequence lower elongation rates, inclusion of alternative exons is greatly increased (Kadener *et al.*, 2001).
3. Trichostatin A (TSA), an inhibitor of histone deacetylation, favours skipping of alternative exons whose inclusion is stimulated by slow elongation (Nogues *et al.*, 2002; Schor *et al.*, 2009).
4. The chromatin-remodelling factor SWI/SNF regulates alternative splicing of the CD44 gene by creating internal roadblocks for Pol II elongation (Batsche *et al.*, 2006).
5. Histone modifications found in the transcribed regions, such as H3K4me3 (tri-methylation of lysine 4 of histone H3), can recruit post-initiation and splicing factors through chromodomain proteins (Sims *et al.*, 2007).
6. The H3K36me3 (tri-methylation of lysine 36 of histone H3) mark displays a remarkable enrichment on exons, as compared with introns within the same gene. Most strikingly, this mark is underrepresented in alternatively spliced exons (Kolasinska-Zwierz *et al.*, 2009), revealing a relation between the chromatin and splicing codes.
7. Membrane depolarization of neuronal cells triggers the skipping of exon 18 from the neural cell adhesion molecule (NCAM) mRNA by causing H3K9Ac (acetylation of lysine 9 of histone H3), restricted to an internal region of the NCAM gene surrounding the alternative exon, with no changes at the promoter. This epigenetic mark is associated to chromatin relaxation and to increase in Pol II processivity, which explains exon 18 skipping (Schor *et al.*, 2009).
8. Exogenously applied ncsRNAs targeting gene sequences close to an alternative exon regulate its splicing by creating heterochromatin marks (H3K9me2 and H3K27me3, di-methylation of lysine 9 of histone H3 and tri-methylation of lysine 27 of histone H3 respectively) at the target site, which reduces Pol II elongation (Alló *et al.*, 2009).
9. An endogenous TGS pathway affects Pol II elongation rate in agreement to the kinetic coupling model in *C. elegans* (Guang *et al.*, 2010). The authors describes the interaction between an Argonaute protein (NRDE3) and factor NRDE2 that triggers TGS, slowing down Pol II and increasing H3K9me3 on target region.

This accumulated evidence prompt us to speculate that any non-coding RNA able to modify the chromatin structure potentially would be able to affect alternative splicing of a neighbouring event. Nevertheless, this kind of modification could be triggered by TGS-AS or by any other similar pathway. Taking into account that ncsRNAs with gene activation capacity have been reported (Li *et al.*, 2006; Janowski *et al.*, 2007), it would not be a surprise that some non-coding RNAs could lead to an opposite effect, producing a local chromatin relaxation, increasing Pol II elongation and favouring exon skipping.

Several non-coding RNAs appear as candidates for endogenous regulators of alternative splicing. The first in line are Piwi RNAs (Kuramochi-Miyagawa *et al.*, 2008; Watanabe *et al.*, 2008; Unhavaithaya *et al.*, 2009) and endogenous siRNAs (esiRNAs) (Zhang *et al.*, 2008a) because both are able to regulate gene expression in mammalian cells through histone modifications and DNA methylation. As far as microRNAs (miRNAs) is concerned, it has been recently shown that they can trigger TGS in the moss *Physcomitrella patens* (Khraiwesh *et al.*,

2010) and mammalian cells (Kim *et al.*, 2008). In a similar way, sense–antisense transcription couples resulting from regions with bidirectional and overlapping transcription could trigger epigenetic silencing (Yu *et al.*, 2008).

On the other hand, a growing number of different categories of non-coding small RNAs have recently been described including PASRs (promoter-associated small RNAs), tiRNAs (transcription initiation RNAs) and spliRNAs (splice-site RNAs). tiRNAs were detected in THP-1 human cells, are 22 nucleotides long and occupy the position –60/+120 (Taft *et al.*, 2009). spliRNAs were found in the same cell line but are localized mainly closer to splicing donor sites (Taft *et al.*, 2010).

Long non-coding RNAs (lncRNAs), which usually contain more than one exon, are longer than 100 bp and are located in antisense orientation to coding genes have been identified as key players of imprinting and X-inactivation. Additionally, it has been shown that they can produce gene silencing both in *cis* (Yap *et al.*, 2010) and in *trans* (Rinn *et al.*, 2007; Huarte *et al.*, 2010) by means of heterochromatin formation mediated by Polycomb (PRC2) group of proteins and the presence of H3K27me3 (Gupta *et al.*, 2010; Tsai *et al.*, 2010). It has been described more than a thousand of long non-coding RNAs, and even some of them are related to transcriptional activation (Orom *et al.*, 2010).

In summary, the regulation of alternative splicing could be seen as the sum of multiple factors of different nature contributing independently or in a concerted way to the thin balance of splicing choices. The dynamic flow of these interactions in the cell produces a very complex network of regulation not only comparable but superimposed to the complexity of transcriptional regulation. Promoter identity and occupation, intron/exon architecture and secondary structure, exon definition, presence of regulatory sequences such as splicing enhancers and silencers and their cognate regulatory factors, recruitment of splicing factors to the transcribing polymerase and changes in transcriptional elongation rates are some examples of this network. Even when we still have a long road ahead on this emerging field, we can expect to find non-coding RNAs, in any version, as novel potential endogenous regulators of splicing in the near future.

References

Adami, G., and Babiss, L.E. (1991). DNA template effect on RNA splicing: two copies of the same gene in the same nucleus are processed differently. EMBO J. *10*, 3457–3465.

Akoulitchev, S., Makela, T.P., Weinberg, R.A., and Reinberg, D. (1995). Requirement for TFIIH kinase activity in transcription by RNA polymerase II. Nature *377*, 557–560.

Alló, M., Buggiano, V., Fededa, J.P., Petrillo, E., Schor, I., de la Mata, M., Agirre, E., Plass, M., Eyras, E., Elela, S.A., *et al.* (2009). Control of alternative splicing through siRNA-mediated transcriptional gene silencing. Nat. Struct. Mol. Biol. *16*, 717–724.

Aravin, A.A., and Bourc'his, D. (2008). Small RNA guides for *de novo* DNA methylation in mammalian germ cells. Genes Dev. *22*, 970–975.

Auboeuf, D., Honig, A., Berget, S.M., and O'Malley, B.W. (2002). Coordinate regulation of transcription and splicing by steroid receptor coregulators. Science *298*, 416–419.

Barash, Y., Calarco, J.A., Gao, W., Pan, Q., Wang, X., Shai, O., Blencowe, B.J., and Frey, B.J. (2010). Deciphering the splicing code. Nature *465*, 53–59.

Batsche, E., Yaniv, M., and Muchardt, C. (2006). The human SWI/SNF subunit Brm is a regulator of alternative splicing. Nat. Struct. Mol. Biol. *13*, 22–29.

Bengal, E., Flores, O., Krauskopf, A., Reinberg, D., and Aloni, Y. (1991). Role of the mammalian transcription factors IIF, IIS, and IIX during elongation by RNA polymerase II. Mol. Cell. Biol. *11*, 1195–1206.

Bentley, D. (1999). Coupling RNA polymerase II transcription with pre-mRNA processing. Curr. Opin. Cell. Biol. *11*, 347–351.

Berget, S.M. (1995). Exon recognition in vertebrate splicing. J. Biol. Chem. *270*, 2411–2414.

Beyer, A.L., and Osheim, Y. N. (1988). Splice site selection, rate of splicing, and alternative splicing on nascent transcripts. Genes Dev. *2*, 754–765.

Black, D.L. (2003). Mechanisms of alternative pre-messenger RNA splicing. Annu. Rev. Biochem. *72*, 291–336.

Bourgeois, C.F., Kim, Y.K., Churcher, M.J., West, M.J., and Karn, J. (2002). Spt5 cooperates with human immunodeficiency virus type 1 Tat by preventing premature RNA release at terminator sequences. Mol. Cell. Biol. *22*, 1079–1093.

Boutz, P.L., Chawla, G., Stoilov, P., and Black, D.L. (2007). MicroRNAs regulate the expression of the alternative splicing factor nPTB during muscle development. Genes Dev. *21*, 71–84.

Bradsher, J.N., Tan, S., McLaury, H.J., Conaway, J.W., and Conaway, R.C. (1993). RNA polymerase II transcription factor SIII. II. Functional properties and role in RNA chain elongation. J. Biol. Chem. *268*, 25594–25603.

Buratti, E., and Baralle, F.E. (2004). Influence of RNA secondary structure on the pre-mRNA splicing process. Mol. Cell. Biol. *24,* 10505–10514.

Caceres, J.F., and Kornblihtt, A.R. (2002). Alternative splicing: multiple control mechanisms and involvement in human disease. Trends Genet. *18,* 186–193.

Cereghini, S., and Yaniv, M. (1984). Assembly of transfected DNA into chromatin: structural changes in the origin-promoter-enhancer region upon replication. EMBO J. *3,* 1243–1253.

Cramer, P., Pesce, C.G., Baralle, F.E., and Kornblihtt, A.R. (1997). Functional association between promoter structure and transcript alternative splicing. Proc. Natl. Acad. Sci. U.S.A. *94,* 11456–11460.

Cramer, P., Caceres, J.F., Cazalla, D., Kadener, S., Muro, A.F., Baralle, F.E., and Kornblihtt, A.R. (1999). Coupling of transcription with alternative splicing: RNA pol II promoters modulate SF2/ASF and 9G8 effects on an exonic splicing enhancer. Mol. Cell *4,* 251–258.

Eperon, L.P., Graham, I.R., Griffiths, A.D., and Eperon, I. C. (1988). Effects of RNA secondary structure on alternative splicing of pre-mRNA: is folding limited to a region behind the transcribing RNA polymerase? Cell *54,* 393–401.

Goldstrohm, A.C., Greenleaf, A.L., and Garcia-Blanco, M. A. (2001). Co-transcriptional splicing of pre-messenger RNAs: considerations for the mechanism of alternative splicing. Gene *277,* 31–47.

Gonzalez, S., Pisano, D.G., and Serrano, M. (2008). Mechanistic principles of chromatin remodeling guided by siRNAs and miRNAs. Cell Cycle *7,* 2601–2608.

Gornemann, J., Kotovic, K.M., Hujer, K., and Neugebauer, K. M. (2005). Cotranscriptional spliceosome assembly occurs in a stepwise fashion and requires the cap binding complex. Mol. Cell *19,* 53–63.

Guang, S., Bochner, A.F., Burkhart, K.B., Burton, N., Pavelec, D.M., and Kennedy, S. (2010). Small regulatory RNAs inhibit RNA polymerase II during the elongation phase of transcription. Nature *465,* 1097–1101.

Guil, S., and Caceres, J.F. (2007). The multifunctional RNA-binding protein hnRNP A1 is required for processing of miR-18a. Nat. Struct. Mol. Biol. *14,* 591–596.

Guillouf, C., Gallais, I., and Moreau-Gachelin, F. (2006). Spi-1/PU.1 oncoprotein affects splicing decisions in a promoter binding-dependent manner. J. Biol. Chem. *281,* 19145–19155.

Gupta, R.A., Shah, N., Wang, K.C., Kim, J., Horlings, H.M., Wong, D.J., Tsai, M.C., Hung, T., Argani, P., Rinn, J.L., *et al.* (2010). Long non-coding RNA HOTAIR reprograms chromatin state to promote cancer metastasis. Nature *464,* 1071–1076.

Hertel, K.J. (2008). Combinatorial control of exon recognition. J. Biol. Chem. *283,* 1211–1215.

Hirose, Y., and Manley, J.L. (2000). RNA polymerase II and the integration of nuclear events. Genes Dev. *14,* 1415–1429.

House, A.E., and Lynch, K.W. (2008). Regulation of alternative splicing: more than just the ABCs. J. Biol. Chem. *283,* 1217–1221.

Howe, K.J., Kane, C.M., and Ares, M., Jr. (2003). Perturbation of transcription elongation influences the fidelity of internal exon inclusion in Saccharomyces cerevisiae. RNA *9,* 993–1006.

Huarte, M., Guttman, M., Feldser, D., Garber, M., Koziol, M.J., Kenzelmann-Broz, D. Khalil, A.M., Zuk, O., Amit, I., Rabani, M., *et al.* (2010). A large intergenic noncoding RNA induced by p53 mediates global gene repression in the p53 response. Cell *142,* 409–419.

Janowski, B.A., Huffman, K.E., Schwartz, J.C., Ram, R., Nordsell, R., Shames, D.S., Minna, J.D., and Corey, D.R. (2006). Involvement of AGO1 and AGO2 in mammalian transcriptional silencing. Nat. Struct. Mol. Biol. *13,* 787–792.

Janowski, B.A., Younger, S.T. Hardy, D.B., Ram, R., Huffman, K.E., and Corey, D.R. (2007). Activating gene expression in mammalian cells with promoter-targeted duplex RNAs. Nat. Chem. Biol. *3,* 166–173.

Jiang, Y., and Gralla, J. D. (1993). Uncoupling of initiation and reinitiation rates during HeLa RNA polymerase II transcription *in vitro.* Mol. Cell. Biol. *13,* 4572–4577.

Jurica, M.S., and Moore, M. J. (2003). Pre-mRNA splicing: awash in a sea of proteins. Mol. Cell *12,* 5–14.

Kadener, S., Cramer, P., Nogues, G., Cazalla, D., de la Mata, M., Fededa, J.P., Werbajh, S.E., Srebrow, A., and Kornblihtt, A.R. (2001). Antagonistic effects of T-Ag and VP16 reveal a role for RNA pol II elongation on alternative splicing. EMBO J. *20,* 5759–5768.

Kadener, S., Fededa, J.P., Rosbash, M., and Kornblihtt, A.R. (2002). Regulation of alternative splicing by a transcriptional enhancer through RNA pol II elongation. Proc. Natl. Acad. Sci. U.S.A. *99,* 8185–8190.

Kalsotra, A., Wang, K., Li, P.F., and Cooper, T.A. (2010). MicroRNAs coordinate an alternative splicing network during mouse postnatal heart development. Genes Dev. *24,* 653–658.

Khraiwesh, B., Arif, M.A., Seumel, G.I., Ossowski, S., Weigel, D., Reski, R., and Frank, W. (2010). Transcriptional control of gene expression by microRNAs. Cell *140,* 111–122.

Kim, D.H., Villeneuve, L.M., Morris, K.V., and Rossi, J. J. (2006). Argonaute-1 directs siRNA-mediated transcriptional gene silencing in human cells. Nat. Struct. Mol. Biol. *13,* 793–797.

Kim, D.H., Saetrom, P., Snove, O., Jr., and Rossi, J.J. (2008). MicroRNA-directed transcriptional gene silencing in mammalian cells. Proc. Natl. Acad. Sci. U.S.A. *105,* 16230–16235.

Klinck, R., Bramard, A., Inkel, L., Dufresne-Martin, G., Gervais-Bird, J., Madden, R., Paquet, E.R., Koh, C., Venables, J.P., Prinos, P., *et al.* (2008). Multiple alternative splicing markers for ovarian cancer. Cancer Res. *68,* 657–663.

Kolasinska-Zwierz, P., Down, T., Latorre, I., Liu, T., Liu, X.S., and Ahringer, J. (2009). Differential chromatin marking of introns and expressed exons by H3K36me3. Nat. Genet. *41,* 376–381.

Kornblihtt, A.R. (2005). Promoter usage and alternative splicing. Curr. Opin. Cell. Biol. *17*, 262–268.

Kornblihtt, A.R. (2007). Coupling transcription and alternative splicing. Adv. Exp. Med. Biol. *623*, 175–189.

Kornblihtt, A.R., de la Mata, M., Fededa, J.P., Munoz, M.J., and Nogues, G. (2004). Multiple links between transcription and splicing. RNA *10*, 1489–1498.

Kotovic, K.M., Lockshon, D., Boric, L., and Neugebauer, K. M. (2003). Cotranscriptional recruitment of the U1 snRNP to intron-containing genes in yeast. Mol. Cell. Biol. *23*, 5768–5779.

Kuramochi-Miyagawa, S., Watanabe, T., Gotoh, K., Totoki, Y., Toyoda, A., Ikawa, M., Asada, N., Kojima, K., Yamaguchi, Y., Ijiri, T.W., *et al.* (2008). DNA methylation of retrotransposon genes is regulated by Piwi family members MILI and MIWI2 in murine fetal testes. Genes Dev. *22*, 908–917.

Lacadie, S.A., and Rosbash, M. (2005). Cotranscriptional spliceosome assembly dynamics and the role of U1 snRNA:5′ss base pairing in yeast. Mol. Cell *19*, 65–75.

Lai, M.C., Teh, B.H., and Tarn, W. Y. (1999). A human papillomavirus E2 transcriptional activator. The interactions with cellular splicing factors and potential function in pre-mRNA processing. J. Biol. Chem. *274*, 11832–11841.

Li, L.C., Okino, S.T., Zhao, H., Pookot, D., Place, R.F., Urakami, S., Enokida, H., and Dahiya, R. (2006). Small dsRNAs induce transcriptional activation in human cells. Proc. Natl. Acad. Sci. U.S.A. *103*, 17337–17342.

Lin, S., Coutinho-Mansfield, G., Wang, D., Pandit, S., and Fu, X.D. (2008). The splicing factor SC35 has an active role in transcriptional elongation. Nat. Struct. Mol. Biol. *15*, 819–826.

Listerman, I., Sapra, A.K., and Neugebauer, K. M. (2006). Cotranscriptional coupling of splicing factor recruitment and precursor messenger RNA splicing in mammalian cells. Nat. Struct. Mol. Biol. *13*, 815–822.

Lorincz, M.C., Dickerson, D.R., Schmitt, M., and Groudine, M. (2004). Intragenic DNA methylation alters chromatin structure and elongation efficiency in mammalian cells. Nat. Struct. Mol. Biol. *11*, 1068–1075.

Luco, R.F., Pan, Q., Tominaga, K., Blencowe, B.J., Pereira-Smith, O.M., and Misteli, T. (2010). Regulation of alternative splicing by histone modifications. Science *327*, 996–1000.

Makeyev, E.V., Zhang, J., Carrasco, M.A., and Maniatis, T. (2007). The MicroRNA miR-124 promotes neuronal differentiation by triggering brain-specific alternative pre-mRNA splicing. Mol. Cell *27*, 435–448.

Mancebo, H.S., Lee, G., Flygare, J., Tomassini, J., Luu, P., Zhu, Y., Peng, J., Blau, C., Hazuda, D., Price, D., and Flores, O. (1997). P-TEFb kinase is required for HIV Tat transcriptional activation *in vivo* and *in vitro*. Genes Dev. *11*, 2633–2644.

Marais, G., Nouvellet, P., Keightley, P.D., and Charlesworth, B. (2005). Intron size and exon evolution in Drosophila. Genetics *170*, 481–485.

Martinez-Contreras, R., Cloutier, P., Shkreta, L., Fisette, J.F., Revil, T., and Chabot, B. (2007). hnRNP proteins and splicing control. Adv. Exp. Med. Biol. *623*, 123–147.

Mason, P.B., and Struhl, K. (2005). Distinction and relationship between elongation rate and processivity of RNA polymerase II *in vivo*. Mol. Cell *17*, 831–840.

de la Mata, M., Alonso, C.R., Kadener, S., Fededa, J.P., Blaustein, M., Pelisch, F., Cramer, P., Bentley, D., and Kornblihtt, A.R. (2003). A slow RNA polymerase II affects alternative splicing *in vivo*. Mol. Cell *12*, 525–532.

de la Mata, M., and Kornblihtt, A.R. (2006). RNA polymerase II C-terminal domain mediates regulation of alternative splicing by SRp20. Nat. Struct. Mol. Biol. *13*, 973–980.

de la Mata, M., Lafaille, C., and Kornblihtt, A.R. (2010). First come, first served revisited: factors affecting the same alternative splicing event have different effects on the relative rates of intron removal. RNA *16*, 904–912.

Michlewski, G., and Caceres, J.F. (2010). Antagonistic role of hnRNP A1 and KSRP in the regulation of let-7a biogenesis. Nat. Struct. Mol. Biol. *17*, 1011–1018.

Michlewski, G., Guil, S., Semple, C.A., and Caceres, J.F. (2008). Posttranscriptional regulation of miRNAs harboring conserved terminal loops. Mol. Cell *32*, 383–393.

Misteli, T., and Spector, D.L. (1999). RNA polymerase II targets pre-mRNA splicing factors to transcription sites *in vivo*. Mol. Cell *3*, 697–705.

Monsalve, M., Wu, Z., Adelmant, G., Puigserver, P., Fan, M., and Spiegelman, B.M. (2000). Direct coupling of transcription and mRNA processing through the thermogenic coactivator PGC-1. Mol. Cell *6*, 307–316.

Morillon, A., Karabetsou, N., O'Sullivan, J., Kent, N., Proudfoot, N., and Mellor, J. (2003). Isw1 chromatin remodeling ATPase coordinates transcription elongation and termination by RNA polymerase II. Cell *115*, 425–435.

Morris, K.V., Chan, S.W., Jacobsen, S.E., and Looney, D.J. (2004). Small interfering RNA-induced transcriptional gene silencing in human cells. Science *305*, 1289–1292.

Muñoz, M.J., Perez Santangelo, M.S., Paronetto, M.P., de la Mata, M., Pelisch, F., Boireau, S., Glover-Cutter, K., Ben-Dov, C., Blaustein, M., Lozano, J.J., *et al.* (2009). DNA damage regulates alternative splicing through inhibition of RNA polymerase II elongation. Cell *137*, 708–720.

Neugebauer, K.M. (2002). On the importance of being co-transcriptional. J. Cell. Sci. *115*, 3865–3871.

Nogues, G., Kadener, S., Cramer, P., Bentley, D., and Kornblihtt, A.R. (2002). Transcriptional activators differ in their abilities to control alternative splicing. J. Biol. Chem. *277*, 43110–43114.

Orom, U.A., Derrien, T., Beringer, M., Gumireddy, K., Gardini, A., Bussotti, G., Lai, F., Zytnicki, M., Notredame, C., Huang, Q., *et al.* (2010). Long noncoding RNAs with enhancer-like function in human cells. Cell *143*, 46–58.

Orphanides, G., and Reinberg, D. (2002). A unified theory of gene expression. Cell *108*, 439–451.

Palancade, B., and Bensaude, O. (2003). Investigating RNA polymerase II carboxyl-terminal domain (CTD) phosphorylation. Eur. J. Biochem. *270*, 3859–3870.

Peng, J., Zhu, Y., Milton, J.T., and Price, D. H. (1998). Identification of multiple cyclin subunits of human P-TEFb. Genes Dev. *12*, 755–762.

Price, D. H. (2000). P-TEFb, a cyclin-dependent kinase controlling elongation by RNA polymerase II. Mol. Cell. Biol. *20*, 2629–2634.

Proudfoot, N. (2000). Connecting transcription to messenger RNA processing. Trends Biochem. Sci. *25*, 290–293.

Proudfoot, N.J., Furger, A., and Dye, M. J. (2002). Integrating mRNA processing with transcription. Cell *108*, 501–512.

Rinn, J.L., Kertesz, M., Wang, J.K., Squazzo, S.L., Xu, X., Brugmann, S.A., Goodnough, L.H., Helms, J.A., Farnham, P.J., Segal, E., and Chang, H.Y. (2007). Functional demarcation of active and silent chromatin domains in human HOX loci by noncoding RNAs. Cell *129*, 1311–1323.

Roberts, G.C., Gooding, C., Mak, H.Y., Proudfoot, N.J., and Smith, C.W. (1998). Co-transcriptional commitment to alternative splice site selection. Nucleic Acids Res. *26*, 5568–5572.

Rondon, A.G., Garcia-Rubio, M., Gonzalez-Barrera, S., and Aguilera, A. (2003). Molecular evidence for a positive role of Spt4 in transcription elongation. EMBO J. *22*, 612–620.

Schor, I.E., Rascovan, N., Pelisch, F., Alló, M., and Kornblihtt, A.R. (2009). Neuronal cell depolarization induces intragenic chromatin modifications affecting NCAM alternative splicing. Proc. Natl. Acad. Sci. U.S.A. *106*, 4325–4330.

Shilatifard, A., Lane, W.S., Jackson, K.W., Conaway, R.C., and Conaway, J.W. (1996). An RNA polymerase II elongation factor encoded by the human ELL gene. Science *271*, 1873–1876.

Sims, R.J., 3rd, Belotserkovskaya, R., and Reinberg, D. (2004). Elongation by RNA polymerase II: the short and long of it. Genes Dev. *18*, 2437–2468.

Sims, R.J., 3rd, Millhouse, S., Chen, C.F., Lewis, B.A., Erdjument-Bromage, H., Tempst, P., Manley, J.L., and Reinberg, D. (2007). Recognition of trimethylated histone H3 lysine 4 facilitates the recruitment of transcription postinitiation factors and pre-mRNA splicing. Mol. Cell *28*, 665–676.

Srebrow, A., and Kornblihtt, A.R. (2006). The connection between splicing and cancer. J. Cell. Sci. *119*, 2635–2641.

Staley, J.P., and Guthrie, C. (1998). Mechanical devices of the spliceosome: motors, clocks, springs, and things. Cell *92*, 315–326.

Taft, R.J., Glazov, E.A., Cloonan, N., Simons, C., Stephen, S., Faulkner, G.J., Lassmann, T., Forrest, A.R., Grimmond, S.M., Schroder, K., *et al.* (2009). Tiny RNAs associated with transcription start sites in animals. Nat. Genet. *41*, 572–578.

Taft, R.J., Simons, C., Nahkuri, S., Oey, H., Korbie, D.J., Mercer, T.R., Holst, J., Ritchie, W., Wong, J.J., Rasko, J.E., *et al.* (2010). Nuclear-localized tiny RNAs are associated with transcription initiation and splice sites in metazoans. Nat. Struct. Mol. Biol. *17*, 1030–1034.

Tennyson, C.N., Klamut, H.J., and Worton, R.G. (1995). The human dystrophin gene requires 16 hours to be transcribed and is cotranscriptionally spliced. Nat. Genet. *9*, 184–190.

Tong, W., Kulaeva, O.I., Clark, D.J., and Lutter, L. C. (2006). Topological analysis of plasmid chromatin from yeast and mammalian cells. J Mol Biol *361*, 813–822.

Tsai, M.C., Manor, O., Wan, Y., Mosammaparast, N., Wang, J.K., Lan, F., Shi, Y., Segal, E., and Chang, H. Y. (2010). Long noncoding RNA as modular scaffold of histone modification complexes. Science *329*, 689–693.

Tufarelli, C., Stanley, J.A., Garrick, D., Sharpe, J.A., Ayyub, H., Wood, W.G., and Higgs, D. R. (2003). Transcription of antisense RNA leading to gene silencing and methylation as a novel cause of human genetic disease. Nat. Genet. *34*, 157–165.

Unhavaithaya, Y., Hao, Y., Beyret, E., Yin, H., Kuramochi-Miyagawa, S., Nakano, T., and Lin, H. (2009). MILI, a PIWI-interacting RNA-binding protein, is required for germ line stem cell self-renewal and appears to positively regulate translation. J. Biol. Chem. *284*, 6507–6519.

Wang, E.T., Sandberg, R., Luo, S., Khrebtukova, I., Zhang, L., Mayr, C., Kingsmore, S.F., Schroth, G.P., and Burge, C.B. (2008). Alternative isoform regulation in human tissue transcriptomes. Nature *456*, 470–476.

Watanabe, T., Totoki, Y., Toyoda, A., Kaneda, M., Kuramochi-Miyagawa, S., Obata, Y., Chiba, H., Kohara, Y., Kono, T., Nakano, T., *et al.* (2008). Endogenous siRNAs from naturally formed dsRNAs regulate transcripts in mouse oocytes. Nature *453*, 539–543.

Weinberg, M.S., Villeneuve, L.M., Ehsani, A., Amarzguioui, M., Aagaard, L., Chen, Z.X., Riggs, A.D., Rossi, J.J., and Morris, K.V. (2006). The antisense strand of small interfering RNAs directs histone methylation and transcriptional gene silencing in human cells. RNA *12*, 256–262.

Yamaguchi, Y., Takagi, T., Wada, T., Yano, K., Furuya, A., Sugimoto, S., Hasegawa, J., and Handa, H. (1999). NELF, a multisubunit complex containing, R.D., cooperates with DSIF to repress RNA polymerase II elongation. Cell 97, 41–51.

Yang, X., Gold, M.O., Tang, D.N., Lewis, D.E., Aguilar-Cordova, E., Rice, A.P., and Herrmann, C.H. (1997). TAK, an HIV Tat-associated kinase, is a member of the cyclin-dependent family of protein kinases and is induced by activation of peripheral blood lymphocytes and differentiation of promonocytic cell lines. Proc. Natl. Acad. Sci. U.S.A. *94*, 12331–12336.

Yap, K.L., Li, S., Munoz-Cabello, A.M., Raguz, S., Zeng, L., Mujtaba, S., Gil, J., Walsh, M.J., and Zhou, M.M. (2010). Molecular interplay of the noncoding RNA ANRIL and methylated histone H3 lysine 27 by polycomb CBX7 in transcriptional silencing of INK4a. Mol. Cell *38*, 662–674.

Yu, W., Gius, D., Onyango, P., Muldoon-Jacobs, K., Karp, J., Feinberg, A.P., and Cui, H. (2008). Epigenetic silencing of tumour-suppressor gene p15 by its antisense RNA. Nature *451*, 202–206.

Yudkovsky, N., Ranish, J.A., and Hahn, S. (2000). A transcription reinitiation intermediate that is stabilized by activator. Nature *408*, 225–229.

Zhang, M.X., Zhang, C., Shen, Y.H., Wang, J., Li, X.N., Chen, L., Zhang, Y., Coselli, J.S., and Wang, X.L. (2008a). Effect of 27nt small RNA on endothelial nitric-oxide synthase expression. Mol. Biol. Cell *19*, 3997–4005.

Zhang, M.X., Zhang, C., Shen, Y.H., Wang, J., Li, X.N., Zhang, Y., Coselli, J., and Wang, X. L. (2008b). Biogenesis of short intronic repeat 27-nucleotide small RNA from endothelial nitric-oxide synthase gene. J. Biol. Chem. *283*, 14685–14693.

6 Natural Antisense Transcripts within Pseudogenes: an EST Survey

Enrique M. Muro and Miguel A. Andrade-Navarro

Abstract

Natural antisense transcripts (NATs) are RNAs that are complementary to other RNAs, being able to hybridize them and modulate their expression. Complementarity is obvious when a NAT is transcribed in *cis* to the target RNA. However, there are also *trans*-NATs, that is, NATs arising from genomic loci different to those of their target RNAs. A question that arises is how a *trans*-NAT can be generated. We hypothesize that the formation of pseudogenes is one possible answer. To support this idea, we present here an analysis of human transcripts expressed in antisense to pseudogenes obtained by a survey of the NCBI's EST database. Our results provide more evidence to the ongoing discovery of pseudogene functionality.

Pseudogenes: not just broken genes

Pseudogenes are genome loci that look like genes but have sequences apparently prevented to produce any functional product owing to genetic defects. However, recent advances in the field of molecular biology urge the revisiting of this definition. In this chapter we will discuss some of those advances.

Pseudogenes are usually the result of the copy of genetic material either by retrotransposition of a transcribed parental gene (resulting in the so called processed pseudogenes, which are intronless) or by genomic DNA duplication of one or more parental genes (resulting in non-processed pseudogenes, which may contain introns). Pseudogenes can also appear by genetic disabling of functional genes (Mighell *et al.*, 2000). In general, pseudogenes are affected by mutations, deletions and insertions that destroy their protein synthesis capability. For this reason, pseudogenes have been broadly considered devoid of function.

The parental genes of pseudogenes can be either currently active genes or genes that were active in an ancient genome but not any more. In the latter case, pseudogenes are considered as genomic relics where a parental gene can be totally erased and the only remaining of its existence are its pseudogenes (Wang *et al.*, 2006; Zhu *et al.*, 2007). A classical example in humans is the presence of pseudogenes that arose before the decrease of our olfactory ability; still there are hundreds of cell surface olfactory receptor genes active in some other mammals that can be traced in humans owing to their remaining pseudogenes (Niimura and Nei, 2003). Particularly interesting are the studies that explain, based on pseudogenes, the loss of primate ability for vitamin C synthesis due to the loss of the enzyme GULO, which is still active for instance in the mouse genome. The list of identified and studied pseudogenes is quite large in many different species; also in humans (UOX, ACYL3, CTF2, CETN4, etc.; Zhu *et al.*, 2007). Until recently pseudogenes were studied as simple reference for comparative studies, for example, in the work reporting the discovery of the loss of haemoglobin by different Antarctic ice fishes (Near *et al.*, 2006). More recently, pseudogenes have been used to demonstrate, for the first time, the existence of gene order within operons in Prokarya by us (Muro *et al.*, 2011).

Pseudogenes look so much like real protein coding genes that sometimes they confuse gene prediction algorithms (Brent, 2008). An approach to avoid this problem consists on masking pseudogenes iteratively with gene prediction (van Baren and Brent, 2006).

After the completion of the genome project, human pseudogenes are genome-wide discovered based on homology methods while using genome annotation information (Torrents *et al.*, 2003; Zhang and Gerstein, 2004); before they were often discovered by PCR. The number of human pseudogenes estimated by different methods varies but it remains around 20,000 (see http://www.pseudogene.org/ for more details).

A pseudogene does not have necessarily all the motifs and other gene regulatory elements that are needed for correct transcription. It is estimated that approximately 5% of the predicted human pseudogenes are transcribed (Harrison *et al.*, 2005). A posterior study on ENCODE regions that was carried out using locus specific RACE microarrays, tiling microarrays and high throughput sequencing showed that the number of transcribed pseudogenes may be higher (Zheng *et al.*, 2007). In that study, tissue and cell line specificity was detected, as well as regulation in the transcribed pseudogene promoter regions. This is not surprising because the different cDNA libraries from GenBank (mRNA and ESTs) point out to tissue specific pseudogene loci transcription (both sense and antisense). The role of these transcripts is not understood. Pseudogene transcription can be as complex as that of genes: for instance alternative splicing in pseudogenes has been already reported (Huang *et al.*, 1997; Cronin *et al.*, 1998; Olsen and Schechter, 1999).

The first studies that showed some pseudogene biological function had not much impact in the scientific community (Ota and Nei, 1995); a very interesting work in a snail *Lymnaea stagnalis* model related the NOS-pseudogene with regulation of the production of NO, which acts as neurotransmitter involved in long-term memory processes and is synthesized by the enzyme nitric oxide synthase (nNOS) (Korneev *et al.*, 1999). The pseudo-NOS transcript, which is partially antisense to its parental gene (nNOS), regulates nNOS expression; the work was later on continued in (Korneev *et al.*, 2005). At that time the state of the art could be seen in Mighell *et al.* (2000). The relevance of these studies was re-discovered after the publication in 2003 of (Hirotsune *et al.*, 2003) where it was claimed that the expression of murine Makorin1 is regulated by its own pseudogene. This result was refuted 3 years later in (Grey *et al.*, 2006) where it was shown that the Makorin1 pseudogene cannot regulate its parental gene because it is methylated in both alleles and cannot be neither expressed nor imprinted. It is probable that this negative result delayed progress in the field of research on pseudogene functionality.

Meanwhile, an important review was published (Balakirev and Ayala, 2003) and since then some other evidence of pseudogene biological relevance have been discovered. For instance in (Duret *et al.*, 2006), where the role of pseudogenes as proto-genes is shown; the *Xist* non-coding gene, involved in the silencing of one of the X chromosomes in placental mammals, is the result of evolution of a pseudogene. More recently in (Tam *et al.*, 2008; Watanabe *et al.*, 2008) it was shown that pseudogenes are the source of dsRNAs that regulate gene expression in mouse oocytes.

In other work (Poliseno *et al.*, 2010) it has been shown that in cancer, a change in the expression of a pseudogene (PTENP1) affects the post-transcriptional regulation of its parental gene (PTEN) through competition for a third molecular agent, PTEN-targeting miRNAs. The same analysis was extended to oncogenic KRAS. A similar idea was explored in (Hawkins and Morris, 2010); in this case the level of transcription of a lncRNA (long non-coding RNA) expressed antisense of an *Oct4* pseudogene affects the levels of transcription of two pseudogenes and this affects the epigenetic regulation of *Oct4* near its promoter.

Some computational studies point to a functional role of the transcripts arising from pseudogenes (Svensson *et al.*, 2006; Guo *et al.*, 2009). In (Svensson *et al.*, 2006), a genome-wide computational screening was made to find biologically functional pseudogenes for the first time. This study selected expressed pseudogenes looking for EST or mRNA evidences within them. Only pseudogenes with an ancient origin and a high degree of conservation (human–mouse) were considered. The two top scored candidates were processed pseudogenes derived from Spinocerebellar ataxia type 1 (ATX1) and from uncharacterized protein Ataxin 7-like 3. To the best of our knowledge, no biological evidence has appeared since then for these candidates. A handicap of the method is, as it is mentioned in the publication, that only a small percentage of transcribed pseudogenes is well conserved in other genomes; just a 5% of transcribed pseudogenes have a potential orthologue in mouse (Harrison *et al.*, 2005). In a posterior work (Guo *et al.*, 2009) antisense transcription arising within rice pseudogenes was analysed, pointing out that those antisense transcripts could be regulating the parental gene in *trans* or that they could be down-regulating the pseudogene transcript in a *cis* manner. In this chapter we will summarize our recent work on antisense transcription arising within pseudogenes, which adds evidence that support the functional role of these antisense transcripts (Muro and Andrade-Navarro, 2010).

In summary, there is experimental and computational evidence of some biological function arising from pseudogene transcription but this evidence is not easy to find. Accordingly, not that many studies have been published on the topic. It seems that if there is pseudogene transcript functionality, it arises in certain tissues, and in certain conditions, with much more specificity than gene expression. Some of this complexity relates to the fact that this function involves non-coding RNAs (ncRNAs), a molecular entity for which novel tools and biological paradigms are still being worked out. A particular type of ncRNAs are natural antisense transcripts (NATs) and these have a special relevance for the study of pseudogene functionality for reasons that we will discuss later. In the next section we review the concept of NATs.

Natural antisense transcripts: a special type of ncRNAs

Natural antisense transcripts (NATs) are RNAs that have at least partial sequence complementarity to some other RNAs (Lavorgna *et al.*, 2004; Werner, 2005; Faghihi and Wahlestedt, 2009). Because of this complementarity they can in principle bind them and modulate their processing. NATs can be protein coding or ncRNA, but the majority are ncRNAs.

NATs are involved in different biological processes such as RNA editing, genome imprinting, chromosome X inactivation, DNA methylation, alternative splicing and RNA masking (Werner, 2005; Parenti *et al.*, 2007). NATs can be alternatively spliced, although usually there is less splicing in the antisense than in their sense counterpart.

There are discrepancies in the percentage of genes with expression in antisense; evaluations in human range from 60% (Yelin *et al.*, 2003) to 35% (Grinchuk *et al.*, 2010) and to 20% (Chen *et al.*, 2004). Only a few NATs have been well characterized, 86 according to Faghihi and Wahlestedt (2009), and not all of them in human or having a human orthologous gene.

A classical example of human NAT is NUDTD6, expressed in antisense to the FGF2 gene (see Fig. 6.1; Chandler *et al.*, 1999). FGF2, human basic fibroblast growth factor 2, is associated with growth and differentiation. Its overexpression is associated with tumour development and malignant progression. The 3′UTR of FGF2 binds the 3′UTR of NUTD6 and the last two exons.

If only exons of the genes coding the NAT and sense transcript overlap this implies that the interaction happens between mature mRNAs already spliced but interaction between pre-mRNAs has also been described. The expression of the pair of sense and antisense transcripts can be positively or negatively correlated in different tissues and cell lines.

The complementarity between a NAT and its sense gene is automatic if the NAT is expressed in *cis* to the sense transcript. However, there are NATs in *trans* to genes. In contrast to *cis*-NATs, *trans*-NATs can evolve separately from their targets.

A question arises however of how a *trans*-NAT can evolve. We presented the hypothesis that pseudogenes could originate *trans*-NATs (Muro and Andrade-Navarro, 2010). The duplication or retrotransposition of a gene with a *cis*-NAT could be followed by pseudogenization, which frees the second NAT to evolve. Alternatively a novel NAT could appear by evolution of a pseudogene

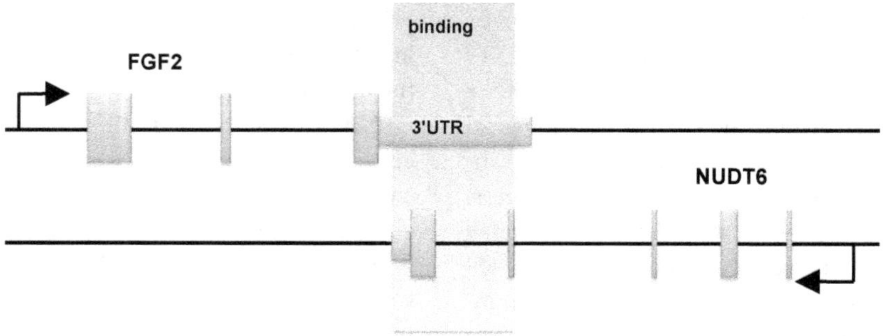

Figure 6.1 Antisense interaction between FGF2 and NUDT6. The 3′UTRs of protein coding genes FGF2 and NUDT6 overlap with each other. The figure represents a region of the human chromosome 4 from positions 123,958,543 to 124,069,307, adapted from the UCSC genome browser representation of the hg18 build of the human genome.

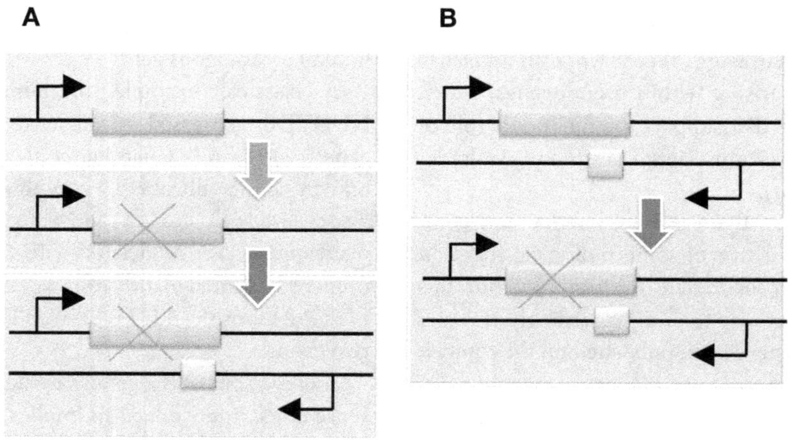

Figure 6.2 Evolution of *trans*-NATs from pseudogenes. (A) A pseudogene (crossed box) appears by duplication of a gene (arrow) followed by deleterious mutations (cross). Further evolution of the pseudogene (arrow) leads to transcription of an antisense transcript that is complementary to the parental gene. (B) If the parental gene contains a cis-NAT then the formation of the pseudogene may carry the cis-NAT, which can be kept as trans-NAT of the parental gene.

without the parental gene having a *cis*-NAT (see Fig. 6.2). We searched for evidence of this process by a survey of abundant public human EST data. In the next paragraphs we will summarize the main details of this analysis, which illustrates how reanalysis of large genomic data sets can result in novel biological insight. Before entering in detailed explanations we will introduce three concepts of importance for our study: polyadenylation signals, EST databases and pseudogene databases.

Transcript polyadenylation

Most eukaryotic protein coding transcripts are polyadenylated. The only known exceptions in metazoan are the histone mRNAs (Dominski and Marzluff, 2007). Non-coding transcripts, including *trans*-NATs, are also polyadenylated and its number in the different databases has increased a lot lately (Claverie, 2005).

Generally, transcript polyadenylation is regulated by a very complex mechanism that is just partially understood (Proudfoot and O'Sullivan,

2002). However, a ubiquitous poly adenylation signal hexamer (PAS) is always located in a range up to 30 nt upstream the transcript cleavage point. The PAS is recognized by the cleavage polyadenylation specificity factor (CPSF) protein complex that is crucial to make RNA polymerase II end the transcription. There are strong ('AAUAAA' and 'AUUAAA') and weak PAS signals (Beaudoing *et al.*, 2000), sequence variations of 'AAUAAA', at least eleven. In our study we used the PAS signal of *trans*-NATs as a comparative location reference, considering only strong PAS signals.

EST database

We take advantage of the EST libraries deposited in NCBI's dbEST database (Sayers *et al.*, 2010) as a resource to search for transcript information: here transcripts in antisense from pseudogenes. Although ESTs report transcripts incompletely, are inaccurately sequenced at some degree, and their sense is sometimes difficult to establish, we have shown that using strict restrictions ESTs can be useful to identify transcript ends with accuracy (Muro *et al.*, 2008). Transcript ends can be localized by EST clustering and direction of transcription can be confirmed by analysis of genomic polyadenylation signals, splicing junctions, and non-trimmed polyadenylated tracks (Fig. 6.3).

Pseudogene database

In order to find a collection of transcripts expressed in antisense of pseudogenes with our analysis we needed a reliable collection of pseudogenes associated to parental genes. A complete set of human pseudogenes (Gerstein and Zheng, 2006; Karro *et al.*, 2007; Zheng and Gerstein, 2007) was obtained from (http://www.pseudogene.org/downloaded on 22/Feb/2008). We took the 20,625 entries (91% of the complete original database) with a unique genomic location disregarding cases like the pseudogene ENSP00000279067.Human.chr7.mb56, which had two entries in different locations of the chromosome 7 both of them related to the same parental gene ENSG00000149656 (Ensembl gene id) coding for a known protein C20orf69 in the chromosome 20. In total, pseudogene.org had 20,197 entries whose parental genes (a total of 4484) were annotated with an Ensembl gene id (ENSG*. See http://www.ensembl.org). From those pseudogenes, 2022 (10%) were duplicated pseudogenes; 10,346 (51%) were processed and 7829 (39%) were ambiguous (Fig. 6.4). Only pseudogenes whose DNA had at least 85% of sequence annotated with A, T, G, C and located in autosomes 1–22 and chromosomes X and Y were considered.

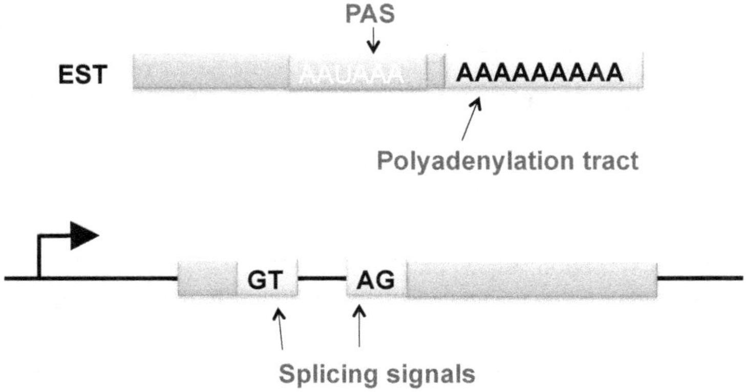

Figure 6.3 Resolving the directionality of ESTs. A spliced EST from the database (top) can be mapped to its originating genomic region (bottom) where exons may be identified. The detection of a polyadenylation signal (PAS) near the possible 3′-end of the EST, the presence of a non-trimmed polyadenylated track in the 3′-end of the EST, and splicing signals in the junctions of the predicted exons are strong evidences for directionality.

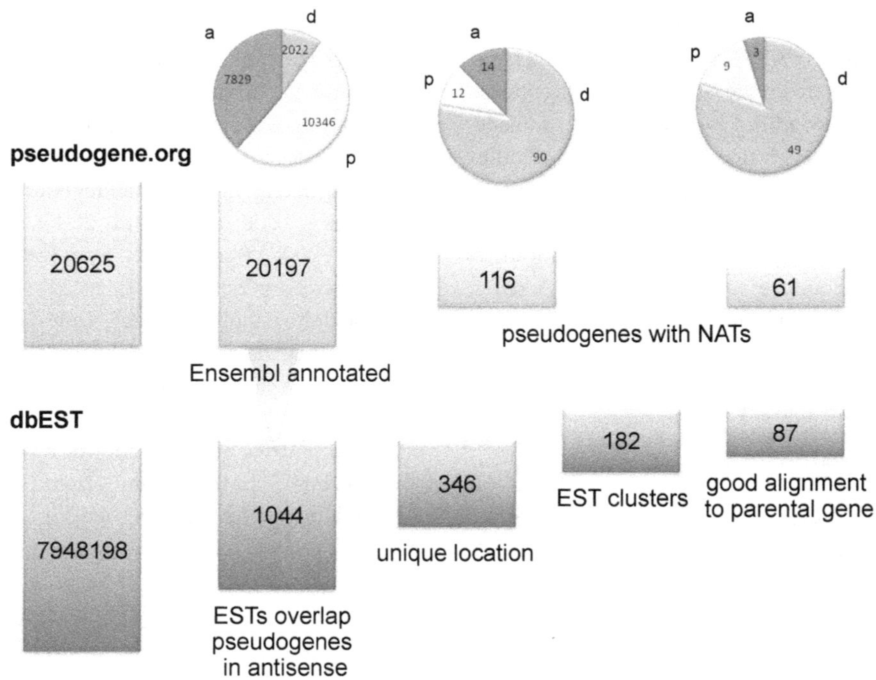

Figure 6.4 The long and winding road to find NATs from pseudogenes. Boxes represent progressive selections of pseudogene data sets from pseudogene.org. Boxes represent progressive selections of human ESTs from the dbEST database. The pie charts identify the composition of the corresponding pseudogene data set with sectors representing (d) duplicated, (p) processed or (a) ambiguous pseudogenes.

Selection of NATs from pseudogenes

We first detected all transcripts overlapping with 20,197 human pseudogenes (Fig. 6.4) and then resolved their directionality to select those in antisense (trans-NAT candidates). Because most ESTs are produced using poly T-primers, most of them map mainly to the 3′UTR end of the transcript and we could identify their probable PAS. In total we obtained 1044 ESTs. In 943 cases we obtained more evidence than the PAS (as described above) for the direction of the transcription, relying only in finding a PAS up to 30 nt upstream the EST end in 101 cases.

Given the ambiguities in EST and pseudogene definition it is important to try to find highly reliable ESTs expressed unambiguously in the pseudogene antisense strand: only those instances with a very good genome alignment were taken into account. From the set of 1044 ESTs, we selected only 349 (33.42%) non-redundant ESTs with an unambiguous genome location. Later on, we excluded three ESTs owing to an overlapping between the pseudogene and the parental gene, reducing the set to 346 ESTs.

Then, we clustered the 346 ESTs using the identified PAS as reference. Each cluster corresponds to a 3′UTR trans-NAT end arising from a pseudogene. For each cluster, the EST with the best alignment to the genome was selected as representative using a score that combined the length of the alignment and the level of identity. The DNA sequence originating the representative EST was obtained and compared with the corresponding parental gene. The resulting set of clusters corresponded to 182 different NAT ends in antisense to a total of 116 different pseudogenes corresponding to 103 different parental genes (see Fig. 6.4).

Regarding the pseudogenes, 90 (78%) were duplicated, 12 (10%) were processed and 14

(12%) were ambiguous. These fractions were very different from those of the total set of pseudogenes: 10% duplicated, 51% processed, and 39% ambiguous. These differences are very interesting considering the way pseudogenes are produced. Basically, this is telling us that although most pseudogenes are produced by processing (retrotranscribed), pseudogenes with antisense transcripts are mostly produced by gene duplication: the latter usually keep exon-intron structures, but also already existing antisense transcripts. This is consistent to the mechanisms of evolution of *trans*-NATs mentioned in Fig. 6.2B.

Characterizing NATs from pseudogenes

We pooled all comparisons between *trans*-NATs within pseudogenes and their parental gene sequences. We computed the resulting cumulative histogram of mutations relative to the distance from the PAS in the trans-NAT. Selective pressure resulting from the interaction of the trans-NAT in the pseudogene with the sense transcript should show a signal in the histogram. Lack of it, on the contrary, should result in all sequence positions having the same mutation probability following neutral evolution.

For each cluster we selected a representative EST, the one with the best alignment to the genome. We obtained the representative DNA sequence for each cluster, the genome sequence that corresponded to each representative EST. Our goal was to compare each trans-NAT pseudogene sequence with that of its parental gene with the purpose of finding the mutations that appeared within the pseudogenes after their formation and the relative positions of these mutations respect to the PAS (our reference) identified at the trans-NAT 3′-end.

From the 182 different NAT ends, we only used the 87 that had a significant alignment using BLAT against the parental gene. Again, we excluded putative good candidates for the sake of a consistent analysis. We required the EST sequence to align both to the pseudogene and to the parental gene to allow a proper analysis.

The 87 NAT ends within pseudogenes corresponded to 61 different pseudogenes (similarly biased as previous: 80% duplicated, 15% processed and 5% ambiguous) for 58 parental genes. Each representative EST was aligned against the pseudogene DNA, and the resulting genomic sequence was then aligned against the parental gene for evaluation of the mutation rates in relation to the distance to the position of the NAT PAS. In ten out of the 87 cases the sequence surrounding the PAS of the trans-NAT did not align to the parental gene, meaning that it had not evolved as part of the formation of the pseudogene and suggesting a more complex mechanism of evolution for these 10 NAT candidates.

The histogram of mutations using the PAS location as reference for the remaining 77 cases indicated a lower level of mutations in a 50 nt region upstream the PAS (Fig. 6.5, right). This degree of conservation suggests that these antisense transcripts are under selective pressure to keep their sequences similar to the parental gene transcript, indicating their functionality as NATs. For a comparison, the histogram of mutations using the 5′-end of the EST as reference did not show such a feature (Fig. 6.5, left).

By examination of the transcripts overlapping the parental genes of these pseudogenes with *trans*-NATs we could find evidence of expression of the equivalent cis-NAT in the majority of those parental genes, suggesting that the mechanism of generation of trans-NAT from pseudogenes proposed in Fig. 6.2B has happened. We also observed that most of the *trans*-NATs were situated in the same chromosome than their targets: this is expected as this property is typical of duplicated pseudogenes.

Conclusion

In the last years, and mainly as a result of a huge improvement in the detection and sequencing of transcripts, the balance of detection evidence between non-coding and protein coding genes is tilting towards the non-coding side. As a result of this, the rate of production of studies on ncRNAs is higher: more instances of functional ncRNAs are discovered while the interest on these molecules increases. Among them, natural antisense transcripts (NATs) seem to interfere

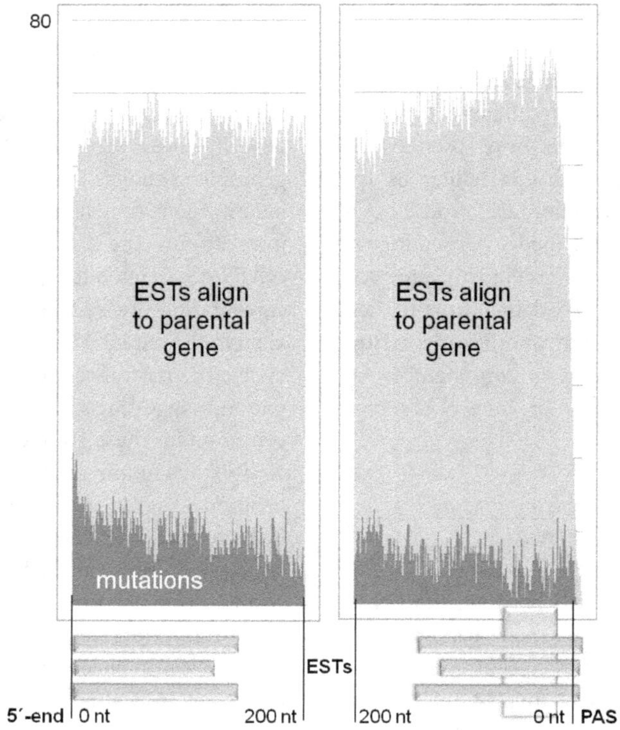

Figure 6.5 Mutation rates in NATs from pseudogenes. Mutation rates in 77 ESTs antisense of pseudogenes respect to their parental genes. Left: the histogram represents how many ESTs align to the parental gene at increasing distance from their 5′-end with grey indicating how many have a mutation. Right: how many ESTs align to the parental gene at decreasing distance to their identified PAS (indicated). The grey box represents a region with low mutation rates.

with the post-transcriptional regulation of sense transcripts because of their complementarity (*cis* or *trans*) to sense transcripts. Formation of *trans*-NATs is a puzzle: it requires the duplication of the genetic material corresponding to the transcript in a different genomic location. In this chapter we explained and presented support for a hypothesis about the origin of *trans*-NATs involving pseudogenes as providers of such duplication of genetic material.

Given the homology of a parental gene to its corresponding pseudogene we proposed that pseudogenes can be a genomic source of *trans*-NATs that bind to their parental gene transcripts as potential regulatory molecules. In Balakirev and Ayala (2003), it was pointed out that 'it is unwarranted to assume that all mutations occurring in pseudogenes are selectively neutral and become fixed in the population with equal probability'. Based on that assumption, we did a survey of human ESTs for the whole transcriptome in antisense within pseudogenes finding potential candidates following a very conservative methodology resulting in a collection of 87 trans-NAT ends within pseudogenes. We used those to determine the mutation rates that appeared within the trans-NAT after the pseudogene formation (see Fig. 6.5). Our observation of a region of high sequence similarity between the parental gene and the *trans*-NATs points out to functionality.

We wondered whether other defined ncRNAs would overlap pseudogenes. In particular, we checked the set of 685 well annotated human microRNAs from miRBase v.11 (Griffiths-Jones *et al.*, 2008). We found six examples of miRNAs that overlapped with the pseudogenes in the human pseudogene database; these in principle are complementary and could regulate the respective parental genes.

Not much is known about the functions of *trans*-NATs within pseudogenes but some works discussed their association to diseases (Runte *et al.*, 2001; Mihalich *et al.*, 2003; Tufarelli *et al.*, 2003). It has been recently discovered that the unbalance in the ratio of pseudogene to parental gene expression could lead to disease in certain tissues or conditions. A clear example is a modification of the levels of expression of an oncogene because of changes in its pseudogene expression (Poliseno *et al.*, 2010). Collectively these publications point out to links between pseudogene expression and diseases but further work is needed to find the mechanisms regulating these transcripts and their mechanism of action.

Here we have increased the evidence towards a functional role of pseudogenes as source of *trans*-NATs, which are potential parental gene regulators. We expect that the next years will see the verification of the functions of many of these. It is probable that many other *trans*-NATs exist with lower levels of similarity to their targets waiting for discovery. Their identification and characterization is challenging and will require the development of novel algorithms. Our work presented cases that might be used towards this goal.

References

Balakirev, E.S., and Ayala, F.J. (2003). Pseudogenes: are they 'junk' or functional DNA? Annu. Rev. Genet. *37*, 123–151.

van Baren, M.J., and Brent, M.R. (2006). Iterative gene prediction and pseudogene removal improves genome annotation. Genome Res. *16*, 678–685.

Beaudoing, E., Freier, S., Wyatt, J.R., Claverie, J.M., and Gautheret, D. (2000). Patterns of variant polyadenylation signal usage in human genes. Genome Res. *10*, 1001–1010.

Brent, M.R. (2008). Steady progress and recent breakthroughs in the accuracy of automated genome annotation. Nat. Rev. Genet. *9*, 62–73.

Chandler, L.A., Sosnowski, B.A., Greenlees, L., Aukerman, S.L., Baird, A., and Pierce, G.F. (1999). Prevalent expression of fibroblast growth factor (FGF). receptors and FGF2 in human tumor cell lines. Int. J. Cancer *81*, 451–458.

Chen, J., Sun, M., Kent, W.J., Huang, X., Xie, H., Wang, W., Zhou, G., Shi, R.Z., and Rowley, J.D. (2004). Over 20% of human transcripts might form sense–antisense pairs. Nucleic Acids Res. *32*, 4812–4820.

Claverie, J.M. (2005). Fewer genes, more noncoding RNA. Science *309*, 1529–1530.

Cronin, C.N., Zhang, X., Thompson, D.A., and McIntire, W.S. (1998). cDNA cloning of two splice variants of a human copper-containing monoamine oxidase pseudogene containing a dimeric Alu repeat sequence. Gene *220*, 71–76.

Dominski, Z., and Marzluff, W.F. (2007). Formation of the 3′ end of histone mRNA: getting closer to the end. Gene *396*, 373–390.

Duret, L., Chureau, C., Samain, S., Weissenbach, J., and Avner, P. (2006). The Xist RNA gene evolved in eutherians by pseudogenization of a protein-coding gene. Science *312*, 1653–1655.

Faghihi, M.A., and Wahlestedt, C. (2009). Regulatory roles of natural antisense transcripts. Nat. Rev. Mol. Cell. Biol. *10*, 637–643.

Gerstein, M., and Zheng, D. (2006). The real life of pseudogenes. Sci. Am. *295*, 43–55.

Gray, T.A., Wilson, A., Fortin, P.J., and Nicholls, R.D. (2006). The putatively functional Mkrn1-p1 pseudogene is neither expressed nor imprinted, nor does it regulate its source gene in trans. Proc. Natl. Acad. Sci. U.S.A. *103*, 12039–12044.

Griffiths-Jones, S., Saini, H.K., van Dongen, S., and Enright, A.J. (2008). miRBase: tools for microRNA genomics. Nucleic Acids Res *36*, D154–158.

Grinchuk, O.V., Jenjaroenpun, P., Orlov, Y.L., Zhou, J., and Kuznetsov, V.A. (2010). Integrative analysis of the human cis-antisense gene pairs, miRNAs and their transcription regulation patterns. Nucleic Acids Res. *38*, 534–547.

Guo, X., Zhang, Z., Gerstein, M.B., and Zheng, D. (2009). Small RNAs originated from pseudogenes: cis- or trans-acting? PLoS Comput. Biol. *5*, e1000449.

Harrison, P.M., Zheng, D., Zhang, Z., Carriero, N., and Gerstein, M. (2005). Transcribed processed pseudogenes in the human genome: an intermediate form of expressed retrosequence lacking protein-coding ability. Nucleic Acids Res. *33*, 2374–2383.

Hawkins, P.G., and Morris, K.V. (2010). Transcriptional regulation of Oct4 by a long non-coding RNA antisense to Oct4-pseudogene 5. Transcription *1*, 165–175.

Hirotsune, S., Yoshida, N., Chen, A., Garrett, L., Sugiyama, F., Takahashi, S., Yagami, K., Wynshaw-Boris, A., and Yoshiki, A. (2003). An expressed pseudogene regulates the messenger-RNA stability of its homologous coding gene. Nature *423*, 91–96.

Huang, Z., Fasco, M.J., and Kaminsky, L.S. (1997). Alternative splicing of CYP2D mRNA in human breast tissue. Arch. Biochem. Biophys. *343*, 101–108.

Karro, J.E., Yan, Y., Zheng, D., Zhang, Z., Carriero, N., Cayting, P., Harrrison, P., and Gerstein, M. (2007). Pseudogene.org: a comprehensive database and comparison platform for pseudogene annotation. Nucleic Acids Res. *35*, D55–60.

Korneev, S.A., Park, J.H., and O'Shea, M. (1999). Neuronal expression of neural nitric oxide synthase (nNOS) protein is suppressed by an antisense RNA transcribed from an NOS pseudogene. J. Neurosci. *19*, 7711–7720.

Korneev, S.A., Straub, V., Kemenes, I., Korneeva, E.I., Ott, S.R., Benjamin, P.R., and O'Shea, M. (2005). Timed and targeted differential regulation of nitric oxide synthase (NOS) and anti-NOS genes by reward conditioning leading to long-term memory formation. J. Neurosci. *25*, 1188–1192.

Lavorgna, G., Dahary, D., Lehner, B., Sorek, R., Sanderson, C.M., and Casari, G. (2004). In search of antisense. Trends Biochem. Sci. *29*, 88–94.

Mighell, A.J., Smith, N.R., Robinson, P.A., and Markham, A.F. (2000). Vertebrate pseudogenes. FEBS Lett. *468*, 109–114.

Mihalich, A., Reina, M., Mangioni, S., Ponti, E., Alberti, L., Vigano, P., Vignali, M., and Di Blasio, A.M. (2003). Different basic fibroblast growth factor and fibroblast growth factor-antisense expression in eutopic endometrial stromal cells derived from women with and without endometriosis. J. Clin. Endocrinol. Metab. *88*, 2853–2859.

Muro, E.M., and Andrade-Navarro, M.A. (2010). Pseudogenes as an alternative source of natural antisense transcripts. BMC Evol. Biol. *10*, 338.

Muro, E.M., Herrington, R., Janmohamed, S., Frelin, C., Andrade-Navarro, M.A., and Iscove, N.N. (2008). Identification of gene 3′ ends by automated EST cluster analysis. Proc. Natl. Acad. Sci. U.S.A. *105*, 20286–20290.

Muro, E.M., Mah, N., Moreno-Hagelsieb, G., and Andrade-Navarro, M.A. (2011). The pseudogenes of Mycobacterium leprae reveal the functional relevance of gene order within operons. Nucleic Acids Res. *39*, 1732–1738.

Near, T.J., Parker, S.K., and Detrich, H.W., 3rd. (2006). A genomic fossil reveals key steps in hemoglobin loss by the antarctic icefishes. Mol. Biol. Evol. *23*, 2008–2016.

Niimura, Y., and Nei, M. (2003). Evolution of olfactory receptor genes in the human genome. Proc. Natl. Acad. Sci. U.S.A. *100*, 12235–12240.

Olsen, M.A., and Schechter, L.E. (1999). Cloning, mRNA localization and evolutionary conservation of a human 5-HT7 receptor pseudogene. Gene *227*, 63–69.

Ota, T., and Nei, M. (1995). Evolution of immunoglobulin VH pseudogenes in chickens. Mol. Biol. Evol. *12*, 94–102.

Parenti, R., Paratore, S., Torrisi, A., and Cavallaro, S. (2007). A natural antisense transcript against Rad*18*, specifically expressed in neurons and up-regulated during beta-amyloid-induced apoptosis. Eur. J. Neurosci. *26*, 2444–2457.

Poliseno, L., Salmena, L., Zhang, J., Carver, B., Haveman, W.J., and Pandolfi, P.P. (2010). A coding-independent function of gene and pseudogene mRNAs regulates tumour biology. Nature *465*, 1033–1038.

Proudfoot, N., and O'Sullivan, J. (2002). Polyadenylation: a tail of two complexes. Curr. Biol. *12*, R855–857.

Runte, M., Huttenhofer, A., Gross, S., Kiefmann, M., Horsthemke, B., and Buiting, K. (2001). The IC-SNURF-SNRPN transcript serves as a host for multiple small nucleolar RNA species and as an antisense RNA for UBE3A. Hum. Mol. Genet. *10*, 2687–2700.

Sayers, E.W., Barrett, T., Benson, D.A., Bolton, E., Bryant, S.H., Canese, K., Chetvernin, V., Church, D.M., Dicuccio, M., Federhen, S., *et al.* (2010). Database resources of the National Center for Biotechnology Information. Nucleic Acids Res. *38*, D5–16.

Svensson, O., Arvestad, L., and Lagergren, J. (2006). Genome-wide survey for biologically functional pseudogenes. PLoS Comput. Biol. *2*, e46.

Tam, O.H., Aravin, A.A., Stein, P., Girard, A., Murchison, E.P., Cheloufi, S., Hodges, E., Anger, M., Sachidanandam, R., Schultz, R.M., *et al.* (2008). Pseudogene-derived small interfering RNAs regulate gene expression in mouse oocytes. Nature *453*, 534–538.

Torrents, D., Suyama, M., Zdobnov, E., and Bork, P. (2003). A genome-wide survey of human pseudogenes. Genome Res. *13*, 2559–2567.

Tufarelli, C., Stanley, J.A., Garrick, D., Sharpe, J.A., Ayyub, H., Wood, W.G., and Higgs, D.R. (2003). Transcription of antisense RNA leading to gene silencing and methylation as a novel cause of human genetic disease. Nat. Genet. *34*, 157–165.

Wang, X., Grus, W.E., and Zhang, J. (2006). Gene losses during human origins. PLoS Biol. *4*, e52.

Watanabe, T., Totoki, Y., Toyoda, A., Kaneda, M., Kuramochi-Miyagawa, S., Obata, Y., Chiba, H., Kohara, Y., Kono, T., Nakano, T., *et al.* (2008). Endogenous siRNAs from naturally formed dsRNAs regulate transcripts in mouse oocytes. Nature *453*, 539–543.

Werner, A. (2005). Natural antisense transcripts. RNA Biol. *2*, 53–62.

Yelin, R., Dahary, D., Sorek, R., Levanon, E.Y., Goldstein, O., Shoshan, A., Diber, A., Biton, S., Tamir, Y., Khosravi, R., *et al.* (2003). Widespread occurrence of antisense transcription in the human genome. Nat. Biotechnol. *21*, 379–386.

Zhang, Z., and Gerstein, M. (2004). Large-scale analysis of pseudogenes in the human genome. Curr. Opin. Genet. Dev. *14*, 328–335.

Zheng, D., and Gerstein, M.B. (2007). The ambiguous boundary between genes and pseudogenes: the dead rise up, or do they? Trends Genet. *23*, 219–224.

Zheng, D., Frankish, A., Baertsch, R., Kapranov, P., Reymond, A., Choo, S.W., Lu, Y., Denoeud, F., Antonarakis, S.E., Snyder, M., *et al.* (2007). Pseudogenes in the ENCODE regions: consensus annotation, analysis of transcription, and evolution. Genome Res. *17*, 839–851.

Zhu, J., Sanborn, J.Z., Diekhans, M., Lowe, C.B., Pringle, T.H., and Haussler, D. (2007). Comparative genomics search for losses of long-established genes on the human lineage. PLoS Comput Biol. *3*, e247.

Non-coding RNAs, Epigenomics and Complexity in Human Cells

7

Fabricio F. Costa

Abstract

During the past two decades, new technologies in molecular biology and human genetics have enabled the discovery of different types of non-coding RNAs. Non-coding RNAs are RNA transcripts that have no apparent protein product. These molecules have been grouped in different classes such as microRNAs, small RNAs and long RNAs (lncRNAs) according to their size and function. LncRNAs have been strongly associated to epigenetic mechanisms in different cell types. Conceivably, they have been described as an essential part of the human epigenome. In this chapter, examples of lncRNAs and their function will be presented. A historical perspective on the impact of lncRNAs in epigenetic mechanisms, human disease and evolution will be also discussed.

Introduction

Growing evidence suggests that the majority of the genomes of complex organisms are transcribed into non-coding RNAs (ncRNAs). Many of these ncRNAs are alternatively spliced and/or processed into smaller products after transcription by different mechanisms, some of which are not completely understood. NcRNA is commonly used for an RNA that does not encode or produce a protein, but this does not mean that such RNAs do not contain information nor have function. For several years, ncRNA genes included highly abundant and functionally important classes of RNAs such as transfer RNA (tRNA) and ribosomal RNA (rRNA). More recently, other RNAs such as small nucleolar RNAs (snoRNAs), microRNAs, small interfering RNAs (siRNAs), piwi-interacting RNAs (piRNAs) and long non-coding RNAs (lncRNAs) were characterized. Examples of lncRNAs have been extensively described and they include XIST, AIR, HOTAIR and others. The advent of different technologies such as tiling arrays, second generation DNA sequencing and novel high-throughput methodologies, have increased our ability to evaluate entire transcriptomes, thereby facilitating the discovery of new classes of ncRNAs.

LncRNAs are a specific group of RNA molecules that have sizes ranging from approximately 200 nucleotides to several thousand kilobases in length. The existence of individual lncRNAs has been known since the early 1990s (see Fig. 7.1 for more details). These lncRNAs were identified by the use of traditional gene mapping approaches and were only discovered to be non-protein-coding after their RNA sequences were analysed in details by specific experimental procedures. However, the majority of these RNAs were generally considered to be artefacts until the advent of technologies that were capable of unbiased high throughput sequencing of all the expressed transcripts in eukaryotic cells. These new methods revealed that lncRNAs, even when detected at very low levels, have widespread expression in mammalian cells. More importantly, several of these examples have functions in epigenetic mechanisms.

Methods such as arrays with high resolution indicated that 60–90% of eukaryotic genomes could be transcribed depending on the cell type

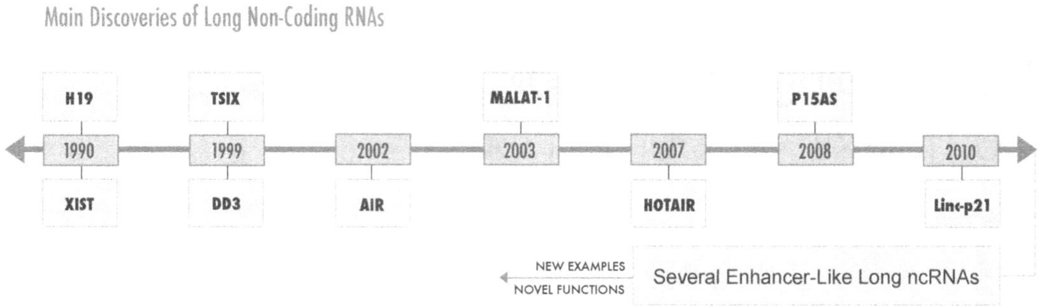

Figure 7.1 Historical timeline showing the main discoveries in the field of lncRNAs. The line with arrows indicates specific time points and the boxes represent the year that important lncRNAs were discovered. New examples will be described in the next years with the advent of novel technologies that have been developed in genomics and transcriptomics to study human cells. Graphic design by Kathryn A. Bischof.

(Nagalakshmi *et al.*, 2008; Affymetrix ENCODE Transcriptome Project, 2009). It is important to note that the fraction of the genome that is analysed by these technologies exclude repetitive elements. This exclusion occurs mainly because of their complexity and difficulties to map them to the human genome. Evidence indicates that eukaryotic cells transcribe high levels of repetitive DNA, although the majority of these non-coding transcripts have unknown function (Falkner *et al.*, 2009).

As discussed above, several lines of evidence indicate an extensive and pervasive transcription of RNAs in a big fraction of mammalian genomes, including humans. Tiling arrays have shown that the human genome contains roughly comparable numbers of protein-coding and non-coding genes that are bound by common transcription factors and probably regulated by common environmental signals (Cawley *et al.*, 2004). Additionally, recent studies reported an association of the deregulated expression of non-coding genes with human diseases and disorders, including cancers (Costa, 2005; Tsai *et al.*, 2011). These discoveries have opened new avenues in areas of research and new possibilities for drug discovery, improvement in therapeutics and the identification of new diagnostic and prognostic markers (Costa, 2009).

LncRNAs have been implicated in several aspects of eukaryotic biology. There are many examples of lncRNAs that can give rise to small RNAs and microRNAs after processing, indicating that some of these might function as primary transcripts intermediating the production of other ncRNAs classes (Costa, 2007). There is also growing evidence that a large fraction of intergenic genomic regions can be transcribed into ncRNAs (Khalil *et al.*, 2009). These lncRNAs have been named LincRNAs for Long Intergenic non-coding RNAs. The expression of lncRNAs can be regulated by transcriptional factors, epigenetic mechanisms and also by other ncRNAs. They are able to act in cooperation with selective ligands, modulating the activity of RNA-binding proteins in response to specific signals, providing a ncRNA-protein-based strategy to integrate transcriptional programmes (Fig. 7.2). Studies have also shown that lncRNAs have roles as activators of the transcription of neighbouring genes (Marques and Ponting, 2009). Some of these events may be a consequence of pervasive transcription, but there is also a hypothesis that widespread non-coding transcription has a function in transcriptional regulation and genome integrity maintenance.

ncRNAs are also implicated in mechanisms of epigenetic inheritance (Zaidi *et al.*, 2011). It is becoming clear that epigenetic modifications, which include different types of ncRNA molecules, are an important factor that affects the phenotype during the lifetime of an organism. A few examples of epigenetic inheritance in plants, mice and even humans have been suggesting that

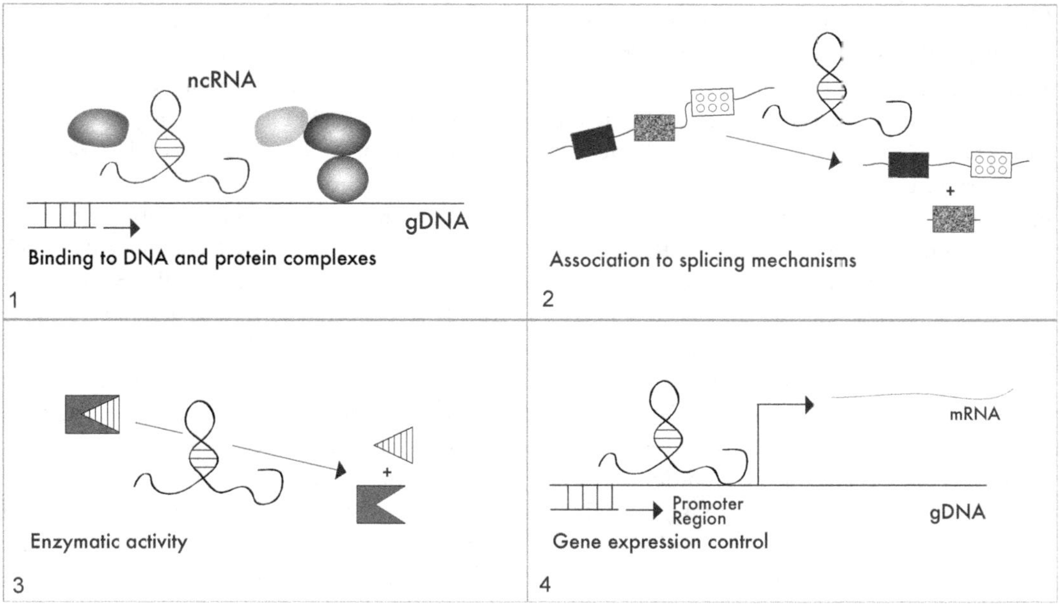

Figure 7.2 LncRNAs and functionality. Different mechanisms that lncRNAs participate are depicted. (1) LncRNAs can bind to DNA and to protein complexes in epigenetic mechanisms that regulate gene expression; (2) LncRNAs might be important players in mechanisms of splicing along with other components of the splicing machinery; (3) LncRNAs could possibly have enzymatic activity; one well-known example is the hTERT catalytic RNA from Telomerase and (4) LncRNAs have been linked to new mechanisms of the gene expression by binding to gene promoters in concert with other factors. *There are other mechanisms in which lncRNAs have been implicated and some of these are not represented in this figure. Graphic design by Kathryn A. Bischof.

epigenetic traits can be passed from generation to generation, and this process might have RNA molecules as intermediaries (Suter and Martin, 2010). As a consequence, evolutionary theories have been re-assessed based on these new discoveries and human molecular evolution has been revisited. lncRNAs linked to human complexity when compared with other organisms has been reported. The expression of ncRNAs, especially lncRNAs, in the mice and human brain has been accessed, indicating that these molecules could be associated to organism complexity. LncRNAs have localized expression in brain cells of higher organisms and this could also indicate that these molecules may be associated to memory by functioning as molecular components of neural circuitries that generate signals in the brain (Mercer *et al.*, 2008). However, more studies in neuroscience will be needed to provide stronger evidence and to support this hypothesis.

The purpose of this book chapter is to give an overview of the burgeoning field of ncRNAs. The following sections will provide information on the history of lncRNAs, including landmark discoveries in this filed with the identification of functional examples. Additionally, the identification of RNAs with no protein-coding potential with the advent of new technologies to study human transcriptomes will be explored. Speculations on the theories behind these discoveries will be also discussed. LncRNA association to human diseases and their potential use as drug targets and biomarkers will be also pinpointed. Finally, the impact that ncRNA discoveries will have in evolutionary theories and on our understanding of human complexity will be hypothesized in the last sections.

A historical perspective of long non-coding RNA discovery and function

Nucleic acids in live cells were first discovered in 1868 by Friedrich Miescher (Dahm, 2005). In the late 1950s, RNA molecules were implicated in protein synthesis by Francis Crick after his studies on possible functions for nucleic acids (Crick, 1958). Some years later, a putative RNA component was identified with functions in protein translation (Warner and Rich, 1964). Following this historical timeline, the first non-coding RNA characterized was an alanine tRNA identified in yeast with its structure published in 1965 (Holley *et al.*, 1965). After the discovery of tRNAs, several other types of RNAs that do not code for proteins have followed; these included rRNAs, snoRNAs, and more recently microRNAs. In the case of lncRNAs, most examples were identified in the beginning of the 1990s (Fig. 7.1).

For many years, RNAs were considered to be just accessory molecules involved in mediating transcription and translation mainly because of the discoveries in the 1950s of tRNAs and rRNAs. However, several pieces of evidence have accumulated in the last two decades demonstrating that this is a simplistic view of how the molecular machinery works in eukaryotic cells. RNA molecules were always very versatile, mainly owing to their chemical properties, allowing them to form complex tertiary structures and bind to proteins, DNA and other RNAs, roles that were thought to be under the exclusive domain of proteins. Additionally, RNAs are capable to form ribocomplexes (Fig. 7.2). RNAs can also associate to specific DNA and/or RNA sequences, being able to control several aspects of gene regulation by affecting the number of molecular networks that happen in eukaryotic cells (Fig. 7.2).

Chromatin structure dynamics is a complex process that involves different epigenetic components. Accumulating evidence indicates that RNA molecules are associated with eukaryotic chromatin and might play an important role in chromatin's structure maintenance. In 1975, in an attempt to isolate the main components associated to eukaryotic heterochromatin and euchromatin, a group has reported that chromatin structure includes several unknown RNAs (Paul and Duerksen, 1975). Several years later, XIST, the first lncRNA with functions in chromatin dynamics was described (see Fig. 7.1 for timeline of discoveries). The lncRNA XIST was discovered because of its specific expression from inactive female X chromosomes in the beginning of the 1990s (Brown *et al.*, 1991). The XIST gene does not encode a protein but rather produces a 17 kilobases functional RNA molecule. XIST lncRNA is only expressed in cells containing at least two X chromosomes and it is not normally expressed in male cells. Higher XIST expression can be seen in cells with more X chromosomes. Research has shown that XIST RNA is both necessary and sufficient for X chromosome inactivation together with other protein complexes (Penny *et al.*, 1996; Wutz and Jaenisch, 2000). It function by recruiting various silencing protein complexes to label and inactivate the X chromosome (Penny *et al.*, 1996; Wutz and Jaenisch, 2000). Increased XIST expression represents a key initiation event in X inactivation, showing the central role of this lncRNA in this mechanism.

Another mechanism involving a different lncRNA in X chromosome inactivation was reported in the late 1990s with the identification of TSIX, an antisense RNA to XIST (Lee *et al.*, 1999). TSIX is 40 kilobases long but it is transcribed in the opposite direction across the entire XIST gene. Like XIST, TSIX only acts on the active chromosome that produces it. Interestingly, there is an inverse relationship between TSIX and XIST in expression. For example, if TSIX transcription is reduced on one X chromosome, XIST expression increases and leads to inactivation of that same X chromosome (Lee and Lu, 1999; Sado *et al.*, 2001). In contrast, overexpression of TSIX prevents increases in XIST expression and blocks inactivation *in cis* on that same X chromosome (Luikenhuis *et al.*, 2001; Stavropoulos *et al.*, 2001).

Genomic imprinting is a phenomenon where expression of an autosomal gene is determined by its parental origin. Imprinted genes are expressed from either one of the alleles: paternal or maternal. Furthermore, the expression pattern is maintained (or imprinted) from one generation to the next with allele-specific DNA methylation and other epigenetic mechanisms that have been strongly implicated in the imprinting process. So far, more than 50 imprinted human and mouse genes have been identified. Approximately half of these are expressed exclusively from the maternal allele while the other half are expressed only from the paternal allele. Imprinted lncRNAs in the human genome include H19, AIR and others.

H19 lncRNA was also discovered in the beginning of the 1990s (Fig. 7.1) and it is 2.3 kb long (Brannan *et al.*, 1990). It is transcribed by RNA polymerase II, spliced and polyadenylated, but it does not appear to be translated (Brannan *et al.*, 1990). After many studies, researchers have concluded that the product of the H19 gene is a lncRNA mainly because the H19 RNA is evolutionarily conserved at the nucleotide level in humans and rodents and there is no known 'open reading frame'. In addition, although the RNA sequence was evolutionarily conserved, at the amino acid level, there was a complete absence of conservation. Furthermore, free energy analysis of the H19 RNA sequence revealed several stable secondary RNA structures, including 16 helices and various hairpin loops. Studies of *in situ* hybridization of the H19 RNA revealed that it localizes in a cytoplasmic ribonucleoprotein particle, suggesting that the H19 RNA functions as a riboregulator (Szymanski *et al.*, 2003). Interestingly, loss of H19 is not lethal in mice (Leighton *et al.*, 1995); however, overexpression of H19 is a dominant and lethal mutation (Brunkow and Tilghman, 1991). H19 is also known as BWS since aberrant H19 expression can be involved in Beckwith-Wiedemann Syndrome, as well as Silver-Russell syndrome, two genetic disorders that cause retardation (Thorvaldsen *et al.*, 1998; Bliek *et al.*, 2006). The H19 gene is expressed exclusively on one parental allele by genomic imprinting and is only transcribed from the maternally inherited allele (Rachmilewitz *et al.*, 1992). Several reports have recently shown that H19 have roles as a tumour suppressor or oncogene in cancers depending on the cancer type (Hao *et al.*, 1993; Hibi *et al.*, 1996).

Similarly to the H19 gene, another example of imprinting lies in a region of mouse chromosome 17, which contains a cluster of 3 imprinted protein-coding genes – Igf2r, Slc22a2, and Slc22a3 – and the lncRNA AIR (Sleutels *et al.*, 2002). Since the lncRNA AIR is conserved in humans and mice, engineered mice containing a polyadenylation site next to the paternal Air allele (called T-Air) were generated and researchers were able to produce a truncated paternal Air RNA (Sleutels *et al.*, 2002). Using different experimental procedures, it was reported that this truncation relieved silencing of the Igf2r, Slc22a2, and Slc22a3 protein-coding paternal alleles (Sleutels *et al.*, 2002). Additionally, AIR lncRNA is able to target repressive histone-modifying activities through molecular interaction with specific chromatin domains to epigenetically silence transcription (Nagano *et al.*, 2008). These data indicate that the non-coding AIR RNA plays key roles in epigenetic mechanisms of gene silencing by imprinting.

Recently, a specific group of lncRNAs was described and named large intergenic non-coding RNAs (lincRNAs) (Woo and Kingston, 2007). LincRNAs are ultraconserved regions of the genome and they are mapped to loci that were first thought to be gene deserts or areas with no identified gene. LincRNAs appear to have been selected for conservation by evolution and are strongly associated with epigenetic modification complexes. They have been implicated in epigenetic mechanisms such as binding to histone modifications and protein complexes, thereby affecting gene transcription. The HOTAIR gene, which stands for HOX antisense intergenic RNA, was the first lincRNA identified (Rinn *et al.*, 2007). HOTAIR is mapped to human chromosome 12 and it encodes for a 2.2 kilobases lncRNA long molecule (Fig. 7.1). HOTAIR controls gene expression on chromosome 2 by a newly described epigenetic mechanism (Rinn *et al.*, 2007). It is the first example of an RNA expressed on one chromosome influencing transcription on another chromosome in *trans*. HOTAIR's mechanism of action includes the interaction with a Polycomb-group protein Polycomb repressive complex 2 (PRC2) and regulation of the chromatin state (Rinn *et al.*, 2007). It is also required for gene silencing of the HOXD locus by PRC2 (Rinn *et al.*, 2007). Importantly, HOTAIR interacts with histone demethylases, another component of the epigenetic machinery (Rinn *et al.*, 2007). HOTAIR is an essential component that affects proper differentiation of the skin over the surface of the body (Rinn *et al.*, 2007). Skin tissues from various anatomical positions are distinct in the human body, e.g. the skin of the eyelid differs markedly from that on the sole of the foot and this is due to differences in the expression of the HOTAIR gene (Rinn *et al.*, 2007). Recently, HOTAIR was associated with diseases such as cancer since it is highly expressed in metastatic tissue compared with primary tumours (Gupta *et al.*, 2010). High levels of expression in primary breast tumours are a significant predictor of subsequent metastasis development and death (Gupta *et al.*, 2010).

DD3 (also known as PCA3) and MALAT-1 (metastasis-associated lung adenocarcinoma transcript-1) lncRNAs were identified by different groups as biomarkers for cancer progression in the late 1990s (Fig. 7.1). DD3 was shown to be a better biomarker than the currently used markers for disease progression in prostate cancers (Bussemaker *et al.*, 1999). MALAT-1 was also identified as a metastasis-associated gene in lung tumours (Ji *et al.*, 2003). Even though the function of DD3 is still unclear, MALAT-1 was implicated as part of the splicing machinery and may have functions in splicing mechanisms (Hutchinson *et al.*, 2007).

Other examples of lncRNAs include p15AS and Linc-p21 that were described as antisense lncRNAs to both p15 and p21 tumour-suppressor genes in 2008 and 2010, respectively (Fig. 7.1). P15AS lncRNA was identified in a cancer study showing that its expression has an inverse correlation to the tumour suppressor p15 gene (Yu *et al.*, 2008). By introducing inducible p15AS expression constructs into mammalian cells, investigators discovered that p15 silencing was induced both in *cis* and in *trans* through heterochromatin formation (Yu *et al.*, 2008). This accompanied changes in histone modifications, but not in DNA methylation (Yu *et al.*, 2008). Another example is the Linc-p21 gene transcription that affects the tumour-suppressor gene p21 through chromatin remodelling in the region of the sense promoter (Huarte *et al.*, 2010). It was shown that the knockdown of the Linc-p21 transcript results in the activation of p21 with a complete loss of suppressive histone epigenetic marks at the p21 promoter (Huarte *et al.*, 2010). Taken together, these studies suggest that p15AS, Linc-p21, and other antisense lncRNAs might have roles in epigenetic silencing.

The first enhancer expressed as an lncRNA was reported in 2006 and this lncRNA had a role in the activation of transcription (Feng *et al.*, 2006). More recently, enhancers that are expressed as lncRNAs were identified and characterized (Ørom *et al.*, 2010). The depletion of a number of these lncRNAs resulted in a decreased expression of their neighbouring protein-coding genes, which included master regulators of important molecular mechanisms (Ørom *et al.*, 2010). Using different assays, it was also demonstrated that these enhancer-like lncRNAs are required for gene expression activation (Ørom *et al.*, 2010). These results revealed unanticipated roles for a class of lncRNAs in the activation of critical regulators of development and differentiation. Based on these discoveries, we can expect the description of several new examples of functional lncRNAs in the future. More details on historical landmark discoveries in the field of lncRNAs are shown in Fig. 7.1.

Human transcriptome, long non-coding RNAs and new definitions for a gene

Although the last reported genome annotation has identified just 20,000–25,000 protein-coding genes in the human genome there are still some questions regarding the number of genes in our genome. This intriguing issue may be explained by the presence of alternative splicing and many ncRNAs, including lncRNAs, in eukaryotic transcriptomes. If we take into account that the majority of the ncRNA fraction of transcriptomes has a function, the number of human genes will be higher than previous estimates. While there is scepticism as to whether novel non-coding transcripts are functional, new paradigms based on ncRNAs are emerging after deep transcriptome analyses and with the identification of functional examples as discussed in the previous sections. Evolving technologies are enabling a better coverage and understanding of eukaryotic transcriptomes, and these new discoveries are changing the *status quo* in genomic science.

Recent data from studies that have analysed both mice and human transcriptomes using tilling arrays and second generation DNA sequencing confirm that eukaryotic transcriptomes express a considerable amount of RNA that do not code for proteins. For example, the members of the FANTOM consortium have reported that almost 70% of the mouse genome is transcribed (Carninci *et al.*, 2005). They have identified around 200,000 independent transcripts, of which half consist of non-coding RNAs (Carninci *et al.*, 2005). Moreover, they found that a high percentage of the mapped transcription units are antisense transcripts, which overlap with a transcript from opposite strands (Katayama *et al.*, 2005). Additionally, the ENCODE Project in human cells have also discovered that a big fraction of human genomes is indeed expressed (ENCODE Project Consortium, 2007). Some of these transcripts have high levels of expression, but on the other hand, the majority has expression levels that are barely detectable. However, low levels of expression do not mean lack of function; one established example is the group of microRNAs that even with low levels of expression have functions in gene regulation. These results could provide a solution to the discrepancy between the number of protein-coding genes and the number of transcripts produced by mammalian cells, especially human cells.

A new classification of transcripts for transcriptomes has emerged with these discoveries; expressed messages will be separated in protein-coding and non-coding. There will also be a need to re-evaluate the number of genes in the genome of several organisms, including humans. Interestingly, a great fraction of the non-coding transcripts are lncRNAs, some of which can give rise to smaller transcriptional units. The mechanisms underlying the control of expression and processing for these lncRNAs in the transcriptome still remains unknown. Some studies are starting to unravel these mechanisms and this will be discussed in the following sections of this book chapter. The discoveries that have been reported until now will help the scientific community to decide whether non-coding RNAs should be considered a separate class of 'genes'. This definition will remain a vocabulary issue for biologists to work out in the years to come.

Long non-coding RNAs in epigenomics, epigenetics and other molecular mechanisms

The way in which multicellular genomes of organisms orchestrate the differentiation of trillions of cells and many organs during development and throughout their lifespan, all from a single fertilized egg, is the subject of intense study. The networks of genes, both protein-coding and non-coding, that are expressed in specific time points can define different cell types in the tissues. This differential expression is regulated at various levels, including epigenetic modifications, which are mainly DNA methylation and histone modifications. These modifications are able to interact with structural and enzymatic proteins, resulting in the activation or silencing of any specific gene or group of genes in the genome at a given developmental time point. Detailed mechanisms are emerging on the role of different chromatin modifications and how these functions are effected at the molecular level, but it is still unclear how their deposition across the epigenomic landscape is regulated in different cells to form tissues. Growing evidence has been accumulating that implicates lncRNAs as new players in these processes. Of special interest are lncRNAs that are able to bind to complexes of both DNA and proteins with functions in gene expression regulation. This section will discuss examples of lncRNAs with functions in epigenomics, epigenetics and other molecular mechanisms that are important for a better understanding of organism development, cell differentiation and disease initiation and progression in humans.

The emergence of the field of epigenomics is improving our knowledge on how eukaryotic cells orchestrate gene transcription. Epigenomics was defined by Callinan and Feinberg as the merged science of epigenetics and genomics and it has arisen as a new discipline with the aim of understanding genetic regulation and its contribution to cellular growth and differentiation, disease, and ageing (Callinan and Feinberg, 2006). At the molecular level, epigenetics is the study of the chemical and physical modifications that occur in the DNA molecule that are able to change its conformation, and consequently, gene expression. There is growing evidence that ncRNAs produced by the genome, especially lncRNAs, can act as important regulators in several aspects of epigenetics, thus representing an additional component of the epigenome.

The first examples of lncRNAs associated to epigenetic mechanisms were XIST, TSIX and H19, which were described as important players in X-chromosome gene silencing and genomic imprinting, respectively. Further, a lncRNA with functions in epigenetic mechanisms that regulates gene expression in tissue differentiation (specifically skin development) named HOTAIR was identified by high-throughput epigenomic analyses (Rinn *et al.*, 2007). Other reports have followed and identified thousands of lncRNAs that may have similar functions to HOTAIR (Khalil *et al.*, 2009). These lncRNA examples were discovered and described as long intergenic ncRNAs or lincRNAs (the definition of lincRNAs was already discussed in a different section). For example, a strong association of epigenetic marks, mainly histone modifications, with lincRNAs in mice and humans was reported (Khalil *et al.*, 2009). LincRNAs were also associated to mechanisms that promote the activation of stem cell phenotypes (Loewer *et al.*, 2010). These lincRNAs are mainly associated to polycomb protein complexes and histone methylation marks having functions in gene expression control of mammalian cells (Khalil *et al.*, 2009). Complexes of proteins were identified within regions in which lincRNAs are localized together with transcriptionally active histone marks in the genome (Khalil *et al.*, 2009). On the other hand, histone marks and lincRNAs associated to gene inactivation were also identified (Khalil *et al.*, 2009). Thus, lincRNAs represent 'flexible scaffolds' that bind to DNA and proteins as important factors in gene expression control. The observation that lncRNAs can bind to an epigenetic regulatory complex of proteins and histone marks indicates that this class of ncRNAs may represent master regulators of gene expression in eukaryotic cells (Costa, 2010a).

Other roles for lncRNAs include the participation in mechanisms of splicing in human cells and possible functions as catalytic RNAs similarly to the hTERT RNA from the human telomerase enzyme (Fig. 7.2). LncRNAs can also control gene expression by physically binding to the DNA and the recruitment of protein complexes controlling gene expression by uncharacterized mechanisms. In the case of splicing, researchers have reported that MALAT-1 lncRNA regulates pre-mRNA splicing which is a critical step in the earliest stage of protein production (Hutchinson *et al.*, 2007). MALAT-1 plays a role in recruiting important proteins, called pre-mRNA splicing factors, to the site of gene transcription in the nucleus (Hutchinson *et al.*, 2007). There is a possibility that other lncRNAs that were not identified yet have similar functions in splicing mechanisms in human cells, since we still know too little about the functions and components of the splicing machinery.

Another recent example includes a group of lncRNAs with enhancer-like functions in gene regulation. As discussed before, a set of lncRNAs with enhancer-like functions were identified in different human cell lines (Ørom *et al.*, 2010). Interestingly, the depletion of a number of these lncRNAs led to a decreased expression of neighbouring protein-coding genes, including regulators of important processes in human cells such as haematopoiesis (Ørom *et al.*, 2010). Using several types of assays, a strong requirement for these lncRNAs in activation of gene expression was demonstrated (Ørom *et al.*, 2010). These results revealed an unanticipated role for a class of lncRNAs with enhancer-like features in the activation of critical regulators of development and differentiation in human cells. A previous report in 2006 has shown that human ultraconserved regions that produce the lncRNA Evf-2 cooperates with Dlx-2 to increase the transcriptional activity of the Dlx-5/6 enhancer in a target and homeodomain-specific manner (Feng *et al.*, 2006). The authors of this study have also hypothesized that a subset of vertebrate ultraconserved regions may function at both the DNA and RNA levels controlling key developmental regulators (Feng *et al.*, 2006). This may explain why ultraconserved sequences exhibit 90% or more conservation even after approximately 500 million years of vertebrate evolution. These discoveries are important to demonstrate that lncRNAs could be strongly associated to organism complexity as will be discussed in the following sections.

Given the increasing number of reports indicating that lncRNAs have important functions in mechanisms of gene regulation, there has been also growing evidence that deregulated expression of lncRNAs is a common feature of human diseases, especially cancers. In some cases, the expression of lncRNAs can predict how aggressive and prone to metastasis a cancer might be. Examples include the MALAT-1 gene, which is a 8 Kilobases long ncRNA that is overexpressed in lung cancers and which has been associated with lung cancer progression (Tano *et al.*, 2010) and DD3, a 2–4 kilobases lng ncRNA that is overexpressed in prostate cancers (Bussemaker *et al.*, 1999). Additionally, new examples of lncRNAs that are deregulated in cancers include HOTAIR that was recently associated to metastasis in breast tumours (Gupta *et al.*, 2010), p15AS in blood tumours (Yu *et al.*, 2008) and Linc-p21 that affects the expression of hundreds of gene targets normally repressed by p53, a well-known tumour-suppressor gene in different cancer types (Huarte *et al.*, 2010). Taken together, these studies indicate that lncRNAs not just have important functions in gene expression control and epigenetic mechanisms, but also that their deregulated expression could be associated to diseases. Other lncRNAs have been also associated to complex disorders. Examples include H19 with cancers and neurological diseases (Thorvaldsen *et al.*, 1998; Bliek *et al.*, 2006), IPW with Prader–Willi and Angelman syndromes (Wevrick *et al.*, 1994), DISC2 with schizophrenia (Millar *et al.*, 2000) and others.

Based on the evidences presented above, it is becoming clear that lncRNAs are major players in different aspects of gene regulation in development, especially in epigenomic and epigenetic mechanisms. Furthermore, lncRNAs with functions in other molecular mechanisms have been also described. The fact that lncRNAs are implicated in mechanisms that are important for proper development and differentiation explain why several of these have their expression deregulated in diseases such as cancers and various types of neurological diseases. We can expect that more examples of lncRNAs will be described in the future with novel functions in human cells and also in complex diseases that still have unknown causes. For example, despite a considerable amount of effort by different groups to evaluate the genetic traits associated with complex diseases by genome-wide association studies (GWAS), just a few regions, that are mainly linked to protein-coding genes, were identified (Costa, 2010b). We can expect that several of these regions map to lncRNAs or other regulatory transcripts (such as enhancer-like lncRNAs) that are still unknown (Costa, 2010b).

Changing our view on evolutionary theories

It has been always accepted that heritable changes are exclusively written in our DNA sequence, however several examples in which 'epigenetic states' can be inherited have been reported in the last years. It is becoming clear that epigenetic modifications are not completely erased after a genome-wide reprogramming during gametogenesis and embryogenesis. Epigenetic 'memories', that is a very intriguing phenomenon, can persist in the following generations as demonstrated in plants (Molinier *et al.*, 2006), flies (Maurange and Paro, 2002), and mammalian cells (Feng *et al.*, 2006). A similar effect such as trans-generational inheritance of RNA interference has been also identified in the worm *C. elegans* more than a decade ago (Fire *et al.*, 1998; Tavernarakis *et al.*, 2000). Most of the mechanisms implicated in the epigenetic 'memory' are associated to paramutation events, which are heritable epigenetic changes in the phenotype of a allele that is paramutable (Chandler, 2007). Paramutation is a mechanism defined as an interaction between two alleles of a single locus that results in heritable changes of one allele that is induced by the other allele. Moreover, the paramutated state can result in heritable changes in the chromatin structure such as changes in DNA methylation and histone modifications affecting gene expression patterns. Paramutation events are of interest to evolutionists mainly because they represent a mechanism that can drive molecular changes in organisms either directly or indirectly via the environment.

The evolutionist Lamarck hypothesized more than 200 years ago that evolution could be driven by specific phenotypic changes that are acquired non-randomly. Lamarck also believed that these changes were inheritable. In contrast, Darwin's theories gave a lot of importance to random changes that could affect natural selection and organism evolution. Both Lamarckian and Darwinian modalities of evolution appear to be important, and reflect different aspects of the interaction between organisms and the environment. In Lamarck's theories, the environment has an important role in influencing the genotype to shape the organism's phenotype. In fact, new studies are showing that epigenetic (not genetic) traits can be passed on to later generations as exemplified by paramutation events in plants and animals (Lolle *et al.*, 2005; Rassoulzadegan *et al.*, 2006). Moreover, it is becoming clear that RNA molecules such as microRNAs and lncRNAs might be the mediators of these newly described mechanisms (Wagner *et al.*, 2008). RNAs seem to be important players in determining how chromatin modifications will be distributed along chromosomes, and RNA might also underlie the complex epigenetic interactions that occur between homologous sequences of the DNA molecule. These newly described epigenetic mechanisms are a type of 'soft-lamarckism', which is a theory of inheritance based in small changes of the old Lamarckism proposed by Lamarck. It maintains the principle that acquired genetic changes (in this case epigenetic changes), can be passed through generations using intermediate molecules such as RNAs (Costa, 2008).

These new discoveries can affect the way we understand and study organism evolution. It could also impact our view on evolutionary theories. Molecular evolution driven by RNAs is not new; however, the advancement and application of new technologies in this field has been helping in a better understanding of the mechanisms that were only perceived by classical genetics. The ncRNA field, especially lncRNAs and their functions, could completely change the way we perceive human complexity at the molecular level.

Long non-coding RNAs and human complexity

Regulatory roles required for the development and function of organisms as complex as vertebrates, including humans, have been associated to lncRNAs. Growing evidence provided by transcriptome analyses have been showing that several examples of lncRNAs are functional and many more might still be hidden in our genome (Washietl *et al.*, 2007). For example, tilling array studies evaluating the expression of long and short RNAs indicated that several ncRNA examples are of biological relevance (Kapranov *et al.*, 2007). LncRNAs might be also associated to epigenetic mechanisms such as chromatin alterations as discussed previously, since several of those are located in the 5′ region of a protein-coding gene.

The expression of lncRNAs in the mouse brain was recently evaluated and it was shown that most of the long ncRNA examples are expressed in a cell-type specific manner, have subcellular localization and many are expressed in regions associated with adult neurogenesis (Lein *et al.*, 2007). Furthermore, the *cis*-antisense pairs of lncRNAs have differences in expression and intronic ncRNAs have different expression levels when compared with the protein-coding sequence in which they reside suggesting that most of these cases are somehow functional (Lein *et al.*, 2007).

The widespread transcription of ncRNA genes by eukaryotic cells and their roles in complex mechanisms of gene regulation indicates that evolution might have selected several of these transcripts for specific functions that can be associated with complexity in molecular networks. A hypothetical model in which ncRNAs might be the key to complexity has already been suggested (Mattick, 2003; Costa, 2007; Taft *et al.*, 2007). According to this theory, ncRNAs would be the main molecular advance in higher organisms and a digital programming system based on their signalling could be responsible for a complex molecular network of interactions in eukaryotic cells (Mattick, 2003; Costa, 2007; Taft *et al.*, 2007). Even though more evidence is required to confirm this hypothesis, some reports have suggested that eukaryotic cells from higher organisms would be like an integrated computer system in which different kinds of ncRNAs, mainly lncRNAs, could be the connectors sending signals into cellular systems (Mattick, 2003; Costa, 2007; Taft *et al.*, 2007). Eukaryotic molecular networks would be mainly controlled by ncRNAs with functions in epigenetic mechanisms. For example, it was already shown that small RNAs can induce transcriptional activation after targeting promoter regions of human genes and these changes are associated with epigenetic modifications (Li *et al.*, 2006). Also, long non-coding interfering transcripts can repress the transcription of genes by targeting their promoter region (Martianov *et al.*, 2007). Evidence indicates that RNAs can be very versatile in activating and turning off genes, strengthening the idea that ncRNAs, especially lncRNAs, are part of a complex digital system in higher eukaryotes.

The hypothesis that lncRNAs are implicated in complexity is supported by studies that have scanned the human genome and have identified human-specific lncRNAs. Examples include a cluster of antisense lncRNAs that display a significant accelerated nucleotide substitution rate in humans (which is consistent with functional significance), indicating that they might be important for brain function (Pollard *et al.*, 2006). These two lncRNAs were named HAR1F and HAR1R. HAR1F is co-expressed with reelin, which is important in the formation of the brain cortex (Pollard *et al.*, 2006). Another ncRNA that appear to have undergone accelerated evolution in humans is snaR-A (Parrott and Mathews, 2007). snaR-A is transcribed by RNA polymerase III from an intragenic promoter and shares sequence identity with Alu RNA and two potential piRNAs (Parrott and Mathews, 2007). Additionally, recent comparisons between the mouse, human and dog genomes were able to reveal several human-specific genes (Clamp *et al.*, 2007). From thousands of computationally selected sequences that are human-specific, approximately 10% are protein-coding but all the remaining sequences do not have any protein-coding potential, thus representing the non-coding pool of transcripts in the transcriptome (Clamp *et al.*, 2007).

The poor conservation of lncRNAs may be the result of recent and rapid adaptive selection. LncRNAs may be more malleable to evolutionary pressures than protein-coding genes, as evidenced by the existence of many lncRNAs that are species-specific, such as XIST and AIR (Pang, 2006). Interestingly, conserved regions of the human genome that are subject to evolutionary change relative to the chimpanzee genome are localized mainly in non-coding regions, many of which are transcribed as non-coding transcripts (Pollard, 2006). This includes the lncRNA, HAR1F, which has undergone rapid evolutionary change in the human brain as discussed above (Pollard, 2006). The fact that many functionally validated lncRNAs are evolving quickly based on different studies (Smith, 2004; Pang, 2006), may result from the fact that these sequences have a more dynamic structure–function constraint. As a consequence, it is expected that some evolutionary innovation occurs in these sequences. This hypothesis is supported by the existence of thousands of sequences in the mammalian genome that show poor conservation at the primary DNA sequence level but have evidences of conserved RNA secondary structures (Torarinsson *et al.*, 2008).

Finally, as represented in Fig. 7.3, it is clear that the proportion of non-coding DNA in the genome of organisms that are lower in the evolutionary scale is very small compared with higher organisms. Studies have been showing that a big part of the non-coding sequences are expressed in more evolved organisms. The model in which the gene pool of humans comprises a small subset of protein-coding genes (2–3%) and a high percentage of ncRNA transcripts (70–90%) is starting to become a reality. Based on this evidence, it is expected that these genomic regions have implications in organism complexity. Although there is still a long way ahead to prove that these non-coding sequences are functional, growing evidence suggests that human-specific non-coding RNA genes (mainly lncRNAs) play roles in a variety of molecular mechanisms. These new findings corroborate to the hypothesis that some clues to organism complexity, especially in humans, might be hidden in the non-coding parts of our genomes.

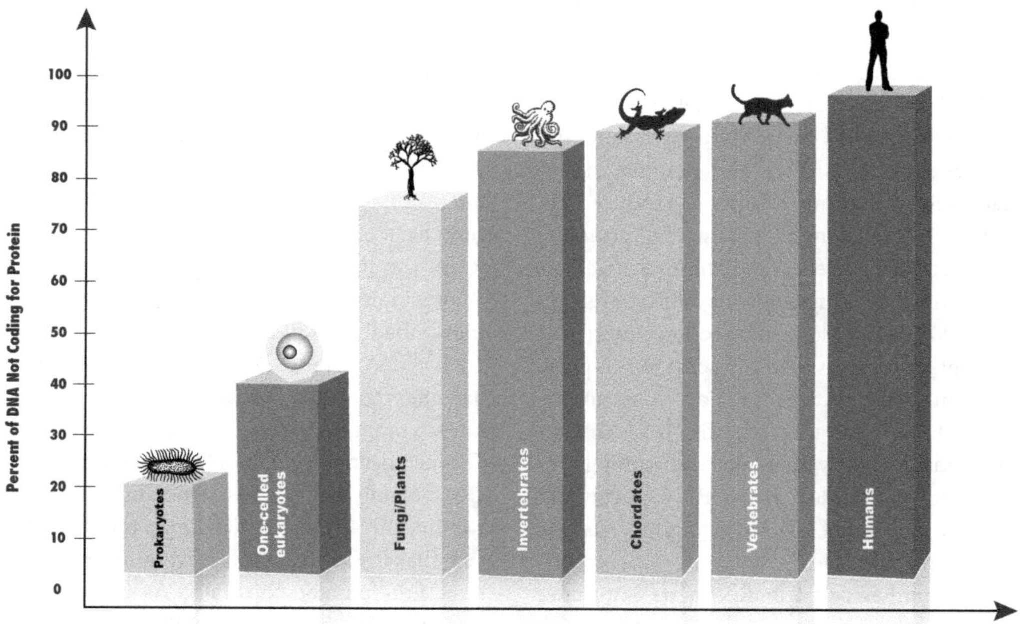

Figure 7.3 Non-coding RNAs as drivers of evolution and complexity? The histogram depicts the percentages of non-coding DNA in different organisms along the evolutionary scale. The Y axis represents the percentage of DNA not coding for protein and the X axis shows histograms for different organisms from prokaryotes to humans. It is clear that in prokaryotes the percentage of DNA not coding for proteins is very small (approximately 20% or less). On the other hand, organisms such as chordates, vertebrates and humans have more than 85% of non-coding DNA in their genomes. Importantly, a big proportion of this non-coding DNA is transcribed into lncRNAs, most of which still have unknown functions. Histograms and chart adapted by Kathryn A. Bischof.

Concluding remarks

The identification of lncRNAs in the beginning of the 1990s and the recent explosion of examples identified in mammalian genomes have shifted some paradigms in molecular biology. Examples of functional ncRNAs, especially lncRNAs, have exponentially grown in the literature. Furthermore, the advent of new technologies in human biology and genomics has helped in the identification of lncRNAs with important functions in epigenetic mechanisms. New studies indicate that these molecules are becoming essential components of the epigenome. In addition, new concepts in evolution have emerged, especially the fact that paramutation events may have RNA molecules as intermediates. Finally, the way we perceive classical genetics has changed with both ncRNAs and proteins being affected in different cellular networks. Since the study of ncRNAs or ncRNOMICS is in its infancy, we might expect more surprises from these new rising stars in the genomics field.

References

Affymetrix ENCODE Transcriptome Project; Cold Spring Harbor Laboratory ENCODE Transcriptome Project. (2009). Post-transcriptional processing generates a diversity of 50-modified long and short RNAs. Nature *457*, 1028–1032.

Bliek, J., Terhal, P., van den Bogaard, M.J., Maas, S., Hamel, B., Salieb-Beugelaar, G., Simon, M., Letteboer, T., van der Smagt, J., Kroes, H., *et al.* (2006). Hypomethylation of the H19 gene causes not only Silver-Russell syndrome (SRS) but also isolated asymmetry or an SRS-like phenotype. Am. J. Hum. Genet. *78*, 604–614.

Brannan, C.I., Dees, E.C., Ingram, R.S., and Tilghman SM. (1990). The product of the H19 gene may function as an RNA. Mol. Cell Biol. *10*, 28–36.

Brown, C.J., Ballabio, A., Rupert, J.L., Lafreniere, R.G., Grompe, M., Tonlorenzi, R., and Willard, H.F. (1991). A gene from the region of the human X inactivation centre is expressed exclusively from the inactive X chromosome. Nature *349*, 38–44.

Brunkow, M.E., and Tilghman, S.M. (1991). Ectopic expression of the H19 gene in mice causes prenatal lethality. Genes Dev. *5*, 1092–1101.

Bussemakers, M.J., van Bokhoven, A., Verhaegh, G.W., Smit, F.P., Karthaus, H.F., Schalken, J.A., Debruyne, F.M., Ru, N., and Isaacs WB. (1999). DD3, a new prostate-specific gene, highly overexpressed in prostate cancer. Cancer Res. *59*, 5975–5979.

Callinan, P.A., and Feinberg, A.P. (2006). The emerging science of epigenomics. Hum. Mol. Genet. *15*, R95–101.

Carninci, P., Kasukawa, T., Katayama, S., Gough, J., Frith, M.C., Maeda, N., Oyama, R., Ravasi, T., Lenhard, B., Wells, C., *et al.* (2005). The transcriptional landscape of the mammalian genome. Science *309*, 1559–1563.

Cawley, S., Bekiranov, S., Ng, H.H., Kapranov, P., Sekinger, E.A., Kampa, D., Piccolboni, A., Sementchenko, V., Cheng, J., and Williams, A.J. (2004). Unbiased mapping of transcription factor binding sites along human chromosomes 21 and 22 points to widespread regulation of noncoding RNAs. Cell *116*, 499–509.

Chandler VL. (2007). Paramutation: from maize to mice. Cell *128*, 641–645.

Clamp, M., Fry, B., Kamal, M., Xie, X., Cuff, J., Lin, M.F., Kellis, M., Lindblad-Toh, K., and Lander, E.S. (2007). Distinguishing protein-coding and noncoding genes in the human genome. Proc. Natl. Acad. Sci. U. S. A. *104*, 19428–19433.

Costa, F.F. (2005). Non-coding RNAs: new players in eukaryotic biology. Gene *357*, 83–94.

Costa, F.F (2007). Non-coding RNAs: lost in translation? Gene *386*, 1–10.

Costa, F.F. (2008). Non-coding RNAs, epigenetics and complexity. Gene *410*, 9–17.

Costa, F.F. (2009). Non-coding RNAs and new opportunities for the private sector. Drug Discov. Today *14*, 446–452.

Costa, F.F. (2010a). Non-coding RNAs: Meet thy masters. Bioessays *32*, 599–608.

Costa, F.F. (2010b). Non-coding RNAs: could they be the answer? Brief Funct. Genomics [Epub ahead of print].

Crick, F.H. (1958). On protein synthesis. Symp. Soc. Exp. Biol. *12*, 138–163.

Dahm, R. (2005). Friedrich Miescher and the discovery of DNA. Dev. Biol. *278*, 274–288.

ENCODE Project Consortium, *et al.* (2007). Identification and analysis of functional elements in 1% of the human genome by the ENCODE pilot project. Nature *447*, 799–816.

Faulkner, G.J., Kimura, Y., Daub, C.O., Wani, S., Plessy, C., Irvine, K.M., Schroder, K., Cloonan, N., Steptoe, A.L., Lassmann, T., *et al.* (2009). The regulated retrotransposon transcriptome of mammalian cells. Nat. Genet. *41*, 563–571.

Feng, J., Bi, C., Clark, B.S., Mady, R., Shah, P., and Kohtz, J.D. (2006). The Evf-2 noncoding RNA is transcribed from the Dlx-5/6 ultraconserved region and functions as a Dlx-2 transcriptional coactivator. Genes Dev. *20*, 1470–1484.

Fire, A., Xu, S., Montgomery, M.K., Kostas, S.A., Driver, S.E., and Mello, C.C. (1998). Potent and specific genetic interference by double-stranded RNA in Caenorhabditis elegans. Nature *391*, 806–811.

Gupta, R.A., Shah, N., Wang, K.C., Kim, J., Horlings, H.M., Wong, D.J., Tsai, M.C., Hung, T., Argani, P., Rinn, J.L., *et al.* (2010). Long non-coding RNA HOTAIR reprograms chromatin state to promote cancer metastasis. Nature *464*, 1071–1076.

Hao, Y., Crenshaw, T., Moulton, T., Newcomb, E., and Tycko, B. (1993). Tumour-suppressor activity of H19 RNA. Nature *365*, 764–767.

Hibi, K., Nakamura, H., Hirai, A., Fujikake, Y., Kasai, Y., Akiyama, S., Ito, K., and Takagi, H. (1996). Loss of H19 imprinting in esophageal cancer. Cancer Res. *56*, 480–482.

Holley, R.W., Apgar, J., Everett, G.A., Madison, J.T., Marquisee, M., Merrill, S.H., Penswick, J.R., and Zamir, A. (1965). Structure of a ribonucleic acid. Science *147*, 1462–1465.

Huarte, M., Guttman, M., Feldser, D., Garber, M., Koziol, M.J., Kenzelmann-Broz, D., Khalil, A.M., Zuk, O., Amit, I., Rabani, M., *et al.* (2009). A large intergenic noncoding RNA induced by p53 mediates global gene repression in the p53 response. Cell *142*, 409–419.

Hutchinson, J.N., Ensminger, A.W., Clemson, C.M., Lynch, C.R., Lawrence, J.B., and Chess, A. (2007). A screen for nuclear transcripts identifies two linked noncoding RNAs associated with SC35 splicing domains. BMC Genomics *8*, 39.

Ji, P., Diederichs, S., Wang, W., Böing, S., Metzger, R., Schneider, P.M., Tidow, N., Brandt, B., Buerger, H., Bulk, E., *et al.* (2003). MALAT-1, a novel noncoding RNA, and thymosin beta4 predict metastasis and survival in early-stage non-small cell lung cancer. Oncogene *22*, 8031–8041.

Kapranov, P., Cheng, J., Dike, S., Nix, D.A., Duttagupta, R., Willingham, A.T., Stadler, P.F., Hertel, J., Hackermüller, J., Hofacker, I.L., *et al.* (2007).RNA maps reveal new RNA classes and a possible function for pervasive transcription. Science *316*, 1484–1488.

Katayama, S., Tomaru, Y., Kasukawa, T., Waki, K., Nakanishi, M., Nakamura, M., Nishida, H., Yap, C.C., Suzuki, M., Kawai, J., *et al.* (2005). Antisense transcription in the mammalian transcriptome. Science *309*, 1564–1566.

Khalil, A.M., Guttman, M., Huarte, M., Garber, M., Raj, A., Rivea Morales, D., Thomas, K., Presser, A., Bernstein, B.E., van Oudenaarden, A., *et al.* (2009). Many human large intergenic noncoding RNAs associate with chromatin-modifying complexes and affect gene expression. Proc. Natl. Acad. Sci. U. S. A. *106*, 11667–16672.

Lee, J.T., and Lu, N. (1999). Targeted mutagenesis of Tsix leads to nonrandom X inactivation. Cell *99*, 47–57.

Lee, J.T., Davidow, L.S., and Warshawsky, D. (1999). Tsix, a gene antisense to Xist at the X-inactivation centre. Nat. Genet. *21*, 400–404.

Leighton, P.A., Saam, J.R., Ingram, R.S., Stewart, C.L., and Tilghman, S.M. (1995). An enhancer deletion affects both H19 and Igf2 expression. Genes Dev. *9*, 2079–2089.

Lein, E.S., Hawrylycz, M.J., Ao, N., Ayres, M., Bensinger, A., Bernard, A., Boe, A.F., Boguski, M.S., Brockway, K.S., Byrnes, E.J., *et al.* (2007). Genome-wide atlas of gene expression in the adult mouse brain. Nature *445*, 168–176.

Li, L.C., Okino, S.T., Zhao, H., Pookot, D., Place, R.F., Urakami, S., Enokida, H., and Dahiya, R. (2006). Small dsRNAs induce transcriptional activation in human cells. Proc. Natl. Acad. Sci. U. S. A. *103*, 17337–17342.

Loewer, S., Cabili, M.N., Guttman, M., Loh, Y.H., Thomas, K., Park, I.H., Garber, M., Curran, M., Onder, T., Agarwal, S., *et al.* (2010). Large intergenic non-coding RNA–RoR modulates reprogramming of human induced pluripotent stem cells. Nat. Genet. *42*, 1113–1117.

Lolle, S.J., Victor, J.L., Young, J.M., and Pruitt RE. (2005). Genome-wide non-mendelian inheritance of extra-genomic information in Arabidopsis. Nature *434*, 505–509.

Luikenhuis, S., Wutz, A., and Jaenisch, R. (2001). Antisense transcription through the Xist locus mediates Tsix function in embryonic stem cells. Mol. Cell Biol. *21*, 8512–8520.

Marques, A.C., and Ponting, C.P. (2009). Catalogues of mammalian long noncoding RNAs: modest conservation and incompleteness. Genome Biol. *10*, R124.

Martianov, I., Ramadass, A., Serra Barros, A., Chow, N., and Akoulitchev, A. (2007). Repression of the human dihydrofolate reductase gene by a non-coding interfering transcript. Nature *445*, 666–670.

Mattick, J/S. (2003). Challenging the dogma: the hidden layer of non-protein-coding RNAs in complex organisms. Bioessays *25*, 930–939.

Maurange, C., and Paro, R. (2002). A cellular memory module conveys epigenetic inheritance of hedgehog expression during Drosophila wing imaginal disc development. Genes Dev. *16*, 2672–2683.

Mercer, T.R., Dinger, M.E., Sunkin, S.M., Mehler, M.F., and Mattick, J.S. (2008). Specific expression of long noncoding RNAs in the mouse brain. Proc. Natl. Acad. Sci. U. S. A. *105*, 716–21.

Millar, J.K., Wilson-Annan, J.C., Anderson, S., Christie, S., Taylor, M.S., Semple, C.A., Devon, R.S., St Clair, D.M., Muir, W.J., Blackwood, D.H., *et al.* (2000). Disruption of two novel genes by a translocation co-segregating with schizophrenia. Hum. Mol. Genet. *9*, 1415–1423.

Molinier, J., Ries, G., Zipfel, C., and Hohn, B. (2006). Transgeneration memory of stress in plants. Nature *442*, 1046–1049.

Nagalakshmi, U., Wang, Z., Waern, K., Shou, C., Raha, D., Gerstein, M., and Snyder, M. (2008).The transcriptional landscape of the yeast genome defined by RNA sequencing. Science *320*, 1344–1349.

Nagano, T., Mitchell, J.A., Sanz, L.A., Pauler, F.M., Ferguson-Smith, A.C., Feil, R., and Fraser, P. (2008). The Air noncoding RNA epigenetically silences transcription by targeting G9a to chromatin. Science *322*, 1717–1720.

Ørom, U.A., Derrien, T., Beringer, M., Gumireddy, K., Gardini, A., Bussotti, G., Lai, F., Zytnicki, M., Notredame, C., Huang, Q., *et al.* (2010). Long noncoding RNAs with enhancer-like function in human cells. Cell *143*, 46–58.

Parrott, A.M., and Mathews, M.B. (2007). Novel rapidly evolving hominid RNAs bind nuclear factor 90 and display tissue-restricted distribution. Nucleic Acids Res. *35*, 6249–6258.

Paul, J., and Duerksen, J.D. (1975). Chromatin-associated RNA content of heterochromatin and euchromatin. Mol. Cell Biochem. *9*, 9–16.

Penny, G.D., Kay, G.F., Sheardown, S.A., Rastan, S., and Brockdorff, N. (1996). Requirement for Xist in X chromosome inactivation. Nature *379*, 131–137.

Pollard, K.S., Salama, S.R., Lambert, N., Lambot, M.A., Coppens, S., Pedersen, J.S., Katzman, S., King, B., Onodera, C., Siepel, A., *et al.* (2006). An RNA gene expressed during cortical development evolved rapidly in humans. Nature *443*, 167–172.

Rachmilewitz, J., Goshen, R., Ariel, I., Schneider, T., de Groot, N., and Hochberg, A. (1992). Parental imprinting of the human H19 gene. FEBS Lett. *309*, 25–28.

Rassoulzadegan, M., Grandjean, V., Gounon, P., Vincent, S., Gillot, I., and Cuzin, F. (2006). RNA-mediated non-mendelian inheritance of an epigenetic change in the mouse. Nature *441*, 469–474.

Rinn, J.L., Kertesz, M., Wang, J.K., Squazzo, S.L., Xu, X., Brugmann, S.A., Goodnough, L.H., Helms, J.A., Farnham, P.J., Segal, E., *et al.* (2007). Functional demarcation of active and silent chromatin domains in human HOX loci by noncoding RNAs. Cell *129*, 1311–1323.

Sado, T., Wang, Z., Sasaki, H., and Li, E. (2001). Regulation of imprinted X-chromosome inactivation in mice by Tsix. Development *8*, 1275–1286.

Sleutels, F., Zwart, R., and Barlow, D.P. (2002). The non-coding Air RNA is required for silencing autosomal imprinted genes. Nature *415*, 810–813.

Stavropoulos, N., Lu, N., and Lee, J.T. (2001). A functional role for Tsix transcription in blocking Xist RNA accumulation but not in X-chromosome choice. Proc. Natl. Acad. Sci. U. S. A. *98*, 10232–10237.

Suter, C.M., and Martin, D.I. (2010). Paramutation: the tip of an epigenetic iceberg? Trends Genet. *26*, 9–14.

Szymanski, M., Erdmann, V.A., and Barciszewski, J. (2003). Eurekah – Riboregulators: an overview. Bioscience Chapter Database. Landes Bioscience. 1–10.

Taft, R.J., Pheasant, M., and Mattick, J.S. (2007). The relationship between non-protein-coding DNA and eukaryotic complexity. Bioessays *29*, 288–299.

Tano, K., Mizuno, R., Okada, T., Rakwal, R., Shibato, J., Masuo, Y., Ijiri, K., and Akimitsu, N. (2010). MALAT-1 enhances cell motility of lung adenocarcinoma cells by influencing the expression of motility-related genes. FEBS Lett. *584*, 4575–4580.

Tavernarakis, N., Wang, S.L., Dorovkov, M., Ryazanov, A., and Driscoll, M. (2000). Heritable and inducible genetic interference by double-stranded RNA encoded by transgenes. Nat. Genet. *24*, 180–183.

Thorvaldsen, J.L., Duran, K.L., and Bartolomei, M.S. (1998). Deletion of the H19 differentially methylated domain results in loss of imprinted expression of H19 and Igf2. Genes Dev. *23*, 3693–3702.

Torarinsson, E., Yao, Z., Wiklund, E.D., Bramsen, J.B., Hansen, C., Kjems, J., Tommerup, N., Ruzzo, W.L., and Gorodkin, J. (2008). Comparative genomics beyond sequence-based alignments: RNA structures in the ENCODE regions. Genome Res. *18*, 242–251.

Tsai, M.C., Spitale, R.C., and Chang, H.Y. (2011). Long intergenic noncoding RNAs: new links in cancer progression. Cancer Res. *71*, 3–7.

Wagner, K.D., Wagner, N., Ghanbarian, H., Grandjean, V., Gounon, P., Cuzin, F., and Rassoulzadegan, M. (2008). RNA induction and inheritance of epigenetic cardiac hypertrophy in the mouse. Dev. Cell. *14*, 962–969.

Warner, J.R., and Rich, A. (1964). The number of soluble rna molecules on reticulocyte polyribosomes. Proc. Natl. Acad. Sci. U. S. A. *51*, 1134–11341.

Washietl, S., Pedersen, J.S., Korbel, J.O., Stocsits, C., Gruber, A.R., Hackermüller, J., Hertel, J., Lindemeyer, M., Reiche, K., Tanzer, A., *et al.* (2007). Structured RNAs in the ENCODE selected regions of the human genome. Genome Res. *17*, 852–864.

Woo, C.J., and Kingston, R.E. (2007). HOTAIR lifts noncoding RNAs to new levels. Cell *129*,n1257–1259.

Wutz, A., and Jaenisch, R. (2000). nA shift from reversible to irreversible X inactivation is triggered during ES cell differentiation. Mol. Cell *4*, 695–705.

Yu, W., Gius, D., Onyango, P., Muldoon-Jacobs, K., Karp, J., Feinberg, A.P., and Cui, H. (2008). Epigenetic silencing of tumour suppressor gene p15 by its antisense RNA. Nature *451*, 202–226.

Zaidi, S.K., Young, D.W., Montecino, M., van Wijnen, A.J., Stein, J.L., Lian, J.B., and Stein, G.S. (2011). Bookmarking the genome: maintenance of epigenetic information. J. Biol. Chem. *286*,18355–18361.

Epigenetic Regulation and *var* Gene Expression in the Malaria Parasite *Plasmodium falciparum*

8

Dacia Kwiatkowski and Kirk Deitsch

Abstract

Much of what is currently know about complex cellular questions, including regulation of gene expression, RNA production and epigenetic influence, has been derived from experiments performed in well-studied model organisms such as yeast, *Drosophila,* mice or human cell lines. In all of these systems, examples have been described of how non-coding RNAs (ncRNAs), which often act in accordance with epigenetic mechanisms, play a vital role in regulating gene expression. These studies have established a strong precedent for the vital role that these molecules can play in complex cellular functions, which is now being extended to other diverse organisms such as viruses and protozoans of distant evolutionary lineages. The focus of this chapter is the role that long ncRNAs play in gene regulation in the protozoan parasite *Plasmodium falciparum,* the causative agent of the most severe form of human malaria. In particular, we will consider ncRNAs that influence the expression of the genes encoding the primary virulence factors expressed by these parasites, the epigenetic marks associated with them, and how this process enables the parasites to avoid the immune response of their human hosts.

Plasmodium

Malaria caused by *Plasmodium falciparum*

Plasmodium is a genus of unicellular protozoan parasites that replicate within the circulating erythrocytes of their vertebrate hosts. These organisms have been found to parasitize a broad range of animals, including birds, rodents, reptiles and mammals. In the case of humans, infection by *Plasmodium* parasites causes malaria, a vector-borne disease transmitted by *Anopheles* mosquitoes. It is endemic to tropical and subtropical geographic regions, putting 40% of the world's population at risk for contracting the disease. Greater than 500 million people worldwide are affected by malaria resulting in up to one million deaths per year, most of which are children under the age of 5 (Snow *et al.*, 1999). There are five *Plasmodium* species known to infect humans including *P. vivax, P. malariae, P. ovale, P. knowlesi,* and *P. falciparum.* However, of these five, *P. falciparum* causes the most severe form of human malaria and is responsible for the vast majority of mortality.

The life cycle of *P. falciparum* includes replication within both the mosquito and human hosts. The parasite differentiates into highly specialized morphological forms as it transitions through these two hosts, including both sexual and asexual replication. The human stage of infection begins when the bite of a female mosquito deposits the sporozoite form of the parasite, stored in the mosquito salivary glands, into the bloodstream. These highly motile parasites travel to the liver where they invade hepatocytes and undergo massive asexual division called schizogeny. This results in a tremendous increase in parasite numbers, which then rupture the hepatocyte and release into the blood stream as invasive merozoites. Once in the bloodstream, the merozoites infect circulating erythrocytes or red blood cells (RBCs), initiating the only stage of the life cycle that causes symptoms in the human host. In the RBCs, the parasites undergo a 48-hour asexual replicative cycle that ultimately ends with bursting of the host cell to release 18–24 new merozoites, each destined to invade additional RBCs (Fig. 8.1). It is this cycle of bursting and reinvasion of RBCs that results in the classical symptoms of malaria; alternating recurrent fevers and chills as well as severe anaemia. Additionally, infected RBCs (iRBCs) display adhesive properties that enable them to adhere to specific cell surface receptors on the endothelium of the post-capillary venules throughout the body. This sequestration and accumulation of iRBCs can lead to obstructed blood flow, thereby damaging the affected tissue and causing potential long-term effects, particularly in children (Miller *et al.*, 1994). The most severe type of sequestration takes place in the brain and causes a syndrome called cerebral malaria, resulting in death in 10–20% of infected children (Holding *et al.*, 1999).

Development of parasites in both the human and mosquito hosts, through both sexual and asexual replication and throughout the frequent cycles of schizogeny, require very specific and tightly controlled regulation of gene expression. Similar to the developmental programs that control cellular differentiation during the development of higher eukaryotic organisms like flies and mammals, malaria parasites also exhibit well defined patterns of gene expression throughout their life cycle which enable them to be successfully transmitted between their vertebrate and insect hosts (Bozdech *et al.*, 2003; Le Roch *et al.*, 2003). Despite the rather vast evolutionary distance between *Plasmodium* and model organisms like Drosophila and mammals, many of the molecular mechanisms that control gene expression and differentiation appear to be shared, including aspects of epigenetic regulation.

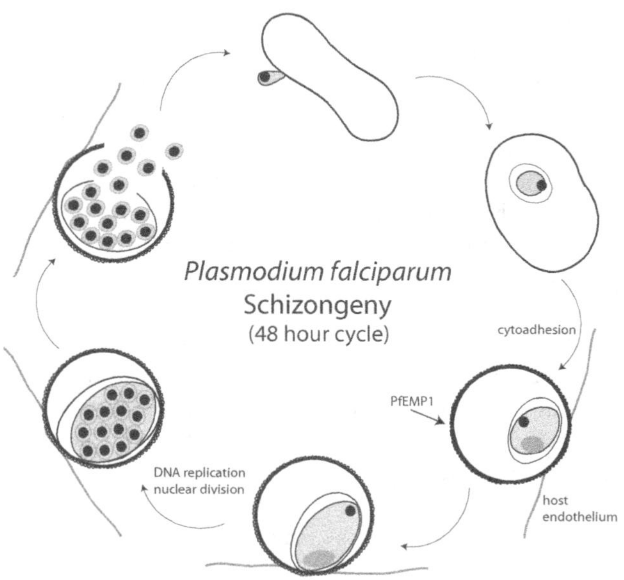

Figure 8.1 Asexual replication of *Plasmodium falciparum* parasites in human red blood cells. Individual parasites, called merozoites, invade circulating RBCs (top). Upon invasion, the parasite becomes surrounded by a vacuole, and begins to modify the cytoskeleton of the host cell, distorting it from its typical bi-concave shape into a more rigid, spherical shape. The host cell membrane also becomes modified through the addition of the adhesive protein *Pf*EMP1, which results in cytoadhesion of the infected cell to the endothelium. Approximately halfway through the cycle the parasite begins multiple rounds of DNA replication and nuclear division, resulting in 18–24 new merozoites, which lyse the RBC and reinvade new host cells.

Evolutionary distance between *Plasmodium* and model organisms

The majority of basic research on eukaryotic gene regulation and genome organization has been focused on the so-called 'crown' group of eukaryotic organisms. This includes plants, yeasts and all animals from fruit flies to humans (Kumar *et al.*, 1996). While we have learned a tremendous amount of functional information through the in depth studies of these organisms, they nonetheless represent a rather narrow slice of the entire eukaryotic lineage. A handful of more distantly related organisms have also been studied to some degree, generally either owing to some special property that makes them particularly interesting or amenable to study, or alternatively because they have an inordinate influence on human health and well being. Malaria parasites are representative of the latter class, causing an enormous public health and economic burden on human populations in the developing world (Sachs *et al.*, 2002).

Malaria parasites (*Plasmodium* spp.) are from the phylum Apicomplexa, a group of invasive protozoan parasites that includes the human pathogens *Toxoplasma*, *Cryptosporidium* and *Babesia*, among others (Morrison, 2009). They are thought to have derived from the dinoflagellates, a group of free-living, photosynthetic protozoans. In fact, most apicomplexan parasites retain chloroplast-like organelles (called plastids), although they are no longer photosynthetic (McFadden *et al.*, 1996; Kohler *et al.*, 1997). Many aspects of gene expression and transcriptional regulation are conserved between the more extensively studied crown eukaryotes, including such basic properties as the basal transcription machinery, mRNA splicing apparatus and chromatin-modifying enzymes (Aravind *et al.*, 2003). However, other components appear highly divergent. For example, apicomplexan organisms appear to be nearly entirely devoid of many standard classes of transcription factors, for example those containing bZip, GATA finger, bHLH or homeodomains. To date, only a single class of specific transcription factors has been identified, a family of proteins that contain what is termed an ApiAP2 domain (Balaji *et al.*, 2005). These appear to be distantly related to the AP2/ERF class of transcription factors in plants. The study of transcriptional regulation, including topics like epigenetic mechanism and ncRNAs, in Apicomplexans (and *Plasmodium* specifically) therefore presents a very distantly related organism that can provide a unique prospective on this rapidly emerging topic of research.

var genes

var genes and cytoadherence

After infection of their vertebrate hosts, malaria parasites undergo asexual replication within the circulating red blood cells, expanding to population densities of up to 10^5 parasites per µl of blood in the case of *P. falciparum* (Miller *et al.*, 1994). Such high numbers of parasites places an obvious stress on the infected individual, leading to a robust and continuous immune assault on the infected cells. Given the hostile environment in which they live, parasites have devoted a remarkably large portion of their genomes specifically to immune evasion, and in particular to the process of antigenic variation. For example, parasites possess highly expanded, hyper-variable families of genes that encode variant surface antigens that are exposed to the immune system of their hosts (Gardner *et al.*, 2002; Lavazec *et al.*, 2006). These large, redundant families of genes must be precisely regulated to enable parasites to maintain persistent infections, and thus to be transmitted to their mosquito vectors. The regulatory pathways that control their expression are complex and are just beginning to be deciphered (Dzikowski *et al.*, 2006). Owing to their importance for both virulence and the persistent nature of malaria infections, these gene families have become the most extensively studied topic in the field of transcriptional regulation in *Plasmodium*. While the details remain elusive, ncRNAs have been implicated in these regulatory pathways.

The most widely studied family of variant surface antigens in *P. falciparum* is known as *Plasmodium falciparum* erythrocyte membrane protein 1 (*Pf*EMP1) and different forms of this protein are encoded by variant members of the *var* gene family (Baruch *et al.*, 1995; Smith *et al.*, 1995; Su *et al.*, 1995). *Pf*EMP1 is present on the exterior of the iRBCs and is thought to be the primary virulence determinant of *P. falciparum* infections.

Each form of *Pf*EMP1 binds to a specific receptor found on the surface of the post-capillary endothelium. These binding properties result in adhesion and sequestration of large numbers of infected cells within specific tissues of the body, leading to blockage of blood flow to the organ and often causing significant tissue damage (Kyes *et al.*, 2001). For example, certain *Pf*EMP1 types are specific for receptors within the brain and result in cerebral malaria, a frequently lethal phenomenon (Kyes *et al.*, 2001). The ability to cytoadhere to blood vessel walls is imperative for the parasites as this property enables them to avoid being cleared from the circulation by the spleen. However, the placement of a parasite-specific protein on the infected cell surface exposes this antigen to the immune system, which readily makes antibodies against it. To avoid complete clearance by these antibodies, *P. falciparum* has evolved a complex transcriptional regulatory pathway that results in sequential expression of different members of the *var* gene family, thus resulting in a continuously changing pattern of *Pf*EMP1 proteins being displayed to the host. This process is referred to as antigenic variation and is key to the parasites ability to maintain a persistent infection. It also provides a fascinating system of gene regulation that is just beginning to be deciphered.

Mutually exclusive expression and antigenic variation

Within a geographical region that is endemic for malaria, many genotypically distinct parasite populations will be in circulation, and each maintains its own unique repertoire of *var* genes (Barry *et al.*, 2007). Within the genome of any given parasite, there are about 60 *var* gene copies located either in subtelomeric regions or in tandem arrays within the central regions of several chromosomes (Gardner *et al.*, 2002). A unique feature of *var* gene regulation is that only one gene is expressed at a time, a property referred to as strict mutually exclusive expression. By only expressing one *var* gene at a time and therefore producing only a single type of *Pf*EMP1 on the surface of the iRBC, the parasite exposes only a small fraction of its entire repertoire of *Pf*EMP1 types to the host immune system. As the host responds to the presence of one *Pf*EMP1 on the iRBC surface by producing antibodies against that particular protein, a subset of parasites arises that has switched expression to a different *var* gene and thereby expresses a different version of *Pf*EMP1. Thus the cycle of competition between host and parasite continues.

This process of antigenic variation is observed clinically by the rising and falling waves of parasitaemia often noted in humans (Fig. 8.2), with each wave representing a population expressing a particular *var* gene (Miller *et al.*, 2002). This process prevents an infected individual from rapidly clearing the parasites, and patients with infections up to 300 days have been reported (Miller *et al.*, 2002). In addition, frequent gene conversion events between *var* genes result in a continuously changing complement of variant forms of *Pf*EMP1 (Frank *et al.*, 2006a; Kraemer *et al.*, 2007) within each parasite's genome. This explains the observation that isolates from the field often have completely different *var* gene repertoires (Kyes *et al.*, 1997; Fowler *et al.*, 2002; Trimnell *et al.*, 2006).

Regulation of *var* genes

The overall regulatory paradigm that underlies *var* gene regulation and antigenic variation in *P. falciparum* is simple: all parasites must express one and only one *var* gene from their repertoire of 60. Further, a switch in which *var* gene is expressed can only result from the coordinated silencing of the currently active gene and activation of a previously silent copy. While a significant amount of information has been gathered in recent years to identify many of the components involved in this regulatory pathway, how all of the parts interact remains largely unknown.

All *var* genes share a similar basic structure consisting of a promoter followed by two exons that are separated by a tightly conserved intron (Fig. 8.3) (Su *et al.*, 1995). The first exon is highly variable and encodes the extracellular portion of *Pf*EMP1 that is displayed on the infected cell surface. It is this part of the protein that accounts for the differences in cytoadherence preference. The second exon is conserved and encodes the part of the protein that is anchored into the RBC membrane. Expression of *var* genes is regulated at the level of transcription such that in any given

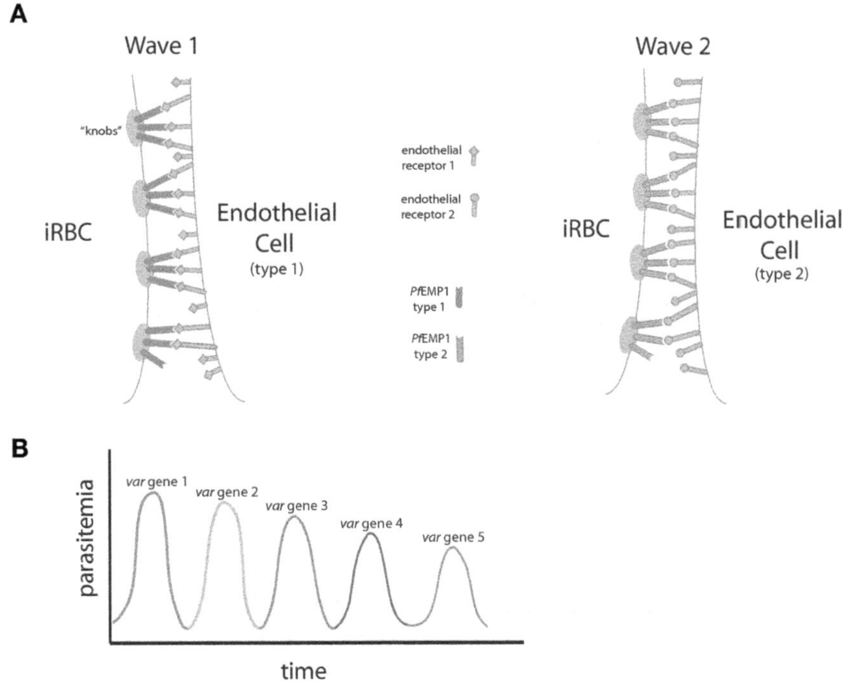

Figure 8.2 Waves of parasitaemia due to antigenic variation. (A) The surface of the infected red blood cell contains electron dense knobs which anchor the parasite-encoded adhesive protein *Pf*EMP1. Any given iRBC only contains a single type of *Pf*EMP1, which binds to a specific host endothelial receptor, thus resulting in adhesion within a specific host tissue (endothelial cell type 1). Over the course of an infection, waves of parasites arise that express different types of *Pf*EMP1 and consequently adhere to alternative receptors in different host tissues (endothelial cell type 2). (B) Different types of *Pf*EMP1 are encoded by different *var* genes. As the immune system generates antibodies against one form of *Pf*EMP1, populations of parasites arise that express alternative *var* genes, thus re-establishing a high parasitaemia and enabling a long-term infection characterized by rising and falling waves of parasites.

parasite, only a single *var* gene is transcribed, spliced and translated into a *Pf*EMP1 that is transported to the exterior of the RBC. The rest of the *var* gene family remain transcriptionally silent, thus preventing production of additional forms of *Pf*EMP1 from being exposed to the host immune system.

Experiments using transgenic parasites and reporter gene constructs have identified two elements that are required for a *var* gene to be recognized by the regulatory pathway that limits expression to a single gene: the promoter/regulatory region found upstream of exon one, and the intron (Frank *et al.*, 2006b; Voss *et al.*, 2006). Interestingly, the intron also contains a bidirectional promoter that when deleted disrupts its regulatory functions (Calderwood *et al.*, 2003). Specifically, a *var* gene without an intron is not recognized by the mutually exclusive expression pathway and is constitutively expressed (Dzikowski *et al.*, 2007, 2008). This results in parasites that have more than one transcriptionally active *var* promoter at a time. Substituting in another promoter for the *var* intron restores the regulatory function suggesting that it is the promoter activity and not the DNA sequence that is important (Dzikowski *et al.*, 2007). It appears that each intron is only capable of regulating a single *var* gene at a time, and it is the pairing of an intron with an upstream promoter/regulatory region that is important for proper *var* gene regulation (Frank *et al.*, 2006b; Swamy *et al.*, 2011). In addition, both the intron and the upstream promoter/regulatory region must be located on the same DNA molecule (*in*

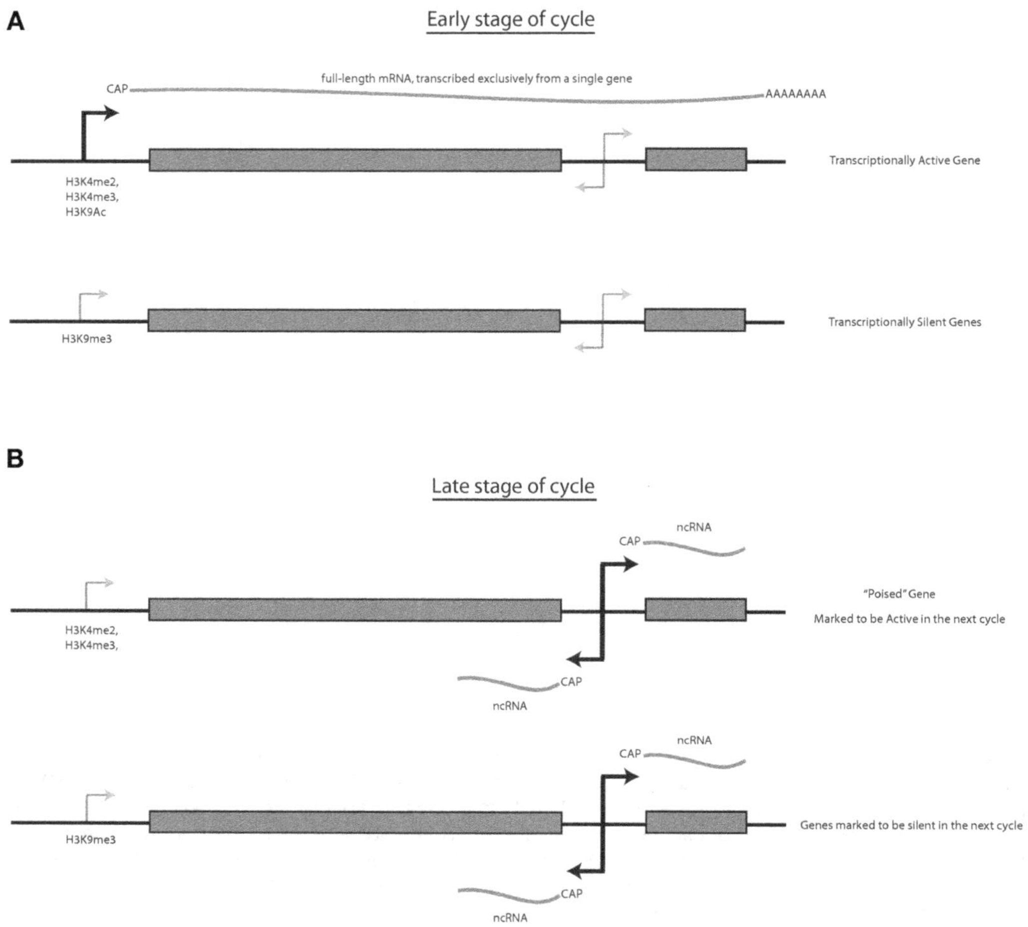

Figure 8.3 Structure of *var* genes. Each *var* gene consists of a two exons (grey boxes) with a uni-directional promoter found upstream of each gene and bi-directional promoter found in each intron (denoted by arrows). (A) Early in each replication cycle, a single *var* gene is actively transcribed into mRNA (top gene) from the upstream promoter and is characterized by chromatin containing the histone modifications H3K4me2, H3K4me3 and H3K9Ac. Silent genes are not transcribed and are associated with histones containing H3K9me3. (B) As the parasites begin to replicate their genome, *var* mRNA is no longer produced and all upstream promoters are silent. The gene that was the source of mRNA earlier in the cycle remains marked by H3K4me2 and H3K4me3, while the silent genes remain associated with H3K9me3. At this time the bi-directional promoters found in *var* introns begin to actively produce ncRNAs from both the silent and active genes. These ncRNAs are capped but not polyadenylated, remain nuclear and associate with chromatin, although the exact nature of their structure and function is unknown.

cis) indicating that *var* gene expression is regulated at the level of individual genes.

Many of the properties described above are reminiscent of several gene regulatory systems described in higher eukaryotes, many that rely on ncRNAs. For example, the concept of 'allelic exclusion' as described for maternal/paternal imprinting or for expression of the immunoglobulin loci in mammals also results in expression of a single gene (Corcoran, 2005; Lewis *et al.*, 2006), albeit in these cases the choice is made between two loci, while in the case of *var* gene expression there are 60 genes that must be coordinated. A similar concept can be applied to the olfactory receptor genes of mammals in which each individual neuron of the olfactory bulb chooses to express a single gene from approximately 1000 different loci (Serizawa *et al.*, 2004). The remaining genes are maintained in a transcriptionally silent state. Perhaps the best studied system is that of X-inactivation in female mammals in which a single X chromosome is chosen to be active while the alternative chromosome is condensed into a heterochromatic state and silenced (Chang *et al.*, 2006). In all of these cases, one gene (or chromosome) is chosen to be active to the exclusion of all others in the genome. It is tempting to think that some of the molecular mechanisms that control these pathways are conserved in *Plasmodium*, even across the rather substantial evolutionary distances that separate these organisms. The presence of ncRNAs, which plays a role in the above-mentioned systems, might be one such conserved aspect.

Basic transcription machinery

As mentioned above, annotation of the *P. falciparum* genome failed to identify many of the transcription factors that are conserved throughout the crown eukaryotic lineage, and the only family of sequence-specific transcription factors that were identified were similar to a family of DNA-binding proteins found in plants (Aravind *et al.*, 2003; Coulson *et al.*, 2004; Balaji *et al.*, 2005). However, these annotations did readily identify most components of the standard chromatin-modifying activities, including histone methyltransferases, acetylases, and the enzymes that reverse several of these epigenetic marks (Iyer *et al.*, 2008). In addition, when adjusted for the number of predicted genes, the *P. falciparum* genome encodes nearly twice as many proteins that are annotated as being involved in RNA processing (Coulson *et al.*, 2004), suggesting that this lineage of obligate parasites might be more dependent on RNA-based mechanisms of gene regulation than members of the crown group of eukaryotes.

One interesting aspect of *Plasmodium* transcriptional machinery centres on the catalytic subunit of RNA polymerase II, the peptide referred to as Rpb1. Like most eukaryotes, *Plasmodium* parasites were shown to rely on RNA pol II for transcription of mRNAs, and this activity is inhibited by alpha-amanatin (Kyes *et al.*, 2007). The C-terminal domain (CTD) of Rpb1, also known as the 'tail', consists of a tandem array of heptapeptides, specifically the amino acid sequence YSPTSPS/K (Allison *et al.*, 1985; Corden, 1990). The number of repeats varies depending on the organism but tends to be longer in organisms of greater complexity and is highly conserved in most evolutionary lineages. For example, the CTD of yeast consists of 26 heptad repeats, while all mammals have 52 heptads in their CTDs (Chapman *et al.*, 2008). This highly flexible 'tail' serves as a docking platform for several factors involved in the process of transcription, including those important for initiation, elongation, and termination of RNA synthesis as well as chromatin-modifiers that can mark histones as the polymerase moves along the chromatin fibre (Carty *et al.*, 2002). Intriguingly, while the length and composition of the CTD is perfectly conserved within all other genera examined, it is highly variable within *Plasmodium* at both the genus and species level (Kishore *et al.*, 2009). Specifically, the CTD repeat length of rodent and bird parasites is the shortest thus far described and consists of only eight repeats. In contrast, primate parasites display an expansion of the CTD to between 12 and 17 heptads, and the exact number is highly variable even within a single species. The implications of this unusual variability in CTD length are unknown, although it has been hypothesized that a longer CTD might play a role in epigenetic control. This hypothesis is supported by the observation that the *yir* gene family of the rodent parasite *P. yoelii*, a multicopy gene family analogous to *var* genes, does not display a phenomenon called epigenetic memory, a trait that is inherent to the *var* gene family of *P. falciparum*.

Epigenetic memory of *var* genes

The form of *Pf*EMP1 expressed on the surface of an iRBC switches over the course of an infection due to changes in which *var* gene is actively transcribed by the majority of parasites within the host (Smith *et al.*, 1995). In order for the parasites to maintain a long-term, persistent infection, the rate of switching must be quite low (estimated at below 1%) (Horrocks *et al.*, 2004; Frank *et al.*, 2007), thereby avoiding a premature exposure of the parasite's entire *Pf*EMP1 repertoire. Thus the *var* gene transcription pattern is maintained for numerous cycles of schizogeny, and only rarely is there a switch in which gene is transcribed (Horrocks *et al.*, 2004; Frank *et al.*, 2007). Switches in *var* gene expression are not accompanied by any changes in DNA sequence or any detectable changes in transcription factors and has been defined as an epigenetic phenomenon (Scherf *et al.*, 1998). The maintenance of expression of a particular *var* gene through multiple cycles of cellular replication has therefore been referred to as a type of 'epigenetic memory'.

The silent or active state of each *var* gene in the genome is reinforced through multiple cycles of replication via the deposition of epigenetic marks at each individual gene. Methylated CpG islands, a major mark of epigenetic gene repression in higher eukaryotic systems, appears to be lacking in the *Plasmodium* genome, however the mark trimethyl histone 3 lysine 9 (H3K9me3), is enriched at silent *var* genes but not on the active/expressed gene (Chookajorn *et al.*, 2007; Lopez-Rubio *et al.*, 2007). Further, *Pf*HP1, a heterochromatin-binding protein known to be enriched in H3K9me3 regions of other organisms, is similarly enriched at silent *var* genes, supporting the observation that silent *var* genes are found within heterochromatin (Flueck *et al.*, 2009; Perez-Toledo *et al.*, 2009). In further support of the role of heterochromatin assembly and maintenance in *var* gene regulation, disruption of the histone deacetylase *Pf*sir2 results in leaky expression of some silent *var* genes (Duraisingh *et al.*, 2005; Tonkin *et al.*, 2009). In addition to maintaining repressive histone marks, histone modifications representative of regions of active transcription, for example di- and trimethyl histone 3 lysine 4 (H3K4me2 and H3K4me3), are enriched in the 5′ upstream regions of the actively transcribed *var* gene. While epigenetic marks appear to play a role in molecular memory of *var* gene transcription, it is unknown what recruits and directs enzymes that catalyse these histone marks to the silent or active *var* genes.

Non-coding RNAs

Centromeric ncRNAs

Given the lack of identifiable sequence-specific transcription factors with homology to those found in animals, plants and fungi, it seems plausible that control of gene expression in *P. falciparum* might be unusually dependent on alternative mechanisms, such as those that utilize ncRNAs. Microarray studies of steady state RNA levels confirmed the presence of transcripts that do not appear to encode proteins (Mourier *et al.*, 2008). ncRNAs have been shown in many systems to function in numerous settings, including chromatin modification, genetic imprinting and RNA-mediated interference pathways (Kelley *et al.*, 2000; Mattick, 2003; Morey *et al.*, 2004). Therefore it is possible that ncRNAs identified in malaria parasites are functioning in a similar manner to exert control over gene expression through epigenetic means or through direct RNA interactions. In addition, the over-representation of CCCH-type zinc finger RNA-binding domains in the parasite's genome provides additional support for the notion that RNA may be a major player in regulating gene expression (Coulson *et al.*, 2004).

One region of each chromosome that is known to be maintained in a unique, heterochromatic structure is the centromere. Centromeres in many organisms have been found to be associated with ncRNAs, and in some cases the ncRNAs have been shown to be processed by proteins from the RNAi pathway (Topp *et al.*, 2004; Pidoux *et al.*, 2005). While centromeres have been annotated in most *P. falciparum* chromosomes, the RNAi pathway appears to be missing (Gardner *et al.*, 2002). Investigation of the *Plasmodium* centromeric regions identified a bidirectional promoter that produced two short ncRNAs of 75 and 175 nt in length that are localized in the nucleus (Li *et al.*, 2008). Additionally, the ncRNAs probably associate directly with the centromeres through a protein-dependent interaction as centromeric RNA signal is lost in florescent *in situ* hybridization (FISH) experiments when cells are treated either with a detergent (which denatures proteins) or with ectopiside (which disrupts centromeric DNA). These ncRNAs represent the first example in *Plasmodium* of ncRNAs associating with chromatin. Given that centromeres are associated with constitutive heterochromatin, it is tempting to speculate that these short ncRNAs might play a role in recruiting chromatin-modifying enzymes to these regions or maintaining the heterochromatin environment.

As mentioned, the centromeric ncRNAs are short in length, appearing as two discrete bands on a Northern blot of 75 and 175 nt in length (Li *et al.*, 2008). The short, discrete length of these ncRNAs suggests that they are being processed in some way, although the genome does not encode orthologues of proteins known to function in this pathway in other organisms. It should be noted, however, that the failure to identify obvious orthologues to proteins involved in ncRNA processing does not necessarily mean that they do not exist. Given the large evolutionary distance between *Plasmodium* and the model organisms in which most work has been done, it is possible that the protein sequences are simply too degenerate to be identified by standard computational analyses. Centromeric RNAs support the hypothesis that ncRNAs may be functioning in gene expression control through epigenetic means as a way to compensate for the apparent lack of transcription factors or other regulatory mechanisms. Another family of ncRNAs were identified as originating in the intron of *var* genes, suggesting that the use of ncRNAs might be a more broadly utilized mechanism to control chromatin assembly and gene expression.

var sterile transcripts

Each cycle of asexual replication in *P. falciparum* takes about 48 hours to complete. It is during the first 18–20 hours of the cycle that the single active *var* gene is transcribed and the encoded PfEMP1 exported to the RBC surface (Kyes *et al.*, 2003). At around 24 hours, *var* mRNA is no longer detectable and the parasite begins DNA replication and nuclear division (Arnot *et al.*, 1998; Kyes *et al.*, 2003). Because *Plasmodium* replicates via schizogeny, DNA synthesis and nuclear division continue repeatedly in the absence of cellular division, ultimately leading to the creation of 18–24 nuclei within a single cytoplasm. The nuclei eventually bud off into individual merozoites, which upon lysis of the host cell, go on to invade other RBCs and reinitiate the cycle. This process requires the repeated replication and division of the genome into individual haploid nuclei. In the case of *var* genes, the epigenetic marks that define individual genes as either active or silent must also be faithfully replicated and transmitted to each daughter nuclei. This is required for the maintenance of epigenetic memory, and the molecular mechanisms that enable the reestablishment of the proper chromatin state at each gene are unknown.

Interestingly, the previously described bidirectional promoter located within each *var* intron becomes activated at the onset of DNA replication, and continues to be active throughout the remainder of the replicative cycle (Kyes *et al.*, 2003). This promoter produces two non-overlapping ncRNAs referred to as 'sterile' transcripts; one is in a sense orientation to the *Pf*EMP1 encoding mRNA while the other is antisense (Su *et al.*, 1995; Epp *et al.*, 2008) (Fig. 8.3). RNA-FISH detects both the sense and antisense sterile transcripts as distinct spots in the nucleus, suggesting that they are tightly localized within discrete regions. The sense transcript is ~2.5 kb but does not terminate in the same region as the mRNA, while the antisense transcript is about 1.7 kb in size. These sterile RNAs are capped but not polyadenylated, and they remain in the nucleus. Chromatin immunoprecipitation experiments found that they are associated with the chromatin fraction, however the exact nature of their association with chromatin is not understood and their function remains unknown.

While the production of *var* mRNA is subject to strict mutually exclusive expression resulting in only one of 60 *var* genes being expressed at a time, the *var* intron derived ncRNAs do not follow this pattern. All *var* introns appear to transcribe ncRNAs during replication regardless of whether the mRNA from the associated upstream mRNA promoter was expressed earlier in the cell cycle (Epp *et al.*, 2008). The intron consists of three distinct regions that can be differentiated based upon the AT richness of the sequence and the asymmetrical nature of their GC content (Calderwood *et al.*, 2003). Specifically, while the central region consists of mostly As and contains the bidirectional promoter, the flanking regions are 'mirror images' of each other in base composition and strand asymmetry. This results in the two ncRNAs (sense and antisense) having very similar 5′ sequences, suggesting they might share a similar function.

Given that the *var* intron is necessary for a gene to be included in the mutually exclusive counting mechanism (Frank *et al.*, 2006b; Dzikowski *et al.*, 2007), it is possible that the sterile RNAs are also involved in this function. Since the ncRNAs are not expressed at the same time as the mRNA it is unlikely that they are acting through an RNA interference type of mechanism. However, given the nuclear localization of the ncRNAs and their association with chromatin, it is possible that they play a role in the recruitment of chromatin-modifying enzymes, similar to how the centromeric ncRNAs have been proposed to function. Considering that the 5′ ends of both the sense and antisense ncRNAs share significant sequence similarity and strand asymmetry, it is tempting to suggest that this portion of the RNAs function by interacting with regulatory proteins, while the rest of the transcript serves to anchor the RNA in its position at the locus. This model would also provide an explanation for why introns can only regulate *var* gene expression when placed *in cis* with the upstream, mRNA promoter.

Speculations/model

Since the *var* gene family was first described in 1995, a significant amount of data has accumulated regarding the epigenetic mechanisms that control mutually exclusive expression and epigenetic memory. However it has been difficult to synthesize all of the data into a single, unified model. While there are many aspects that remain unclear, a basic framework can be designed for how *var* gene expression is controlled as a parasite transitions through a single cycle of schizogeny. ChIP experiments have shown that early in the cycle, when a single *var* gene is actively being transcribed into mRNA, the promoter of this gene is bound by nucleosomes carrying the histone marks H3K4me2, H3K4me3 and H3K9Ac. As the cell transitions into the second half of the cycle and *var* mRNA production ceases, the active gene is no longer marked by H3K9Ac, but the H3K4me2 and H3K4me3 marks remain, a state referred to as 'poised'. If these marks persist through the multiple rounds of genome replication during schizogeny, the active gene will remain appropriately marked in the following cycle, thus achieving epigenetic memory. In contrast, the silent *var* genes are marked by H3K9me3 throughout the entire cycle (Lopez-Rubio *et al.*, 2007). It is also during the replication phase that transcription of the *var* ncRNAs occurs. Given that deletion of the *var* intron, and thereby abrogation of these transcripts, disrupts proper *var* gene regulation, it seems probable that the ncRNAs play a vital, though ill-defined role in this process. These data provide a relatively straightforward mechanism for maintaining a particular *var* gene expression pattern through multiple rounds of replication, however they do not provide an explanation for how switches in expression are achieved, and results from more recent experiments suggest additional complexity.

By using a technique called promoter titration, it is possible to transiently prevent the 'on' *var* gene from being actively transcribed by RNA pol II, thus creating parasites that do not transcribe any *var* mRNA (Dzikowski *et al.*, 2008). Interestingly, when parasites were released from the titration and allowed to resume *var* gene transcription, they did not return to transcription of the previously active *var* gene, but rather randomly activated any *var* gene in the genome. Thus epigenetic memory was lost. This suggests that the movement of RNA pol II through the active locus during the early portion of the cycle might be required to maintain the epigenetic marks necessary for memory. In higher eukaryotes, RNA pol II has been shown to directly recruit histone modifiers, specifically the histone methyltransferase SET2, through recruitment by the CTD, thus providing a link between active transcription and histone modification (Kizer *et al.*, 2005). Consistent with the potential for this mechanism to be important for *var* gene epigenetic memory, *P. falciparum* has an extended CTD when compared with rodent parasites, and also possesses SET2 and its cognate demethylase JmjC1, while rodent parasites are missing the genes encoding these two proteins. The fact that rodent parasites do not appear to utilize epigenetic memory for antigenic variation (Cunningham *et al.*, 2009) makes it tempting to speculate that this pathway might be important for regulating *var* gene expression.

The observations described above provide a model whereby parasites could enter S-phase of the cycle with a single *var* gene marked as the active gene while all others in the family carry epigenetics marks typical of silent chromatin. The details for how these marks are perpetuated through the multiple rounds of DNA replication and nuclear division that occur during schizogeny remains unclear; however, evidence suggests that the *var* intron derived ncRNAs play a role in this process. Ongoing work will further define the nature of the proteins that bind these ncRNAs, their structure in the nucleus, and the part they play in coordinating expression of this large, tightly regulated gene family.

References

Allison, L.A., Moyle, M., Shales, M., and Ingles, C.J. (1985). Extensive homology among the largest subunits of eukaryotic and prokaryotic RNA polymerases. Cell *42*, 599–610.

Aravind, L., Iyer, L.M., Wellems, T.E., and Miller, L.H. (2003). *Plasmodium* biology: Genomic gleanings. Cell *115*, 771–785.

Arnot, D.E., and Gull, K. (1998). The *Plasmodium* cell-cycle: facts and questions. Ann. Trop. Med. Parasitol. *92*, 361–365.

Balaji, S., Babu, M.M., Iyer, L.M., and Aravind, L. (2005). Discovery of the principal specific transcription factors of Apicomplexa and their implication for the evolution of the AP2-integrase DNA binding domains. Nucleic Acids Res. *33*, 3994–4006.

Barry, A.E., Leliwa-Sytek, A., Tavul, L., Imrie, H., Migot-Nabias, F., Brown, S.M., McVean, G.A., and Day, K.P. (2007). Population genomics of the immune evasion (*var*) genes of *Plasmodium falciparum*. PLoS. Pathog. *3*, e34

Baruch, D.I., Pasloske, B.L., Singh, H.B., Bi, X., Ma, X.C., Feldman, M., Taraschi, T.F., and Howard, R.J. (1995). Cloning the *P. falciparum* gene encoding PfEMP*1*, a malarial variant antigen and adherence receptor on the surface of parasitized human erythrocytes. Cell *82*, 77–87.

Bozdech, Z., Llinas, M., Pulliam, B.L., Wong, E.D., Zhu, J., and Derisi, J.L. (2003). The transcriptome of the intraerythrocytic developmental cycle of *Plasmodium falciparum*. PLoS. Biol. *1*, E5

Calderwood, M.S., Gannoun-Zaki, L., Wellems, T.E., and Deitsch, K.W. (2003). *Plasmodium falciparum var* genes are regulated by two regions with separate promoters, one upstream of the coding region and a second within the intron. J. Biol. Chem. *278*, 34125–34132.

Carty, S.M. and Greenleaf, A.L. (2002). Hyperphosphorylated C-terminal repeat domain-associating proteins in the nuclear proteome link transcription to DNA/chromatin modification and RNA processing. Mol. Cell Proteomics *1*, 598–610.

Chang, S.C., Tucker, T., Thorogood, N.P., and Brown, C.J. (2006). Mechanisms of X-chromosome inactivation. Front. Biosci. *11*, 852–866.

Chapman, R.D., Heidemann, M., Hintermair, C., and Eick, D. (2008). Molecular evolution of the RNA polymerase II CTD. Trends Genet. *24*, 289–296.

Chookajorn, T., Dzikowski, R., Frank, M., Li, F., Jiwani, A.Z., Hartl, D.L., and Deitsch, K.W. (2007). Epigenetic memory at malaria virulence genes. Proc. Natl. Acad. Sci. U.S.A. *104*, 899–902.

Corcoran, A.E. (2005). Immunoglobulin locus silencing and allelic exclusion. Semin. Immunol. *17*, 141–154.

Corden, J.L. (1990). Tails of RNA polymerase II. Trends Biochem. Sci. *15*, 383–387.

Coulson, R.M., Hall, N., and Ouzounis, C.A. (2004). Comparative genomics of transcriptional control in the human malaria parasite *Plasmodium falciparum*. Genome Res. *14*, 1548–1554.

Cunningham, D., Fonager, J., Jarra, W., Carret, C., Preiser, P., and Langhorne, J. (2009). Rapid changes in transcription profiles of the *Plasmodium* yoelii yir multigene family in clonal populations: lack of epigenetic memory? PLoS ONE *4*, e4285

Duraisingh, M.T., Voss, T.S., Marty, A.J., Duffy, M.F., Good, R.T., Thompson, J.K., Freitas-Junior, L.H., Scherf, A., Crabb, B.S., and Cowman, A.F. (2005). Heterochromatin silencing and locus repositioning linked to regulation of virulence genes in *Plasmodium falciparum*. Cell *121*, 13–24.

Dzikowski, R., and Deitsch, K.W. (2008). Active transcription is required for maintenance of epigenetic memory in the malaria parasite *Plasmodium falciparum*. J. Mol. Biol. *382*, 288–297.

Dzikowski, R., Templeton, T.J., and Deitsch, K. (2006). Variant antigen gene expression in malaria. Cell Microbiol. *8*, 1371–1381.

Dzikowski, R., Li, F., Amulic, B., Eisberg, A., Frank, M., Patel, S., Wellems, T.E., and Deitsch, K.W. (2007). Mechanisms underlying mutually exclusive expression of virulence genes by malaria parasites. EMBO Rep. *8*, 959–965.

Epp, C., Li, F., Howitt, C.A., Chookajorn, T., and Deitsch, K.W. (2008). Chromatin associated sense and antisense noncoding RNAs are transcribed from the *var* gene family of virulence genes of the malaria parasite *Plasmodium falciparum*. RNA *15*, 116–127.

Flueck, C., Bartfai, R., Volz, J., Niederwieser, I., Salcedo-Amaya, A.M., Alako, B.T., Ehlgen, F., Ralph, S.A., Cowman, A.F., Bozdech, Z., *et al.* (2009). *Plasmodium falciparum* heterochromatin protein 1 marks genomic loci linked to phenotypic variation of exported virulence factors. PLoS Pathog. *5*, e1000569

Fowler, E.V., Peters, J.M., Gatton, M.L., Chen, N., and Cheng, Q. (2002). Genetic diversity of the DBLalpha region in *Plasmodium falciparum var* genes among Asia-Pacific isolates. Mol. Biochem. Parasitol. *120*, 117–126.

Frank, M., and Deitsch, K. (2006). Activation, silencing and mutually exclusive expression within the *var* gene family of *Plasmodium falciparum*. Int. J. Parasitol. *36*, 975–985.

Frank, M., Dzikowski, R., Constantini, D., Amulic, B., Burdougo, E., and Deitsch, K. (2006). Strict pairing of var promoters and introns is required for *var* gene silencing in the malaria parasite *Plasmodium falciparum*. J. Biol. Chem. *281*, 9942–9952.

Frank, M., Dzikowski, R., Amulic, B., and Deitsch, K. (2007). Variable switching rates of malaria virulence genes are associated with chromosomal position. Mol. Microbiol. *64*, 1486–1498.

Gardner, M.J., Hall, N., Fung, E., White, O., Berriman, M., Hyman, R.W., Carlton, J.M., Pain, A., Nelson, K.E., Bowman, S., *et al.* (2002). Genome sequence of the human malaria parasite *Plasmodium falciparum*. Nature *419*, 498–511.

Holding, P.A., Stevenson, J., Peshu, N., and Marsh, K. (1999). Cognitive sequelae of severe malaria with impaired consciousness. Trans. Royal Soc. Trop. Med. Hygiene *93*, 529–534.

Horrocks, P., Pinches, R., Christodoulou, Z., Kyes, S.A., and Newbold, C.I. (2004). Variable *var* transition rates underlie antigenic variation in malaria. Proc. Natl. Acad. Sci. U.S.A. *101*, 11129–11134.

Iyer, L.M., Anantharaman, V., Wolf, M.Y., and Aravind, L. (2008). Comparative genomics of transcription factors and chromatin proteins in parasitic protists and other eukaryotes. Int. J. Parasitol. *38*, 1–31.

Kelley, R.L., and Kuroda, M.I. (2000). Noncoding RNA genes in dosage compensation and imprinting. Cell *103*, 9–12.

Kishore, S.P., Perkins, S.L., Templeton, T.J., and Deitsch, K.W. (2009). An unusual recent expansion of the C-terminal domain of RNA polymerase II in primate malaria parasites features a motif otherwise found only in mammalian polymerases. J. Mol. Evol. *68*, 706–714.

Kizer, K.O., Phatnani, H.P., Shibata, Y., Hall, H., Greenleaf, A.L., and Strahl, B.D. (2005). A novel domain in Set2 mediates RNA polymerase II interaction and couples histone H3 K36 methylation with transcript elongation. Mol. Cell. Biol. *25*, 3305–3316.

Kohler, S., Delwiche, C.F., Denny, P.W., Tilney, L.G., Webster, P., Wilson, R.J., Palmer, J.D., and Roos, D.S. (1997). A plastid of probable green algal origin in Apicomplexan parasites. Science *275*, 1485–1489.

Kraemer, S.M., Kyes, S.A., Aggarwal, G., Springer, A.L., Nelson, S.O., Christodoulou, Z., Smith, L.M., Wang, W., Levin, E., Newbold, C.I., *et al.* (2007). Patterns of gene recombination shape *var* gene repertoires in *Plasmodium falciparum*: comparisons of geographically diverse isolates. BMC Genomics *8*, 45

Kumar, S., and Rzhetsky, A. (1996). Evolutionary relationships of eukaryotic kingdoms. J. Mol. Evol. *42*, 183–193.

Kyes, S., Taylor, H., Craig, A.G., Marsh, K., and Newbold, C.I. (1997). Genomic representation of *var* gene sequences in *Plasmodium falciparum* field isolates from different geographic regions. Molecular and Biochemical Parasitology *87*, 235–238.

Kyes, S., Horrocks, P., and Newbold, C. (2001). Antigenic variation at the infected red cell surface in malaria. Annu. Rev. Microbiol. *55*, 673–707.

Kyes, S.A., Christodoulou, Z., Raza, A., Horrocks, P., Pinches, R., Rowe, J.A., and Newbold, C.I. (2003). A well-conserved *Plasmodium falciparum var* gene shows an unusual stage-specific transcript pattern. Mol. Microbiol. *48*, 1339–1348.

Kyes, S., Christodoulou, Z., Pinches, R., Kriek, N., Horrocks, P., and Newbold, C. (2007). *Plasmodium falciparum var* gene expression is developmentally controlled at the level of RNA polymerase II-mediated transcription initiation. Mol. Microbiol. *63*, 1237–1247.

Lavazec, C., Sanyal, S., and Templeton, T.J. (2006). Hypervariability within the *Rifin*, *Stevor* and *Pfmc-2TM* superfamilies in *Plasmodium falciparum*. Nucleic Acids Res. *34*, 6696–6707.

Le Roch, K.G., Zhou, Y.Y., Blair, P.L., Grainger, M., Moch, J.K., Haynes, J.D., De la Vega, P., Holder, A.A., Batalov, S., Carucci, D.J., *et al.* (2003). Discovery of gene function by expression profiling of the malaria parasite life cycle. Science *301*, 1503–1508.

Lewis, A., and Reik, W. (2006). How imprinting centres work. Cytogenet. Genome Res. *113*, 81–89.

Li, F., Sonbuchner, L., Kyes, S.A., Epp, C., and Deitsch, K.W. (2008). Nuclear non-coding RNAs are transcribed from the centromeres of *Plasmodium falciparum* and are associated with centromeric chromatin. J. Biol. Chem. *283*, 5692–5698.

Lopez-Rubio, J.J., Gontijo, A.M., Nunes, M.C., Issar, N., Hernandez, R.R., and Scherf, A. (2007). 5′ flanking region of *var* genes nucleate histone modification patterns linked to phenotypic inheritance of virulence traits in malaria parasites. Mol. Microbiol. *66*, 1296–1305.

Mattick, J.S. (2003). Challenging the dogma: the hidden layer of non-protein-coding RNAs in complex organisms. Bioessays *25*, 930–939.

McFadden, G.I., Reith, M.E., Munholland, J., and Lang-Unnasch, N. (1996). Plastid in human parasites. Nature *381*, 482

Miller, L.H., Good, M.F., and Milon, G. (1994). Malaria pathogenesis. Science *264*, 1878–1883.

Miller, L.H., Baruch, D.I., Marsh, K., and Doumbo, O.K. (2002). The pathogenic basis of malaria. Nature *415*, 673–679.

Morey, C., and Avner, P. (2004). Employment opportunities for non-coding RNAs. Febs Lett. *567*, 27–34.

Morrison, D.A. (2009). Evolution of the Apicomplexa: where are we now? Trends Parasitol. *25*, 375–382.

Mourier, T., Carret, C., Kyes, S., Christodoulou, Z., Gardner, P.P., Jeffares, D.C., Pinches, R., Barrell, B., Berriman, M., Griffiths-Jones, S., *et al.* (2008). Genome-wide discovery and verification of novel structured RNAs in *Plasmodium falciparum*. Genome Res. *18*, 281–292.

Perez-Toledo, K., Rojas-Meza, A.P., Mancio-Silva, L., Hernandez-Cuevas, N.A., Delgadillo, D.M., Vargas, M., Martinez-Calvillo, S., Scherf, A., and Hernandez-Rivas, R. (2009). *Plasmodium falciparum* heterochromatin protein 1 binds to tri-methylated histone 3 lysine 9 and is linked to mutually exclusive expression of var genes. Nucleic Acids Res. *37*, 2596–2606.

Pidoux, A.L., and Allshire, R.C. (2005). The role of heterochromatin in centromere function. Philos. Trans. R. Soc. Lond. B. Biol. Sci. *360*, 569–579.

Sachs, J., and Malaney, P. (2002). The economic and social burden of malaria. Nature *415*, 680–685.

Scherf, A., Hernandez-Rivas, R., Buffet, P., Bottius, E., Benatar, C., Pouvelle, B., Gysin, J., and Lanzer, M. (1998). Antigenic variation in malaria: *in situ* switching, relaxed and mutually exclusive transcription of *var* genes during intra-erythrocytic development in *Plasmodium falciparum*. EMBO J. *17*, 5418–5426.

Serizawa, S., Miyamichi, K., and Sakano, H. (2004). One neuron-one receptor rule in the mouse olfactory system. Trends Genet. *20*, 648–653.

Smith, J.D., Chitnis, C.E., Craig, A.G., Roberts, D.J., Hudson-Taylor, D.E., Peterson, D.S., Pinches, R., Newbold, C.I., and Miller, L.H. (1995). Switches in expression of *Plasmodium falciparum var* genes correlate with changes in antigenic and cytoadherent phenotypes of infected erythrocytes. Cell *82*, 101–110.

Snow, R.W., Craig, M., Deichmann, U., and Marsh, K. (1999). Estimating mortality, morbidity and disability due to malaria among Africa's non-pregnant population. Bull. World Health Organ. *77*, 624–640.

Su, X., Heatwole, V.M., Wertheimer, S.P., Guinet, F., Herrfeldt, J.V., Peterson, D.S., Ravetch, J.V., and Wellems, T.E. (1995). A large and diverse gene family (*var*) encodes 200–350 kD proteins implicated in the antigenic variation and cytoadherence of *Plasmodium falciparum*-infected erythrocytes. Cell *82*, 89–100.

Swamy, L., Amulic, B., and Deitsch, K.W. (2011). *Plasmodium falciparum var* gene silencing is determined by *cis* DNA elements that form stable and heritable interactions. Eukaryot. Cell *10*, 530–539.

Tonkin, C.J., Carret, C.K., Duraisingh, M.T., Voss, T.S., Ralph, S.A., Hommel, M., Duffy, M.F., Silva, L.M., Scherf, A., Ivens, A., *et al.* (2009). Sir2 paralogues cooperate to regulate virulence genes and antigenic variation in *Plasmodium falciparum*. PLoS. Biol. *7*, e84

Topp, C.N., Zhong, C.X., and Dawe, R.K. (2004). Centromere-encoded RNAs are integral components of the maize kinetochore. Proc. Natl. Acad. Sci. U.S.A. *101*, 15986–15991.

Trimnell, A.R., Kraemer, S.M., Mukherjee, S., Phippard, D.J., Janes, J.H., Flamoe, E., Su, X.Z., Awadalla, P., and Smith, J.D. (2006). Global genetic diversity and evolution of *var* genes associated with placental and severe childhood malaria. Mol. Biochem. Parasitol. *148*, 169–180.

Voss, T.S., Healer, J., Marty, A.J., Duffy, M.F., Thompson, J.K., Beeson, J.G., Reeder, J.C., Crabb, B.S., and Cowman, A.F. (2006). A *var* gene promoter controls allelic exclusion of virulence genes in *Plasmodium falciparum* malaria. Nature *439*, 1004–1008.

Part III

Non-coding RNAs: Disease and Therapeutics

Long Non-coding RNAs (lncRNAs) and Cancer

9

Jessica M. Silva and David I. Smith

Abstract

The sequencing of the human genome revealed that less than 1.2% of the genome actually codes for protein while another 45% comprises highly repetitive sequences and approximately 54% of the genome is neither repetitive sequences nor codes for proteins (Consortium, 2004). This portion of the genome contains a number of important housekeeping transcripts such as ribosomal RNAs and transfer RNAs. There are also a number of small non-coding transcripts such as the microRNAs, piwiRNAs and snoRNAs. In addition to these small regulatory transcripts there are a number of longer non-coding RNAs (lncRNAs) which have also been found to play important roles within the genome. LncRNAs have been found to be aberrantly expressed in a number of different diseases including psoriasis, mental disorders, autism and cancer. The list of lncRNAs associated with cancer is rapidly increasing. These lncRNAs can been differentiated from each other based upon where they are derived from within the genome. For instance, there are intronic lncRNAs such as AK024366 that are transcribed from between the exons of genes, intergenic lncRNAs including NEAT1, MALAT1 and LSINCTs that are transcribed in the space between genes, and lncRNAs that are more complex as they overlap both intron and exon of a coding gene such as DLEU2 and CDKN2BAS/ANRIL. These lncRNAs may also be transcribed in the sense or in the antisense direction. Although cancer-associated lncRNAs are being discovered, relatively little is still known about the precise functions of most of these lncRNAs. These long non-coding transcripts function in many different ways within cells and a number of them promote cancer development and progression. We will review what is currently known about the role that some of the lncRNAs play in cancer.

Introduction

Even before the sequencing of the human genome it was known by re-association studies that a significant portion of the genome comprised highly repetitive DNA sequences. However, it was not until the human genome was sequenced and analysed that it was determined that only 1.2% of the genome actually codes for protein (Consortium, 2004). The technology of whole genome tiling arrays made it possible to interrogate the entire non-repetitive portion of the genome for transcriptional activity. These initial studies revealed that there were considerably more transcripts that previously expected and that much of the non-coding non-repetitive portion of the genome was transcriptionally active.

There are numerous non-coding transcripts in the genome. The housekeeping transcripts, ribosomal RNAs and transfer RNAs, that participate in the processes of protein production within the cells. These include the large number of small RNAs such as microRNAs, piwiRNAs and snoRNAs. In addition to these relatively small transcripts there are also considerably longer non-coding transcripts that also play important roles within the genome. These long non-coding RNAs (lncRNAs) can be differentiated from each other based upon where they are derived from within the genome. For instance, there are intronic lncRNAs (transcribed between exons of genes), intergenic lncRNAs (transcribed in the space between genes), and lncRNAs that are more complex as they overlap both intron and exon of a coding gene. Each of these lncRNAs may also be transcribed in the sense or in the antisense direction.

The list of lncRNAs that are associated with diseases is rapidly increasing. Some lncRNAs have been found to be aberrantly expressed in various diseases including psoriasis, mental disorders, autism, and also cancer (Barciszewski *et al.*, 2002; Szegedi *et al.*, 2010). The list of cancer-associated lncRNAs are also increasing. However, relatively little is known about the precise functions of most cancer-associated lncRNAs, but of what is known suggest that these long non-coding transcripts function in many different ways within cells and promote cancer development and progression. We will review what is known about the role that some of the lncRNAs play in cancer (Table 9.1).

Intronic long non-coding RNAs

Long ncRNAs transcribed from intronic regions have generally been found to show tissue-specific expression, to be localized predominantly within the nucleus, and to be responsive to physiological stimuli (Louro *et al.*, 2009). In 2004, Verjovski-Almeida's group hybridized RNA from 27 prostate cancer patient tumour samples with Gleason Scores ranging from 5–10 onto a custom made DNA microarray containing intronic transcript sequences (Reis *et al.*, 2004). They identified three novel antisense intronic transcripts produced from introns of the *RASSF1* gene (~1.1 kb), *PPHCN1* gene (~0.6 kb), *AP1GBP1* gene (~0.8 kb), and also an altered splice form within the intron of the *TLE3* gene (~5.53 kb). The antisense intronic *RASSF1* intron was further assessed and found to have decreased expression in tumour tissues with a high Gleason score, demonstrating a correlation between the expression of an lncRNA and prostate cancer progression. Further analysis of prostate cancer intronic lncRNAs was conducted by the same group identifying multiple intronic ncRNAs which were associated with androgen responsiveness in prostate cancer (Louro *et al.*, 2007). The expression levels of the intronic RNAs correlated with levels corresponding to the protein coding genes surrounding those lncRNAs and a fraction of the naturally transcribed intronic non-coding RNAs (ncRNAs) identified were found to be regulated by common physiological signals such as hormones.

This same group utilized a custom cDNA microarray platform on six patients with clear cell renal carcinoma (RCC) and identified both protein-coding transcripts as well as intronic ncRNAs which were down-regulated in RCC (Costa Brito *et al.*, 2008). Among these down-regulated RCC transcripts, 49 mapped to both translated and untranslated regions, and of those untranslated regions seven were intronic transcripts. These intronic transcripts down-regulated in RCC were derived from within intron #1 of the *SLC2A1* gene, intron #14 of the *SND1* gene, intron #8 of the *ACTN4* gene, intron #18 of the *NIBP* gene, intron #2 of the *TACC1* gene, intron #3 of the *HDAC5* gene, and intron #6 of the *H2AF4* gene. Down regulation of the transcripts derived from introns within the *ACTN4* and *SND1* genes were validated by qPCR. This analysis suggested a potential role for both coding and intronic non-coding tumour-suppressor gene candidates which may play a role in RCC.

Three putative intronic lncRNAs were also identified by de la Chappelle's group using genome wide linkage analysis on a Affymetrix chip 50K SNP array and fine mapping microsatellite markers in families with papillary thyroid cancer (PTC) and melanoma (He *et al.*, 2009). These studies identified a susceptibility locus for PTC on chromosome 8q24. Contained within this region were two known coding genes, *TG* (thyroid globulin) and SLA (Src-like adaptor), and one previously uncharacterized cDNA, *AK023948*. The lncRNAs identified had homology to a previously annotated cDNA *AK023948* and two other cDNAs *AK023852* and *AK024366* (2.1 kb to 2.8 kb) which are located in the introns of the SLA gene. The intronic lncRNAs *AK023948* and *AK023852* gene expression levels were 5-fold lower in PTC tumour versus normal thyroid tissue in 50% of the samples they tested with no significant changes in expression of either the *TG* or *SLA* genes. Deceased expression of *AK023948* was also detected in tumour samples of families with sporadic PTC. This work demonstrates examples of intronic lncRNAs found within a susceptibility locus for PTC. These non-coding transcripts could actually be the targets of alteration in PTC, but this will need to be functionally proven.

Finally, a stress induced lncRNA, *LSINCT10* (long stress induced non-coding transcript 10) was identified by an entire human genome tiling array analysis of stress on NHBE (normal human bronchial epithelial) cells induced with a tobacco carcinogen. This lncRNA is located within the intron of the SDC5 gene (Silva *et al.*, 2010). *LSINCT10* seems to be a spliced transcript with two isoforms (750bp and 2.5kb) by Northern blot analysis. It also has increased expression in certain breast and cervical cancer cell lines (Silva *et al.*, 2010).

Although many intronic lncRNAs have been identified there is a lack of detailed characterization of these transcripts or of their functions. Hence, the majority of these are known as transcripts of unknown function (TUFs).

Long intergenic non-coding RNAs (lincRNAs)

Intergenic long non-coding RNAs also known as lincRNAs are the ncRNAs which have been more fully characterized. Several lincRNAs have known functions. Many lincRNAs have been implicated in epigenetic regulation by modifying chromosomes thereby affecting gene expression (Khalil *et al.*, 2009). However, a number of the lincRNAs currently being characterized have been found to play more central roles in cellular regulation. It is thus not too surprising that cancer-specific alterations to a number of these transcripts are observed. Cancer-associated lincRNAs have been found to be associated with X-inactivation, nuclear speckle enrichment, imprinting, tissue specificity, stress, and have been found to encode or to be regulated by smaller ncRNAs.

X-inactivation ncRNA

One of the most well-known and well-characterized lincRNA is *XIST* (*X*-inactive specific transcript), which is involved in chromatin remodelling events associated with X-chromosome inactivation and dosage compensation in eukaryotes. Copies of this 8 kb transcript are responsible for coating the inactive X chromosome (Akhtar *et al.*, 2003; Chow *et al.*, 2005; Payer *et al.*, 2008).Recently, a link between X inactivation and tumorigenesis has been made when it was observed that there was a gain or loss of the X chromosome in different tumour cells including testicular germ cell tumours, lymphoma, recurrent ovarian cancer, breast cancer, and cervical cancer (Wutz *et al.*, 2009). There is also evidence to suggest that *BRCA1* is involved in *XIST* regulation on the inactive X chromosome. *BRCA1* expression leads to higher levels of *XIST* expression thereby allowing *XIST* to be a potential marker for BRCA1-associated breast cancers compared with sporadic basal-like cancers (Sirchia *et al.*, 2009).

Nuclear-associated ncRNAs

Two adjacent lincRNAs *NEAT* (nuclear paraspeckle assembly transcript) 1 and 2 also known as *tncRNA* (trophoblast derived ncRNA) and *MALAT1* (metastasis-associated lung adenocarcinoma transcript 1) have been demonstrated to be enriched in nuclear speckles and play roles in cancer development (Geirsson *et al.*, 2004; Hutchinson *et al.*, 2007; Clemson *et al.*, 2009). *NEAT1* is a 4.5 kb transcript associated with the suppression of the MHC class II antigen promoter in T cells, thus suppressing the display of fetal histocompatibility antigens to allow fetal evasion of the maternal immune system (Geirsson *et al.*, 2003, 2004, 2007). *NEAT1* is also required for the formation and maintenance of nuclear paraspeckle formation (Fox *et al.*, 2002). In addition, *NEAT1* has also been found to have increased expression in HELA cells (Souquere *et al.*, 2010). *NEAT2* is the well studied 8 kb nuclear speckle enriched lincRNA found to be expressed in numerous tissues and has been found to be associated with synapse formation, maintenance of paraspeckles (Ji *et al.*, 2003; Sasaki *et al.*, 2009; Sunwoo *et al.*, 2009; Bernard *et al.*, 2010), and be a potential regulator of several genes (Tano *et al.*, 2010). *NEAT2* is overexpressed in a variety of different cancers and is a marker for metastasis in early stage non-small cell lung cancer patients (Ji

Table 9.1 lncRNAs associated with cancer

Category	Name	Length	Cancer	Reference
Intronic	**Intronic antisense transcripts (RASSF1, PPHCN1, APLGBP1)**	**600–1.1 kb**	**Prostate cancer**	Reis (2004)
	Androgen responsive intronic ncRNAs	500–5 kb	Prostate cancer	Louro (2007)
	NC intronic RNAs down-regulated in RCC	165–705 bp	Clear cell renal carcinoma	Brito (2008)
	AK024366,023948,023952	2.1–2.8 kb	Papillary thyroid cancer	He (2009)
	LSINCT10	750 bp/2.5 kb	Breast and cervical cancer	Silva (2010)
Intergenic	**XIST (X-inactive specific transcript)**[a]	143 kb	**Breast cancer**	Akhtar (2003), Chow (2005), Payer (2008), Sirchi, (2009), Wutz (2009)
	NEAT1 (tncRNA)	4–>17 kb	Cervical cancer	Fox (2002), Geirsson (2003, 2004, 2007), Hutchinson (2007), Clemson (2009), Souquere (2010)
	NEAT 2 (MALAT-1)	>8 kb	Human hepatocellular carcinomas and hepatoblastomas, endometrial stromal sarcoma of the uterus, lung adenocarcinomas, cervical cancer, neuroblastoma	Ji (2003), Luo (2006), Yamada (2006), Hutchinson (2007), Lin (2007), Clemson (2009), Sasaki (2009), Sunwoo (2009), Bernard (2010), Guo (2010), Koshimizu (2010),Tano (2010)
	PCGEM1 (prostate specific transcript 1)	530–1.6 kb	Prostate cancer	Srikantan (2000), Fu (2006)
	UCA1 (urothelial cancer-associated 1)	1442 kb	Bladder carcinoma	Wang (2006, 2008)
	H19[b]	**2640 bp**	**Hepatocellular carcinoma, colorectal, prostate cancer, breast cancer, choriocarcinoma, testicular cancers, bladder carcinoma**	Hao (1993), Lustig-Yariv (1997), Verkerk (1997), Adriaenss (1998), Ariel (2000), Manoharan (2003), Cai and Cullen (2007), Matouk (2007), Berteaux (2008), Paradowska (2009), Riccio (2009), Gabory (2010), Tsang (2010), Yoshimizu (2008)
	KCNQ1OT1 (LIT1) imprinted gene antisense[a]	**60 kb**	**Breast and colorectal**	Mitsuya, (1999), Smilinich (1999), Gallagher (2006), Nakano (2006), Murakami (2007), Faghihi and Morgan (2008), Mohamma (2008, 2010), Redrup (2009)
	LSINCTs (long stress induced non-coding transcripts)	1–5 kb	Breast, cervix, lung, ovarian	Silva (2010)

	HOTAIR[a] (HOX transcript antisense RNA)	**2158 bps**	**Breast cancer**	Rinn (2007), Gupta (2010)
	GAS5 (growth arrest specific 5)	**4087 bps**	**Breast cancer, B-cell lymphoma**	Schneider (1988), Smith and Steitz (1998), Mourtada-Maarabouni (2008, 2009, 2010), Nakamura (2008), Kino (2010)
	HULC (highly up-regulated in liver cancer)	1637 bps	Liver cancer, hepatic colorectal cancer, hepatocellular carcinoma	Matouk (2007, 2009), Panzitt (2007), Wang (2010)
	SRA (steroid receptor RNA activator ncRNA)	**700–850 bps and 1.3–1.5 kb**	**Breast cancer, uterine, ovarian cancer**	Lanz (1999, 2003), Leygue, (1999a,b), Watanabe (2001), Chooniedass-Kothari (2006), Hube (2006, 2010), Cooper (2009), Foulds (2010)
Complex	B2 (Homo sapiens NSCLC non-coding RNA)	13 251 bps	Non-small cell lung cancer	Jacquot (2004), Espinoza (2007), Malleter (2010)
	DLEU2 (Homo sapiens deleted in lymphocytic leukaemia 2 non-coding RNA)	**2768 bps**	**Chronic lymphocytic leukaemia**	Corcoran (2004); Lerner (2009); Mertens (2009)
	CDKN2BAS/ANRIL (CDKN2B antisense RNA)	**3834 bps**	**Melanoma-NST syndrome, breast cancer, nasopharyngeal carcinoma, basal cell carcinoma, and glioma**	McPherson (2007), Pasmant (2007), Scott (2007), Bilguvar (2008), Bei (2010), Rodriguez, (2010), Sato (2010), Shete (2009), Stacey, 2009, Turnbull (2010), Yasuno (2010)

Antisense lncRNAs are in bold font.

[a]Epigenetically regulated.

[b]Endo-siRNA.

et al., 2003; Luo et al., 2006; Yamada et al., 2006; Lin et al., 2007; Koshimizu et al., 2010).

Tissue-specific ncRNAs

PCGEM1 (prostate-specific transcript 1) was identified by differential display analysis on a panel of African American patients with prostate cancer (Srikantan *et al.*, 2000). *PCGEM1* expression is found exclusively in prostate tissue and detected only in androgen receptor positive cell lines such as LNCAP. *PCGEM1* expression has been found to increase colony formation in both LNCAP cancer cells and in NIH3T3 mouse fibroblast cells. However, it also has been shown that overexpression in LNCAP cells leads to inhibition of apoptosis by PARP cleavage when induced with doxorubicin, etoposide and sodium selenite treatments (Fu *et al.*, 2006).

Another interesting tissue specific lincRNA is *UCA1* (urothelial carcinoma-associated 1). *UCA1* was identified as a candidate urine marker for bladder transitional cell carcinoma (TCC) and is one of the most TCC-specific genes identified (Wang *et al.*, 2006). It has been shown to be associated with embryonic development and upregulated in bladder cancer affecting proliferation, migration, invasion, and drug resistance as well as inducing tumorgenesis in nude mice (Wang *et al.*, 2008).

Imprinting ncRNAs

Imprinted genes are genes whose expression is determined by the parent that contributed them. One well studied imprinting gene is the non-coding RNA *H19*. *H19* is a 2.6 kb, capped and polyadenylated lincRNA (Gabory *et al.*, 2010). *H19* is located in an imprinted region of chromosome 11p15.5, 90 kb downstream of the (insulin-like growth factor 2) *IGF2* imprinted gene. *H19* is only expressed from the maternally inherited chromosome and *IGF2* is only expressed from the paternally inherited chromosome. The *H19* maternal and *IGF2* paternal monoallelic expression is regulated upstream by methylation of the differentially methylated region (DMR) and imprinting control region (ICR), two regulatory elements required for the imprinting of these genes located on the paternal chromosome. Mutations in this gene were first associated with Beckwith–Wiedemann syndrome (BWS) and Wilms tumorigenesis (Riccio *et al.*, 2009). *H19* further has been found to act both as a tumour suppressor and an oncogene (Hao *et al.*, 1993; Lustig-Yariv *et al.*, 1997; Verkerk *et al.*, 1997; Adriaenssens *et al.*, 1998; Ariel *et al.*, 2000; Matouk *et al.*, 2007). Dandolo's group utilized *in vivo* murine models of tumorigenesis in teratocarcinoma, colorectal, and hepatocarcinoma, and found *H19* to display a clear tumour suppressor effect in mice (Manoharan *et al.*, 2003; Yoshimizu *et al.*, 2008). In prostate cancer, the *IGF2/H19* gene was shown to contain differences in DNA methylation and histone methylation status of the ICR and DMR regions providing insight into early steps of carcinogenesis (Paradowska *et al.*, 2009). In addition to these roles, the 120 kb antisense transcript of *H19*, known as *91H* has recently been identified to also regulate *IGF2* and further been shown to be overexpressed in breast cancer and spontaneous hepatocarcinogenesis (Adriaenssens *et al.*, 1998; Manoharan *et al.*, 2003; Berteaux *et al.*, 2008). In 2007, this important lincRNA was found to be a precursor for a miRNA, miR-675 and this year the oncofetal H19-derived miR-675 was found to regulate the tumour suppressor *RB* in colorectal cancer (Cai and Cullen *et al.*, 2007; Tsang *et al.*, 2010).

Another imprinted lincRNA located on the same chromosomal region as IGF2/H19 which is also associated with cancer is the *Kcnq1ot1/Lit1* (long QT intronic transcript 1) ncRNA. *Kcnq1ot1/Lit1* is a RNA polymerase II transcribed, nuclear, unspliced, maternally expressed lncRNA (Smilinich *et al.*, 1999; Mohammad *et al.*, 2008; Redrup *et al.*, 2009). This lincRNA was identified in 1999 by Oshimura's group to be an imprinted antisense transcript encoded within the KvLQT1 locus, which was associated with multiple balanced chromosomal rearrangements in BWS and an additional breakpoint in embryonal rhabdoid tumours (Mitsuya *et al.*, 1999). They further found methylation of the intronic CpG island on the silent maternal allele and also showed four of 13 BWS patients had complete loss of maternal methylation at the CpG island, suggesting that antisense regulation is involved in the development of human disease. *Kcnq1ot1/Lit1* has also been identified to regulate H3K9 trimethylation in the perinucleaur compartment during S-phase (Faghihi and Morgan *et al.*, 2008) and has been shown to be important for the transcriptional silencing of the imprinted domain (Murakami *et al.*, 2007). In mouse *Kcnq1ot1/Lit1* recruits DNMT1 to somatic DMRs by interacting with DNMT1 for allele specific methylation (Mohammad *et al.*, 2010). *Kcnq1ot1/Lit1* is expressed preferentially from the paternal allele and produced in most human tissues. *Kcnq1ot1/Lit1* is located on the IGF2 locus and was also found to be discordantly regulated in cancer from this region (Mitsuya *et al.*, 1999). Finally, *Kcnq1ot1/Lit1* has been found to be associated *CDKN1C* expression; however, loss of imprinting of *Kcnq1ot1/Lit1* via epigenetic disruption was found to be important in colorectal cancer (Gallagher *et al.*, 2006; Nakano *et al.*, 2006; Murakami *et al.*, 2007).

Stress and cancer-associated lncRNAs

The role of stress in cancer development has been well studied. Environmental stresses directly linked to cancer initiation and/or promotion include radiation exposure, DNA damaging chemicals such as inhaled asbestos, certain dioxins, and carcinogens within tobacco smoke. Long stress induced non-coding transcripts (*LSINCTs*), are a class of lncRNAs which are associated with both stress and cancer. *LSINCTs* were identified by utilizing an entire human genome tiling array with normal human bronchial epithelial cells (NHBE) induced with a potent tobacco carcinogen NNK (4-(methylnitrosamino)-1-(3-pyridyl)-1-butanone) (Silva *et al.*, 2010). The tobacco carcinogen NNK causes DNA mutations that dysregulate various genes affecting viability, cell movement, cell cycle, and cell proliferation (Hecht *et al.*, 1988; Jorquera *et al.*, 1994; Lonardo *et al.*, 2002; Chuang *et al.*, 2006; Abdel-Aziz *et al.*, 2007). Twelve NNK-induced *LSINCTs* were identified throughout the entire genome. In this analysis, *NEAT 1* and *NEAT 2* were also identified as long stress-induced transcripts. *LSINCTs* range in size from 1.2 to 5 kbs. They have increased expression in more proliferative normal tissues such as spleen and decreased expression in less proliferative tissues such as brain. In addition, *LSINCTs* had increased expression in certain lung and breast cancer cell lines.

Hox loci ncRNAs

The *HOX* genes are a group of transcriptional factors, which are master regulators of embryonic development and stem cell differentiation and their mis-regulation leads to human disease and cancer (Rastegar *et al.*, 2010). *HOX* genes encode highly conserved transcription factors that organize the anterior-posterior body axis during embryogenesis. *HOX* gene expression during development is tightly regulated in a spatiotemporal manner, partly by chromatin structure and epigenetic modifications (Soshnikova and Duboule *et al.*, 2009).

In mammals, 39 *HOX* transcription factors are clustered on four chromosomal loci, termed *HOXA* though *HOXD*. The four *HOX* loci was analysed and found to have considerable transcription within the intergenic regions (Rinn *et al.*, 2007). The ncRNAs identified in the intergenic regions were found to be conserved and contain distinct expression patterns. One antisense lincRNA identified from this analysis was termed *HOTAIR* (Hox Antisense intergenic RNA). This 2158 nt lncRNA is a spliced and polyadenylated ncRNA. It is produced within one cluster of *HOX* genes and is responsible for regulating another *HOX* gene cluster on a different chromosome by PRC2-mediated silencing of chromatin (Rinn *et al.*, 2007). Another study, analysing RNA derived from normal human breast epithelia, primary breast carcinomas, and distant metastases to an ultra-dense *HOX* tiling array identified 233 differentially expressed transcribed regions consisting of 170 ncRNAs and 60 HOX exons (Gupta *et al.*, 2010). *HOTAIR* was one of these differentially expressed ncRNAs found to be increased in primary breast tumours and their metastases and is also suggested to be a potential novel predictor for metastasis and death of breast cancer (Gupta *et al.*, 2010).

microRNA-regulated ncRNAs

One interesting lincRNA, *GAS5* (growth arrested specific transcript 5), was identified from a cDNA library enriched for RNA sequences preferentially expressed in growth-arrested NIH3T3 cells

(Schneider *et al.*, 1988). *GAS5* was found to be a ncRNA which codes for multiple conserved small nucleolar RNA (snoRNAs), specifically nine mouse box C/D snoRNAs or ten human snoRNAs and was also found be a 5′ terminal oligo pyramidine (5′TOP) gene (Smith and Steitz *et al.*, 1998; Mourtada-Maarabouni *et al.*, 2008). *GAS5* plays an essential role in normal growth arrest in T-cells lines and non-transformed lymphocytes and when overexpressed was found to increase apoptosis and decrease the rate of progression through the cell cycle. However, down regulation of endogenous *GAS5* also lead to apoptosis and maintained a more rapid cell cycle (Mourtada-Maarabouni *et al.*, 2008). This suggests that *GAS5* is necessary and sufficient for normal growth arrest of T-Cells. Chrousos's group identified *GAS5* to be a riborepressor of the glucocorticoid receptor which influenced cell survival and metabolic activities (Kino *et al.* 2010). In addition, a patient with diffuse large B cell lymphoma was found to have the *GAS5* gene fused to the BCL6 gene t(1;3) (q25;q27) (Nakamura *et al.*, 2008). A result of the translocation allowed the *GAS5/BCL6* chimeric transcripts to be expressed, leading *GAS5* to be the second *BCL6* partner as *SNHG5* is the first. In breast cancer, *GAS5* has been found to have decreased expression in certain breast cancer cell lines in breast ductal carcinoma and adjacent normal tissue (Mourtada-Maarabouni *et al.*, 2009). Moreover, overexpression of *GAS5* in MCF7 cancer cells inhibited cell growth and induced apoptosis. All together, the recent findings on *GAS5* have proven it to be a novel snoRNAs encoding lincRNA which is associated with many processes.

microRNA regulated long ncRNAs

Recently, lincRNAs have been found to bind and be regulated by microRNAs (miRNAs). miRNAs are small non-coding RNAs which bind and inhibit or degrade mRNAs (Garzon *et al.*, 2006). miRNAs have also been found to be associated with cancer (Yu *et al.*, 2007) so their regulation of cancer-associated lincRNAs is not surprising. One lincRNA that has been found to be regulated by a miRNA is *HULC* (highly up-regulated in liver cancer). *HULC* was identified as a specific marker for hepatocellular carcinomas (HCC). *HULC* is a capped, spliced, polyadenylated, RNA polymerase III transcribed lincRNA located in the cytoplasm where it copurifies with ribosomes and resembles the mammalian LTR transposon 1A (Panzitt *et al.*, 2007; Matouk *et al.*, 2009). Recently, *HULC* has been found to not only be up-regulated in HCCs but also to be up-regulated in colorectal carcinomas which metastasized to the liver. *HULC* was not found to be expressed in normal colon, primary colorectal carcinomas and colorectal carcinomas which metastasized to the lymph nodes (Matouk *et al.*, 2009). In order to identify a possible mechanism for *HULC*, Fan's group analysed the one cAMP response element binding protein (*CREB*) site within the *HULC* proximal promoter region and found that it can specifically bind phospho-*CREB* transcription factors through the *PKA* pathway and that mutations in this area diminishes *HULC* activity in liver cancer cells (Wang *et al.*, 2010). They further demonstrated that *HULC* has the ability to inhibit several miRNAs, in particular *miR-372*. The inhibition of *miR-372* by *HULC* leads to the induction of the *PRKACB* gene which then induces phosphorylation of *CREB* (Wang *et al.*, 2010).

Complex lncRNAs

In addition to purely intronic and purely intergenic lncRNAs there are also complex lncRNAs. These complex lncRNAs may be transcribed within the encoding regions of genes such as in exons or in the antisense direction of the entire gene or gene cluster but still do not have putative open reading frames.

The steroid receptor RNA activator (*SRA*) encodes both a protein-coding transcript that produces, *SRAP*, and a non-protein coding RNA, *SRA* (Leygue *et al.*, 1999; Hube *et al.*, 2006, 2010). *SRA* was the first identified lncRNA that functioned to regulate genes in *trans* (Lanz *et al.*, 1999). *SRA* is a spliced, polyadenylated transcript which enhances ligand activity on reporter genes and was also found to be present in steroid receptor coactivator *SRC-1/SRC-2* coactivator complexes (Watanabe *et al.*, 2001). *SRA* is overexpressed in breast, uterine and ovarian tumours, and affects the growth of certain hormone-dependent breast and prostate cancer cell lines (Lanz *et al.*, 1999;

Leygue *et al.*, 1999; Chooniedass-Kothari *et al.*, 2006; Hube *et al.*, 2006; Cooper *et al.*, 2009). SRA knockdown was shown to coactivate certain target promoters including *DIO2*, *THBS1*, and *SLC23* gene promoters and also enhance activity of some co-regulator proteins such as *PGC-1a* and *SRC-1* (Foulds *et al.*, 2010) in MCF7 breast cancer cells.

Another complex lncRNA, *B2* RNA is located in 6p24-p25 beginning 10 kb upstream of exon 1 of the *HEF1* gene on the 5′ end to exon 4 of *HEF1* on the 3′ end (Jacquot *et al.*, 2004). *B2* RNA does not have any open reading frames and contains only one exon. *B2* RNA does have homology with the genomic DNA of the *HEF1/NEDD9/Cas-L* gene, which was found to regulate apoptosis and cell proliferation by modulating the *HEF1* in non-small cell lung cancer (Malleter *et al.*, 2010). Proliferation arrest was also detected using *B2* antisense oligos and VT1 treatment in NSCLC-N6 cells.

DLEU2 (deleted in lymphocytic leukaemia 2) encodes a conserved lncRNA that overlaps with the first exon of *RFP2/LEU5* in the antisense direction. *DLEU2* is a very large RNA which has 15 exons that cover 143 kb in humans and 80 kb in mouse. *DLEU2* shown by Northern blot assay has been shown to have various sized transcripts when using different probes, in exon 11 (1.2 kb, 2.4 kb) and in exon 1b and 8 (2 kb, 1.8 kb, and 1.4 kb) (Corcoran *et al.*, 2004). *DLEU2* is located in the 10 kb minimally deleted region (MDR) on chromosome 13q14.3 frequently lost in chronic lymphocytic leukaemia (CLL) in addition to several other genes and miRNAs (Mertens *et al.*, 2009). *DLEU2* functions as a critical host gene of *miR15/16-1*, by regulating in G1 Phase cyclins E1 and D1 through *miR15a/16-1* (Lerner *et al.*, 2009). The overexpression of *DLEU2* in MCF7, U2Os, and HEK293 was also found to block cell proliferation and inhibit the colony forming ability of these tumour cells in a *miR15a/16-1*-dependent manner(Lerner *et al.*, 2009).

Finally, *CDKN2AS/ANRIL* (antisense non-coding RNA in the INK4 locus) which is transcribed in the antisense direction of the *p15/CDKN2B-p16/CDKN2A-p14/ARf* gene cluster was identified when characterizing a germ line deletion in a melanoma-neural system tumour family (Pasmant *et al.*, 2007a,b). *CDKN2AS/ANRIL* was found to be expressed in both normal human tissues such as normal ovary and in certain breast cancer cell lines and NF1-associated tumours. *ANRIL* is associated with the coordinated trans-regulation of its nearby genes *CDKN2A* and *CDKN2B* however its regulation and function are not yet fully understood. It was shown that *ANRIL* is a target of *E2F1* and may play a role in regulating nuclear genes (Sato *et al.*, 2010). Recently, *CTCF* (CCCTC-binding factor) binding sites mapped in the CpG island overlapping the *ANRIL-p14/ARF* promoters gave new insights into co-regulation of *ANRIL* and *p14/ARF* transcription (Rodriguez *et al.*, 2010). More recently, GWAS identified *ANRIL* as a genetic susceptibility locus associated with coronary disease, intracranial aneurysm, type 2 diabetes and several cancers (McPherson *et al.*, 2007; Scott *et al.*, 2007; Bilguvar *et al.*, 2008; Yasuno *et al.*, 2010) including breast cancer, nasopharyngeal carcinoma, basal cell carcinoma, and glioma (Shete *et al.*, 2009; Stacey *et al.*, 2009; Bei *et al.*, 2010; Turnbull *et al.*, 2010).

Conclusions

In this review, we have summarized what is currently known about the various types of lncRNAs found to be associated with cancer development. These cancer-associated lncRNAs include ones encoded from intronic regions. These are tissue-specific transcripts, localized predominantly within the nucleus, and found to be responsive to physiological stimuli. Other cancer-associated lncRNAs include lincRNAs and complex lncRNAs that are associated with X-inactivation, nuclear speckle enrichment, imprinting, tissue specificity, *Hox* loci, stress, and have been found to encode or to be regulated by smaller ncRNAs.

Only a few of the lncRNAs have been fully characterized and have known functions and associations with regulating gene expression. These exceptions include the well-studied *XIST* gene and its role in X-inactivation. However, the precise role that alterations in *XIST* play in cancer development is unknown. Similarly many years of studying the imprinted *IGF2R/H19* locus have revealed the role these two imprinted genes play in both normal cells and how they function as putative tumour suppressors in multiple cancers. Although the lncRNA *H19* is well studied the precise manner in how it performs its' regulatory functions is still unknown. The recent demonstration that this lncRNA is a precursor for an important regulatory miRNA helps to explain some but not all of how this locus participates in cancer development. For instance, does *H19* regulate genes with its full length sense transcript, its larger antisense transcript, or through utilizing its smaller miRNA, or all three mechanisms. Finally, the detailed characterization of the *HOTAIR* locus reveals yet another important way lncRNAs may function and that is to regulate another locus in *trans* (in this case it is another group of *HOX* genes on another chromosome). Hence, lncRNAs can function to regulate a gene or genes in the immediate vicinity of where they are encoded, while others, like *HOTAIR*, function in *trans*.

There is little information on the majority of the known lncRNAs and how they could possibly function. The demonstration that a particular lncRNA has increased or decreased expression in a specific cancer does little to help understand how those lncRNAs function and how their altered expression contributes to cancer development. Instead, detailed functional characterization of each lncRNA is required to begin to uncover how that particular lncRNA functions both within normal and in cancer cells.

Invariably, we are still drawn back to the important question of how much of the human genome is actually being transcribed and how many distinct transcripts are encoded by our genome. Then it will be possible to begin to further characterize all the distinct functional groups of lncRNAs and then to determine which of these are important targets or players involved in cancer development and progression. While we currently know only a small amount of the ncRNA world it is clear that there is a significant amount of potential cellular regulation made possible by these types of transcripts. Thus, it will not be too surprising to find that a number of these regulatory transcripts, such as those listed here, play important roles in cancer development and thus make suitable targets for cancer treatment.

References

Abdel-Aziz, H.O., Takasaki, I., Tabuchi, Y., Nomoto, K., Murai, Y., Tsuneyama, K., and Takano Y. (2007). High-density oligonucleotide microarrays and functional network analysis reveal extended lung carcinogenesis pathway maps and multiple interacting genes in NNK [4-(methylnitrosamino)-1-(3-pyridyle)-1-butanone] induced CD1 mouse lung tumor. J. Cancer Res. Clin. Oncol. *133*, 107–115.

Adriaenssens, E., Dumont, L., Lottin, S., Bolle, D., Lepretre, A., Delobelle, A., Bouali, F., Dugimont, T., Coll, J., and Curgy, J.J. (1998). H19 overexpression in breast adenocarcinoma stromal cells is associated with tumor values and steroid receptor status but independent of p53 and Ki-67 expression. Am. J. Pathol. *153*, 1597–1607.

Akhtar, A. (2003). Dosage compensation: an intertwined world of RNA and chromatin remodelling. Curr. Opin. Genet. Dev. *13*, 161–169.

Ariel, I., Sughayer, M., Fellig, Y., Pizov, G., Ayesh, S., Podeh, D., Libdeh, B.A., Levy, C. Birman, T., Tykocinski, M.L., *et al.* (2000). The imprinted H19 gene is a marker of early recurrence in human bladder carcinoma. Mol. Pathol. *53*, 320–323.

Barciszewski, S.A. (2002). Beyond the proteome: non-coding regulatory RNAs. Gen. Biol. *3*, e1–8.

Bei, J.-X., Li, Y., Jia, W.-H., Feng, B.-J., Zhou, G., Chen, L.-Z., Feng, Q.-S., Low, H.-Q., Zhang, H., He, F., *et al.* (2010). A genome-wide association study of nasopharyngeal carcinoma identifies three new susceptibility loci. Nat. Genet. *42*, 599–603.

Bernard, D., Prasanth, K.V., Tripathi, V., Colasse, S., Nakamura, T., Xuan, Z., Zhang, M.Q., Sedel, F., Jourdren, L., Coulpier, F., *et al.* (2010). A long nuclear-retained non-coding RNA regulates synaptogenesis by modulating gene expression. EMBO J. *29*, 3082–3093.

Berteaux, N., Aptel, N., Cathala, G., Genton, C., Cull, J., Daccache, A., Spruyt, N., Hondermarck, H., Dugimont, T., Curgy, J.J., *et al.* (2008). A novel H19 antisense RNA overexpressed in breast cancer contributes to paternal IGF2 expression. Mol. Cell. Biol. *28*, 6731–6745.

Bilguvar, K., Yasuno, K., Niemela, M., Ruigrok, Y.M., von UndZu Fraunberg, M., van Duijn, C.M., van den Berg, L.H., Hane, S., Mason, C.E., *et al.* (2008). Susceptibility loci for intracranial aneurysm in European and Japanese populations. Nat. Genet. *40*, 1472–1477.

Brito, G.C., Fachel, A.A., Vettore, A.L., Vignal, G.M., Gimba, E.R., Campos, F.S., Barcinski, M.A., Verjovski-Almeida, S., and Reis, E.M. (2008). Identification of protein-coding and intronic noncoding RNAs down-regulated in clear cell renal carcinoma. Mol. Carcinogenesis *47*, 727–767.

Cai, X., and Cullen, B.R. (2007). The imprinted H19 noncoding RNA is a primary microRNA precursor. RNA *13*, 313–316.

Carmen Rodriguez, J.B., Court, F., Cathala, G., Forné, T., and Piette, J. (2010). CTCF is a DNA methylation-sensitive positive regulator of the INK/ARF locus. Biochem. Biophys. Res. Commun. *392*, 129–134.

Chooniedass-Kothari, S., Hamedani, M., Troup, S., Hubé, F., and Leygue, E. (2006). The steroid receptor RNA activator protein is expressed in breast tumor tissues. Int. J. Cancer *118*, 1054–1059.

Chow, J.C., Yen, Z., Ziesche, S.M., and Brown, C.J. (2005). Silencing the mammalian X chromosome. Annu. Rev. Genet. Hu. Genet. *6*, 69–92.

Chuang, C.H., and Hu, M.L. (2006). Synergistic DNA damage and lipid peroxidation in cultured human white blood cells exposed to 4-(methyl-nitrosamino)–1-(3-pyridyl)–1-butanone and ultraviolet A. Environ. Mol. Mutagen. *47*, 73–81.

Clemson, C.M., Hutchinson, J.N., Sara, S.A., Ensminger, A.W., Fox, A.H., Chess, A., and Lawrence, J.B. (2009). An architectural role for a nuclear noncoding RNA: NEAT1 RNA is essential for the structure of paraspeckles. Mol. Cell *33*, 717–726.

Consortium, E.P. (2004). The ENCODE (Encyclopedia Of DNA Elements) Project. Science *306*, 636–640.

Cooper, C., Guo, J., Yan, Y., Chooniedass-Kothari, S., Hube, F., Hamedani, M.K., Murphy, L.C., Myal, Y., and Leygue, E. (2009). Increasing the relative expression of endogenous non-coding Steroid Receptor RNA Activator (SRA) in human breast cancer cells using modified oligonucleotides. Nucleic Acids Res. *37*, 4518–4531.

Corcoran, M.M., Hammarsund, M., Zhu, C., Lerner, M., Kapanadze, B., Wilson, B., Larsson, C., Forsberg, L., Ibbotson, R.E., Einhorn, S., *et al.* (2004). DLEU2 encodes an antisense RNA for the putative bicistronic RFP2/LEU5 gene in humans and mouse. Genes Chromosomes Cancer *40*, 285–297.

Costa Brito, G., Fachel, A.A., Vettore, A.L., Vignal, G.M., Gimba, E.R., Campos, F.S., Barcinski, M.A., Verjovski-Almeida, S., and Reis, E.M. (2008). Identification of protein-coding and intronic noncoding RNAs down-regulated in clear cell renal carcinoma. Mol. Carcinogenesis *47*, 727–767.

Espinoza, C.A., G.J., Kugel JF. (2007). Characterization of the structure, function, and mechanism of B2 RNA, an ncRNA repressor of RNA polymerase II transcription. RNA *13*, 583–596.

Faghihi, M.A., Mohammad, F., Khalil, A.M., Wood, D.E., Sahagan, B.G., Morgan, T.E, Finch, C.E., St. Laurent, G., Kenny, P.J., and Wahlestedt, C. (2008). Expression of a noncoding RNA is elevated in Alzheimer's disease and drives rapid feed-forward regulation of b-secretase. Nat. Med. *14*, 723–730.

Foulds, C.E., Tsimelzon, A., Long, W., Le, A., Tsai, S.Y., Tsai, M.J., and O'Malley, B.W. (2010). Research resource: expression profiling reveals unexpected targets and functions of the human steroid receptor RNA activator (SRA) gene. Mol. Endo. *24*, 1090–1105.

Fox, A.H., Lam, Y.W., Leung, A.K., Lyon, C.E., Andersen, J., Mann, M., and Lamond, A.I. (2002). Paraspeckles: a novel nuclear domain. Curr. Biol. *12*, 13–25.

Fu, X., Ravindranath, L., Tran, N., Petrovics, G., and Srivastava, S. (2006). Regulation of apoptosis by a prostate-specific and prostate cancer-associated noncoding gene, PCGEM1. DNA Cell Biol. *25*, 135–141.

Gabory, A., Jammes, H., and Dandolo, L. (2010). The H19 locus: role of an imprinted non-coding RNA in growth and development. Bioessays *32*, 473–480.

Gallagher, E., McGoldrick, A., Chung, W.Y., McCormack, O., Harrison, M., Kerin, M., Dervan, P.A., and McCann, A. (2006). Gain of imprinting of SLC22A18 sense and antisense transcripts in human breast cancer. Genomics *88*, 12–17.

Garzon, R., Fabbri, M., Cimmino, A., Calin, G.A., and Croce, C.M. (2006). MicroRNA expression and function in cancer. Trends Mol. Med. *12*, 580–587.

Geirsson, A., Lynch, R.J., Paliwal, I., Bothwell, A.L., and Hammond, G.L. (2003). Human trophoblast noncoding RNA suppresses CIITA promoter III activity in murine B-lymphocytes. Biochem. Biophys. Res. Commun. *301*, 718–724

Geirsson, A., Bothwell, A.L., and Hammond, G.L. (2004). Inhibition of alloresponse by a human trophoblast non-coding RNA suppressing class II transactivator promoter III and major histocompatibility class II expression in murine B-lymphocytes. J. Heart Lung Transplant. *23*, 1077–1081.

Guo, F., Liu, Y., Wang, J., Li, Y., and Li, G. (2010). Inhibition of metastasis-associated lung adenocarcinoma transcript 1 in CaSki human cervical cancer cells suppresses cell proliferation and invasion. Acta Biochem. Biophys. Sin. *42*, 224–229.

Gupta, R.A., Shah, N., Wang, K.C., Kim, J., Horlings, H.M., Wong, D.J., Tsai, M.-C., Hung, T., Argani, P., Rinn, J.L., *et al.* (2010). Long non-coding RNA HOTAIR reprograms chromatin state to promote cancer metastasis. Nature *464*, 1071–1076.

Hao, Y., Crenshaw, T., Moulton, T., Newcomb, E., and Tycko, B. (1993). Tumour-suppressor activity of H19 RNA. Nature *365*, 764–767.

He, H., Nagy, R., Liyanarachchi, S., Jiao, H., Li, W., Suster, S., Kere, J., and de la Chapelle, A. (2009). A susceptibility locus for papillary thyroid carcinoma on chromosome 8q24. Cancer Res. *69*, 625–631.

Hecht, S.S., and Hoffmann, D. (1988). Tobacco-specific nitrosamines, an important group of carcinogens in tobacco and tobacco smoke. Carcinogenesis *9*, 875–884.

Hube, F., G.J., Chooniedass-Kothari, S., Cooper, C., Hamedani, M.K., Dibrov, A.A., Blanchard, A.A., Wang, X., Deng, G., Myal, Y., and Leygue, E. (2006). Alternative splicing of the first intron of the steroid receptor RNA activator (SRA) participates in the generation of coding and noncoding RNA isoforms in breast cancer cell lines. DNA Cell Biol. *25*, 418–428.

Hubé, F., Velasco, G., Rollin, J., Furling, D., and Francastel, C. (2010). Steroid receptor RNA activator protein binds to and counteracts SRA RNA-mediated activation of MyoD and muscle differentiation. Nucleic Acids Res. *39*, 513–525.

Hutchinson, J.N., Ensminger, A.W., Clemson, C.M., Lynch, C.R., Lawrence, J.B., and Chess, A. (2007). A screen for nuclear transcripts identifies two linked noncoding RNAs associated with SC35 splicing domains. BMC Genomics *8*, 39.

Jacquot, C., Carbonella, D., Tomasoni, C., Papaconstadinou, A., Roussis, V., and Roussakis, C. (2004). Identification of a novel putative non-coding RNA involved in proliferation arrest of a non-small cell lung carcinoma cell line treated with an original chemical substance, methyl-4-methoxy-3-(3-methyl-2-butanoyl) benzoate. Int. J. Oncol. *25*, 519–527.

Ji, P., Diederichs, S., Wang, W., Boing, S., Metzger, R., Schneider, P.M., Tidow, N., Brandt, B., Buerger, H. Bulk, E., *et al.* (2003). MALAT-*1*, a novel noncoding RNA, and thymosin beta4 predict metastasis and survival in early stage non-small cell lung cancer. Oncogene *22*, 8031–8041.

Jiaya Wang, Q.F. (2010). CREB up-regulates long non-coding RNA, HULC expression through interaction with microRNA-372 in liver cancer. Nucleic Acids Res. *38*, 5366–5383.

Jorquera, R., Castonguay, A., and Schuller, H.M. (1994). DNA single-strand breaks and toxicity induced by 4-(methyl-nitrosamino)-1-(3- pyridyl)-1-butanone or N-nitrosodimethylamine in hamster and rat liver. Carcinogenesis *15*, 389–394.

Khalil, A.M., Guttman, M., Huarte, M., Garber, M., Raj, A., Rivea Morales, D., Thomas, K., Presser, A., Bernstein, B.E., van Oudenaarden, A., *et al.* (2009). Many human large intergenic noncoding RNAs associate with chromatin-modifying complexes and affect gene expression. Proc. Natl. Acad. Sci. U.S.A. *106*, 11667–11672.

Kino, T., Hurt, D.E., Ichijo, T., Nader, N., and Chrousos, G.P. (2010). Noncoding RNA gas5 is a growth arrest- and starvation-associated repressor of the glucocorticoid receptor. Sci. Signal *3*, ra8.

Koshimizu, T.A., Fujiwara, Y., Sakai, N., Shibata, K., and Tsuchiya, H. (2010). Oxytocin stimulates expression of a noncoding RNA tumor marker in a human neuroblastoma cell line. Life Sci. *86*, 455–460.

Lanz, R.B., Chua, S., Barron, N., Soder, B.M., DeMayo, F., and O'Malley, B.W. (2003). Steroid Receptor RNA activator stimulates proliferation as well as apoptosis *in vivo*. Mol. Cell. Biol. *23*, 7163–7176.

Lanz, R.B., McKenna, N.J., Onate, S.A., Albrecht, U., Wong, J., Tsai, S.Y., Tsai, M.J., and O'Malley, B.W. (1999). A steroid receptor coactivator, SRA, functions as an RNA and is present in an SRC-1 complex. Cell *97*, 17–27.

Lerner, M., Harada, M., Lovén, J., Castro, J., Davis, Z., Oscier, D., Henriksson, M., Sangfelt, O., Grandér, D., and Corcoran, M.M. (2009). DLEU2, frequently deleted in malignancy, functions as a critical host gene of the cell cycle inhibitory microRNAs miR-15a and miR-16-1. Exp. Cell Res. *315*, 2941–2952.

Leygue, E., Dotzlaw, H., Watson, P.H., and Murphy, L.C. (1999). Expression of the steroid receptor RNA activator in human breast tumors. Cancer Res. *59*, 4190–4193.

Leygue E., Dotzlaw, H., Watson, P.H., and Murphy, L.C. (1999). Expression of the Steriod RNA activator in human breat tumors. Cancer Res. *59*, 4190–4193.

Lin, R., Maeda, S., Liu, C., Karin, M., and Edgington, T.S. (2007). A large noncoding RNA is a marker for murine hepatocellular carcinomas and a spectrum of human carcinomas. Oncogene *26*, 851–858.

Lonardo, F., Dragnev, K.H., Freemantle, S.J., Ma, Y., Memoli, N., Sekula, D., Knauth, E.A., Beebe, J.S., and Dmitrovsky, E. (2002). Evidence for the epidermal growth factor receptor as a target for lung cancer prevention. Clin. Cancer Res. *8*, 54–60.

Louro, R., Nakaya, H.I., Amaral, P.P., Festa, F., Sogayar, M.C., da Silva, A.M., Verjovski-Almeida, S., and Reis, E.M. (2007). Androgen responsive intronic non-coding RNAs. BMC Biol. *5*, 4.

Louro, R., Smirnova, A.S., and Verjovski-Almeida, S. (2009). Long intronic noncoding RNA transcription: expression noise or expression choice? Genomics *93*, 291–298.

Luo, J.H., Ren, B., Keryanov, S., Tseng, G.C., Rao, U.N., Monga, S.P., Strom, S., Demetris, A.J., Nalesnik, M., Yu, Y.P., *et al.* (2006). Transcriptomic and genomic analysis of human hepatocellular carcinomas and hepatoblastomas. Hepatology *44*, 1012–1024.

Lustig-Yariv, O., Schulze, E., Komitowski, D., Erdmann, V., Schneider, T., de Groot, N., and Hochberg, A. (1997). The expression of the imprinted genes H19 and IGF-2 in choriocarcinoma cell lines. Is H19 a tumor suppressor gene? Oncogene *15*, 169–177.

Malleter, M., Jacquot, C., Moreau, D., Tomasoni, C., Tsvetanova, M., Chinou, I., Juge, M., Pineau, A., Le Pape, P., and Roussakis C. (2010). A novel large regulator RNA, B2, partially overlaps the HEF1/NEDD9/Cas-L gene. Int. J. Oncol. *25*, 897–903.

Manoharan, H., Babcock, K., Willi, J., and Pitot, H.C. (2003). Biallelic expression of the H19 gene during spontaneous hepatocarcinogenesis in the albumin SV40 T antigen transgenic rat. Mol. Carcinog. *38*, 40–47.

Matouk, I.J., DeGroot, N., Mezan, S., Ayesh, S., Abu-lail, R., Hochberg, A., and Galun, E. (2007). The H19 non-coding RNA is essential for human tumor growth. PLoS One *2*, e845.

Matouk, I.J., Abbasi, I., Hochberg, A., Galun, E., Dweik, H., and Akkawi, M. (2009). Highly up-regulated in liver cancer noncoding RNA is overexpressed in hepatic colorectal metastasis. Eur. J. Gastroenterol. Hepatol. *21*, 688–692.

McPherson, R., Pertsemlidis, A., Kavaslar, N., Stewart, A., Roberts, R., Cox, D.R., Hinds, D.A., Pennacchio, L.A., Tybjaerg-Hansen, A., Folsom, A.R., *et al.* (2007). A common allele on chromosome 9 associated with coronary heart disease. Science *316*, 1488–1491.

Mertens, D., Philippen, A., Ruppel, M., Allegra, D., Bhattacharya, N., Tschuch, C., Wolf, S., Idler, I., Zenz, T., and Stilgenbauer, S. (2009). Chronic lymphocytic leukemia and 13q*14*, miRs and more. Leuk. Lymphoma *3*, 502–505.

Mitsuya, K., Meguro, M., Lee, M.P., Katoh, M., Schulz, T.C., Kugoh, H., Yoshida, M.A., Niikawa, N., Feinberg, A.P., and Oshimura, M. (1999). LIT*1*, an imprinted antisense RNA in the human KvLQT1 locus identified by screening for differentially expressed transcripts using monochromosomal hybrids. Hum. Mol. Genet. *8*, 1209–1217.

Mohammad, F., Mondal, T., Guseva, N., Pandey, G.K., and Kanduri C. (2010). Kcnq1ot1 noncoding RNA mediates transcriptional gene silencing by interacting with Dnmt1. Dev. Biol. *137*, 2493–2499.

Mohammad, F., Pandley, R.R., Nagano, T., Chakalova, L., Mondal, T., Fraser, P., and Kanduri C. (2008). Kcnq1ot1/Lit1 noncoding RNA mediates transcriptional silencing by targeting to the perinucleolar region. Mol. Cell. Biol. *28*, 3713–3728.

Mourtada-Maarabouni, M., Hedge, V.L., Kirkham, L., Farzaneh, F., and Williams, G.T. (2008). Growth arrest in human T-cells is controlled by the non-coding RNA growth-arrest-specific transcript 5 (GAS5). J. Cell. Sci. *121*, 939–946.

Mourtada-Maarabouni, M., Pickard, M.R., Hedge, V.L., Farzaneh, F., and Williams, G.T. (2009). GAS5, a non-protein-coding RNA, controls apoptosis and is down-regulated in breast cancer. Oncogene *28*, 195–208.

Mourtada-Maarabouni, M., Hasan, A.M., Farzaneh, F., and Williams, G.T. (2010). Inhibition of human T-cell proliferation by mTOR antagonists requires non-coding RNA growth-arrest-specific transcript 5 (GAS5). Mol. Pharm. *78*, 19–28.

Murakami, K., Oshimura, M., and Kugoh, H. (2007). Suggestive evidence for chromosomal localization of non-coding RNA from imprinted LIT1. J. Hum. Genet. *52*, 926–933.

Nakamura, Y., Takahashi, N., Kakegawa, E., Yoshida, K., Ito, Y., Kayano, H., Niitsu, N., Jinnai, I., and Bessho, M. (2008). The GAS5 (growth arrest-specific transcript 5) gene fuses to BCL6 as a result of t(1;3) (q25;q27) in a patient with B-cell lymphoma. Cancer Genet. Cytogenet. *182*, 144–149.

Nakano, S., Murakami, K., Meguro, M., Soejima, H., Higashimoto, K., Urano, T., Kugoh, H., Mukai, T., Ikeguchi, M., and Oshimura*1*, M. (2006). Expression profile of LIT1/KCNQ1OT1 and epigenetic status at the KvDMR1 in colorectal cancers. Cancer Sci. *97*, 1147–1154.

Panzitt, K., Tschernatsch, M., Guelly, C., Moustafa, T., Stradner, M., Strohmaier, H.M., Buck, C.R., Denk, H., Schroeder, R., Trauner, M., and Zatloukal K. (2007). Characterization of HULC, a novel gene with striking up-regulation in hepatocellular carcinoma, as noncoding RNA. Gastroenterology *132*, 330–342.

Paradowska, A., Fenic, I., Konrad, L., Sturm, K., Wagenlehner, F., Weidner, W., and Steger, K. (2009). Aberrant epigenetic modifications in the CTCF binding domain of the IGF2/H19 gene in prostate cancer compared with benign prostate hyperplasia. Int. J. Oncol. *35*, 87–96.

Pasmant, E., Laurendeau, I., Heron, D., Vidaud, M., Vidaud, D., and Bieche, I. (2007). Characterization of a germ-line deletion, including the entire INK4/ARF locus, in a melanoma-neural system tumor family: identification of ANRIL, an antisense noncoding RNA whose expression coclusters with ARF. Cancer Res. *67*, 3963–3969.

Payer, B. (2008). X chromosome dosage compensation: how mammals keep the balance. Annu. Rev. Genet. *42*, 733–772.

Rastegar, B.B.A.M. (2010). Epigenetic control of Hox genes during neurogenesis, development, and disease. Ann. Anat. *192*, 261–274.

Redrup, L., Branco, M.R., Perdeaux, E.R., Krueger, C., Lewis, A., Santos, F., Nagano, T., Cobb, B.S., Fraser, P., and Reik W. (2009). The long noncoding RNA Kcnq1ot1 organises a lineage-specific nuclear domain for epigenetic gene silencing. Dev. Biol. *136*, 525–530.

Reis, E.M., Nakaya, H.I., Louro, R., Canavez, F.C. Flatschart, A.V., Almeida, G.T., Egidio, C.M., Paquola, A.C., Machado, A.A., Festa, F., *et al.* (2004). Antisense intronic non-coding RNA levels correlate to the degree of tumor differentiation in prostate cancer. Oncogene *23*, 6684–6692.

Riccio, A., Sparago, A., Verde, G., De Crescenzo, A., Citro, V., Cubellis, M.V., Ferrero, G.B., Silengo, M.C., Russo, S., Larizza, L., *et al.* (2009). Inherited and sporadic epimutations at the IGF2-H19 locus in Beckwith–Wiedemann syndrome and Wilms' tumor. Endocr. Dev. *14*, 1–9.

Rinn, J.L., Kertesz, K., Wang, J.K., Squazzo, S.L., Xu, X., Brugmann, S.A., Goodnough, H., Helms, J., Farnham, P., Segal E., and Chang, H.Y. (2007). Functional demarcation of active and silent chromatin domains in human HOX loci by noncoding RNAs. Cell *129*, 1311–1323.

Sasaki, Y.T., Ideue, T., Sano, M., Mituyama, T., and Hirose, T. (2009). MENepsilon/beta noncoding RNAs are essential for structural integrity of nuclear paraspeckles. Proc. Natl. Acad. Sci. U.S.A. *106*, 2525–2530.

Sato, K., Nakagawa, H., Tajima, A., Yoshida, K., and Inoue, I. (2010). ANRIL is implicated in the regulation of nucleus and potential transcriptional target of E2F1. Oncol. Rep. *24*, 701–707.

Schneider, C., King, R.M., and Philipson, L. (1988). Genes specifically expressed at growth arrest of mammalian cells. Cell *54*, 787–793.

Scott, L.J., Mohlke, K.L., Bonnycastle, L.L., Willer, C.J., Li, Y., Duren, W.L., Erdos, M.R., Stringham, H.M., Chines, P.S., Jackson, A.U., *et al.* (2007). A genome-wide association study of type 2 diabetes in Finns detects multiple susceptibility variants. Science *316*, 1341–1345.

Shete, S., Hosking, F.J., Robertson, L.B., Dobbins, S.E., *et al.* (2009). Genome-wide association study identifies five susceptibility loci for glioma. Nat. Genet. *41*, 899–904.

Silva, J., Perez, D., Pritchett, J., Halling, M., Tang, H., Smith, D. (2010). Identification of long stress-induced non-coding transcripts that have altered expression in cancer. Genomics 95, 355–362.

Sirchia, S.M., Tabano, S., Monti, L., Recalcati, M.P., Gariboldi, M., Grati, F.R., Porta, G., Finelli, P., Radice, P., and Miozzo, M. (2009). Misbehaviour of XIST RNA in breast cancer cells. PLoS One *4*, e5559.

Smilinich, N.J., Day, C.D., Galina, V., Fitzpatrick, G.M., Caldwell, A.C., Lossie, P.R., Cooper, A.C., Smallwood, J.A., Schofield, P.N., Reik, W., Nicholls, R.D., *et al.*

(1999). A maternally methylated CpG island in KvLQT1 is associated with an antisense paternal transcript and loss of imprinting in Beckwith–Wiedemann syndrome. PNAS *96*, 8064–8069.

Smith, C.M., and Steitz, J.A. (1998). Classification of gas5 as a multismall-nucleolar-RNA (snoRNA) host gene and a member of the 5′-terminal oligopyrimidine gene family reveals common features of snoRNA host genes. Mol. Cell. Biol. *18*, 6897–6909.

Soshnikova, N., and Duboule, D. (2009). Epigenetic temporal control of mouse Hox genes *in vivo*. Science *324*, 1320–1323.

Souquere, S., Beauclair, G., Harper, F., Fox, A., and Pierron, G. (2010). Highly ordered spatial organization of the structural long non-coding NEAT1 RNAs within paraspeckle nuclear bodies. Mol. Biol. Cell *21*, 4020–4027.

Srikantan, V., Zou, Z., Petrovics, G., Xu, L., Augustus, M., Davis, L., Livezey, J.R., Connell, T., Sesterhenn, I.A., Yoshino, K., Buzard, G.S., Mostofi, F.K., *et al.* (2000). PCGEM*1*, a prostate-specific gene, is overexpressed inprostate cancer. PNAS *97*, 12216–12221.

Stacey, S.N., Sulem, P., Masson, G., Gudjonsson, S.A., Thorleifsson, G., Jakobsdottir, M., Sigurdsson, A., Gudbjartsson, D.F., Sigurgeirsson, B., Benediktsdottir, K.R., *et al.* (2009). New common variants affecting susceptibility to basal cell carcinoma. Nat. Genet. *41*, 909–914.

Sunwoo, H., Dinger, M.E., Wilusz, J.E., Amaral, P.P., Mattick, J.S., and Spector, D.L. (2009). MEN epsilon/beta nuclear-retained non-coding RNAs are up-regulated upon muscle differentiation and are essential components of paraspeckles. Genome Res. *19*, 347–359.

Szegedi, K., S.E., Nagy, N., Németh, I.B., Bata-Csörgo, Z., Kemény, L., Dobozy, A., and Széll, M. (2010). The anti-apoptotic protein G1P3 is overexpressed in psoriasis and regulated by the non-coding RNA, PRINS. Exp. Dermatol. *19*, 269–278.

Tano, K., Mizuno, R., Okada, T., Rakwal, R., Shibato, J., Masuo, Y., Ijiri, K., and Akimitsu, N. (2010). MALAT-1 enhances cell motility of lung adenocarcinoma cells by influencing the expression of motility-related genes. FEBS Lett. *584*, 4575–4580.

Tsang, W.P., Ng, E.K, Ng, S.M., Jin, H., Sung, J., Tak, T., and Kwok, T.T. (2010). Oncofetal H19-derived miR-675 regulates tumor suppressor RB in human colorectalcancer. Carcinogenesis *31*, 350–358.

Turnbull, C., Ahmed, S., Morrison, J., Pernet, D., Renwick, A., Maranian, M., Seal, S., Ghoussaini, M., Hines, S., Healey, C., *et al.* (2010). Genome-wide association study identifies five new breast cancer susceptibility loci. Nat. Genet. *42*, 504–507.

Verkerk, A.J., Ariel, I., Dekker, M.C., Schneider, T., van Gurp, R.J., de Groot, N., Gillis, A.J., Oosterhuis, J.W., Hochberg, A.A., and Looijenga, L.H. (1997). Unique expression patterns of H19 in human testicular cancers of different etiology. Oncogene *14*, 95–107.

Wang, F., Li, X., Xie, X., Zhao, L., and Chen, W. (2008). UCA*1*, a non-protein-coding RNA up-regulated in bladder carcinoma and embryo, influencing cell growth and promoting invasion. FEBS Lett. *582*, 1919–1927.

Wang, J., Li, X., Wu, H., Ni, P., Gu, Z., Qiao, Y., Chen, N., Sun, F., and Fan, Q. (2010). CREB up-regulates long non-coding RNA, HULC expression through interaction with microRNA-372 in liver cancer. Nucleic Acids Res. *12*, 1–18.

Wang, X.S., Zhang, Z., Wang, H.C., Cai, J.L., Xu, Q.W., Li, M.Q., Chen, Y.C., Qian, X.P., Lu, T.J., Yu, L.Z., *et al.* (2006). Rapid identification of UCA1 as a very sensitive and specific unique marker for human bladder carcinoma. Clin Cancer Res. *12*, 4851–4858.

Watanabe, M., Yanagisawa, J., Kitagawa, H., Takeyama, K., Ogawa, S., Arao, Y., Suzawa, M., Kobayashi, Y., Yano, T., Yoshikawa, H., *et al.* (2001). A subfamily of RNA-binding DEAD-box proteins acts as an estrogen receptor alpha coactivator through the N-terminal activation domain (AF-1) with an RNA coactivator, SRA. EMBO J. *20*, 1341–1352.

Wutz, R.A.A.A. (2009). ConteXt of change-X inactivation and disease. EMBO Mol. Med. *2*, 6–15.

Yamada, K., Kano, J., Tsunoda, H., Yoshikawa, H., Okubo, C., Ishiyama, T., and Noguchi, M. (2006). Phenotypic characterization of endometrial stromal sarcoma of the uterus. Cancer Sci. *97*, 106–112.

Yasuno K., Bilguvar, K., Bijlenga P., Low S.K., Krischek, B., Auburger, G., Simon, M., Krex, D., Arlier, Z., Nayak, N., Ruigrok, Y.M., *et al.* (2010). Genome–wide association study of intracranial aneurysm identifies three new risk loci. Nat. Genet. *42*, 420–425.

Yoshimizu, T., Miroglio, A., Ripoche, M.A., Gabory, A., Vernucci, M., Riccio, A., Colnot, S., Godard, C., Terris, B., Jammes, H., *et al.* (2008). The H19 locus acts *in vivo* as a tumor suppressor. Proc. Natl. Acad. Sci. U.S.A. *105*, 12417–12422.

Yu, S.L., Chen, H.Y., Yang, P.C., and Chen, J.J. (2007). Unique MicroRNA signature and clinical outcome of cancers. DNA Cell Biol. *26*, 283–292.

Antisense RNAs and Modulation of Tumour-suppressor Genes

10

Hengmi Cui, Isabelle Cui and Xi Yang

Abstract

Antisense RNA is the first RNA molecule identified to function as a regulator in the cell. With advances in biological science, antisense RNA is now recognized to be involved in not only post-transcriptional regulation, but also the transcriptional regulation of various important genes including tumour-suppressor genes (TSGs). Recent studies indicate that the modulation of TSGs by antisense RNA may be through either up- or down-regulation, occurring either transcriptionally, post-transcriptionally or both, dependent on features of sense and antisense. Antisense RNA is the main contributor of TSG epigenetic silencing in tumorigenesis. While a general picture of the pathways involved in antisense RNA mediated gene regulation has emerged, many questions remain unaddressed: Why does some antisense RNAs function as down-regulators but others as up-regulators in different TSGs? Are the molecular mechanisms of antisense RNA regulation highly gene- and species-specific? Are they spatially and temporally restricted? Is there an antisense RNA-induced silencing complex (ARISC)? What factor(s) cause aberrant expression of antisense RNA in tumour cells? There is the potential for using antisense RNA as a biomarker for the early diagnosis of tumours and developing new therapeutic strategies for tumour treatment by targeting and controlling antisense RNA.

Antisense RNAs

Antisense RNA is a type of endogenous single-stranded transcript complementary to an annotated transcript. Thus, it is also called a natural antisense transcript (NAT). While categorized as long coding or non-coding RNAs, distinct from short non-coding RNAs like miRNA, siRNA and piRNA in size, antisense RNAs function as emerging regulatory molecules in biological processes within cells. Transcription of antisense RNA is a common phenomenon in the mammalian transcriptome (Katayama *et al.*, 2005), although it's biological significance has not been completely elucidated yet. Antisense RNA regulation was the first identified process mediated by RNA molecules, preceding the discovery of RNA interference (RNAi), one of the most significant findings of modern biology. Interestingly, RNAi was discovered accidentally when scientists were researching the function of oligo antisense RNA (Fire *et al.*, 1998). With advance in biological science, we are increasingly recognizing the regulatory function of antisense RNA, extending from the repression of mRNA translation to transcriptional or epigenetic regulation of important genes in eukaryotic cells. Although the mechanism whereby antisense RNA regulates gene expression has not been completely revealed yet, we infer that antisense RNA-directed regulation should be multilayered, acting at both the transcriptional and post-transcriptional levels, and occurring in both the cytoplasm and nucleus to more efficiently control cellular processes. Antisense RNA regulation may result in the natural selection and subsequent evolution and might play an important role in development and differentiation like other ncRNAs (Mercer *et al.*, 2009). Meanwhile, aberrant expression of antisense RNA may contribute to human diseases.

To comprehend the function and mechanisms of antisense RNA, it is necessary to know the history behind antisense RNA research, as summarized by Cui and Cui (2010) (see Table 10.1). The phenomenon of antisense RNA repression was first uncovered in 1963 by Singer *et al.* (1963) in an *in vitro* experiment in which antisense RNAs shut off protein synthesis in cell-free extracts. This result was confirmed by two other groups more than 10 years later, suggesting that the translation of mRNAs could be blocked by complementary RNAs (Paterson *et al.*, 1977; Hastie and Held, 1978). The *in vivo* repressive function of natural antisense RNA was first reported in plasmid ColEI of bacteria in the early 1980s (Lacatena and Cesareni, 1981; Tomizawa *et al.*, 1981). Following this discovery, other systems in which antisense RNA inhibits gene expression and regulates plasmid replication, transposition, and conjugation, were characterized in prokaryotes (Simons and Kleckner, 1988). Izant and Weintraub were the first to demonstrate the function of antisense gene constructs in eukaryotic cells; they modified a vector by excising the HSV thymidine kinase gene (*TK*) and reinserting it back into the vector in reverse orientation. When co-injecting the modified and unmodified vectors with wild-type *TK* into *TK*- mouse L cells, they found that *TK* activity dropped 4-fold in the modified vectors (Izant and Weintraub, 1984). A year later, through the introduction of antisense plasmid constructs into eukaryotic cells via transfection or microinjection, these same investigators found that antisense transcripts complementary to as few as 52 nucleotides of 5′ region of untranslated genes in target mRNA could suppress gene activity, further confirming that suppression by antisense RNA is conserved in eukaryotic cells (Izant and Weintraub, 1985). Meanwhile, two other research groups also demonstrated that RNA antisense repression occurs in higher organisms. Rosenberg *et al.* (1985) applied a SP6 vector *in vitro* transcription system to produce a *Kr* antisense RNA. The *Kr* antisense RNA was then injected into wild-type *Drosophila* embryos, leading to inactivation of the Kr gene. These *Drosophila* embryos developed into phenocopies of *Kr* mutant embryos. Crowley *et al.* (1985) used an antisense construct of the Discoidin gene and transfected it into *Dictyostelium*. They found that transformants exhibited more than a 90% reduction in Discoidin mRNA and protein leading to phenocopies of these mutants. Subsequently, numerous natural antisense RNAs were identified

Table 10.1 History of antisense RNA research

Discovery	Model	References
Antisense RNA shut off protein synthesis	Cell-free extracts	Singer *et al.* (1963), Paterson *et al.* (1977), Hastie *et al.* (1978)
In vivo repression of natural antisense RNA in plasmid ColEI	Bacteria	Lacatena *et al.* (1981),Tomizawa *et al.* (1981)
Antisense RNA repression function in eukaryotic cells	Mouse	Izant *et al.* (1984)
Identification of natural antisense RNA in eukaryotic cells	Xenopus	Kimelman *et al.* (1989), Volk *et al.* (1989)
Prediction of sense–antisense pairs by computational tool	Human	Yelin (2003)
Deletion-induced antisense transcription leads to epigenetic silencing and disease	Human	Tufarelli *et al.* (2003)
Genome-wide Identification of endogenous antisense RNAs	Human	Rosok *et al.* (2004)
Epigenetic silencing of tumour-suppressor genes by antisense RNAs	Human	Yu *et al.* (2008), Morris *et al.* (2008)
Global identification of antisense RNA by ASSAGE	Human	He *et al.* (2008)
Genetic deletion gives rise to antisense transcription and epigenetic silencing of tumour-suppressor gene *MSH2*	Human	Ligtenberg *et al.* (2009)

in eukaryotic systems. Two research groups identified naturally occurring endogenous antisense RNA of the bFGF gene in *Xenopus*; this antisense transcript is thought to be involved in regulating the stability of the *bFGF* mRNA (Kimelman and Kirschner, 1989; Volk *et al.*, 1989). Kumar and Carmichael (1997) discovered that nuclear antisense RNA induces adenosine deamination, nuclear retention of transcripts, and transcript degradation. Nowadays, many naturally occurring antisense RNAs with repressive functions have been described in a variety of organisms (Werner and Berdal, 2005; Beiter *et al.*, 2009). In 2003, the Higgs group showed that a deletion in *LUC7L* results in juxtaposition to a neighbouring HBA2 gene, producing antisense RNA that leads to the epigenetic silencing of *HBA2* in patients with α-thalassemia. This was the first reported case of a genetic alteration giving rise to epigenetic silencing that causes human disease (Tufarelli *et al.*, 2003). With advancements in biotechnology, antisense transcription on a genome-wide level has been addressed in a number of animal and plant species, including humans (Yelin *et al.*, 2003; Chen *et al.*, 2004; Rosok and Sioud, 2004), mice (Katayama *et al.*, 2005), rice (Osato *et al.*, 2003; Zhou *et al.*, 2009), and *Arabidopsis thaliana* (Wang *et al.*, 2005). Powerful computational tools were utilized to predict sense–antisense transcript pairs based on known public genomic databases. In the mouse genome, 2431 (20.1%) sense–antisense pairs were predicted (Wang *et al.*, 2005), whereas in the human genome, 2667 and 2940 sense–antisense pairs were reported by the Chen and Yelin groups, respectively. These two groups applied different criteria to scrutinize the orientation of cDNAs. The latter required the transcripts to be spliced and have a polyadenylation signal, while the former chose mRNAs with an open reading frame, a poly(A) tail and/or polyadenylation signals (Yelin *et al.*, 2003; Chen *et al.*, 2004). Computational approaches predicted that over 20% of sense–antisense transcript pairs exist in both mouse and human genomes. However, the number of antisense transcripts predicted is largely underestimated since they only represent processed cytoplasm mRNAs while neglecting unprocessed or nuclear RNAs. The first systematic identification of genome-wide sense–antisense transcripts by molecular biological approaches was performed in 2004. Rosok and Sioud developed an experimental strategy that allowed genome-wide identification of endogenous mRNA with long sequences complementary to other transcripts in normal human mammary epithelial and breast cancer cells (Rosok and Sioud, 2004). He *et al.* developed a technique called asymmetrical strand-specific analysis of gene expression (ASSAGE) that identified sense and antisense transcripts from expressed genes at a global level in five different human cell types (He *et al.*, 2008). It had been previously thought that gene regulation by antisense RNA is a post-transcriptional phenomenon and that antisense RNA functions only as a repressor of mRNA translation. However, a recent study showed that antisense RNA might silence sense genes at the transcriptional level, affecting DNA methylation and histone modifications, and leading to epigenetic silencing of the tumour-suppressor gene *p15* (Yu *et al.*, 2008). It was also suggested that the *p15* antisense RNA might be a trigger initiating the epigenetic silencing of *p15* during tumorigenesis (Yu *et al.*, 2008). This result was later confirmed by Morris *et al.*, who found that the tumour-suppressor gene *p21* is epigenetically down-regulated by its antisense transcript, suggesting that the phenomenon may be the result of an imbalance in bidirectional transcriptional levels (Morris *et al.*, 2008). In general, as the first type of regulatory RNA identified, antisense RNA possesses a vital regulatory role not only in the repression of mRNA translation but also in the regulation of transcription, giving rise to epigenetic changes in gene expression in cells. In this chapter, we will focus on reviewing the regulation of tumour-suppressor genes by antisense RNAs.

Down-regulation of tumour-suppressor genes by antisense RNA

Tumour-suppressor genes (TSGs) encode a group of proteins that guard the fidelity of the genome via multiple mechanisms including, but not limited to, the activation of downstream pathways that inhibit cell proliferation. Loss of one or more of these TSGs contributes to carcinogenesis through a variety of pathways affecting critical cellular processes such as mitosis, cell adhesion, apoptosis, DNA replication and DNA repair. (Levine, 1990; Hollingsworth and Lee, 1991; Marshall, 1991; Weinberg, 1991). It has been widely recognized that TSGs are frequently silenced epigenetically in tumour cells without the presence of genetic mutations (Jones and Baylin, 2002; Feinberg and Tycko, 2004). Although DNA hypermethylation is often associated with the silencing of tumour-suppressor genes, the mechanisms whereby epigenetic silencing is initiated during tumorigenesis were not elucidated until *p15* antisense was reported to be involved in epigenetic silencing.

p15 antisense (*p15AS*)

A group at the Johns Hopkins University was the first to address how the epigenetic silencing of a tumour-suppressor gene is initiated in cancer cells (Yu *et al.*, 2008). The p15 gene, encoding a cyclin-dependent kinase inhibitor, is a well documented tumour-suppressor gene that is frequently genetically deleted or epigenetically silenced in a wide variety of tumours including leukaemia, melanomas, gliomas, and lung cancers. Up to 60% of leukaemias show epigenetic silencing of *p15* (Lubbert, 2003). Researchers at Johns Hopkins identified the *p15* antisense RNA responsible for the epigenetic silencing of *p15* (Yu *et al.*, 2008) (Fig. 10.1). They found an inverse relationship between *p15* antisense and sense expression with a higher expression of *p15* antisense and lower expression of *p15* sense in leukaemia samples compared with normal lymphocytes using real-time RT-PCR (Fig. 10.2). The hypothesis that *p15AS* plays a mechanistic role in the silencing of *p15* in tumour cells was confirmed by transfecting engineered reporter constructs into cells. Fluorescence-activated cell sorting (FACS) analysis showed that cells stably transfected with pP15-AS have significantly decreased percentages of GFP-positive cells (12%), whereas cells transfected with pP15 and pP15-ASΔ have much higher GFP expression (51% and 59%), suggesting that *p15AS* transcript induces partner sense gene silencing. Real-time RT-PCR further showed that the expressed *p15AS* transcript down-regulates *p15* both exogenously and endogenously (Yu *et al.*, 2008). To determine whether antisense-induced silencing acts post-transcriptionally or transcriptionally, Chromatin-immunoprecipitation (ChIP) experiments were performed. Results show a marked increase in the dimethylation of H3K9 and a decrease in the dimethylation of H3K4 in cells with *p15AS* transcripts. This finding suggests that antisense RNA may trigger heterochromatin formation, leading to the transcriptional silencing of sense *p15* although the possibility of a post-transcriptional regulatory mechanism cannot be completely excluded. Altered chromatin modifications affect both the exogenous and endogenous *p15* promoters, suggesting that antisense RNA has both *cis*- and *trans*-acting functions in heterochromatin formation. The silenced *p15* promoter could be reactivated with methylation and heterochromatin inhibitors, 5-aza-deoxytidine and trichostatin, further confirming that antisense RNA-induced silencing is an epigenetic or transcriptionally down-regulated process rather than only a post-transcriptional event. Interestingly, *p15AS* in mouse embryonic stem cells was observed to cause exogenous *p15* silencing and increased growth through heterochromatin formation rather than DNA methylation alteration.

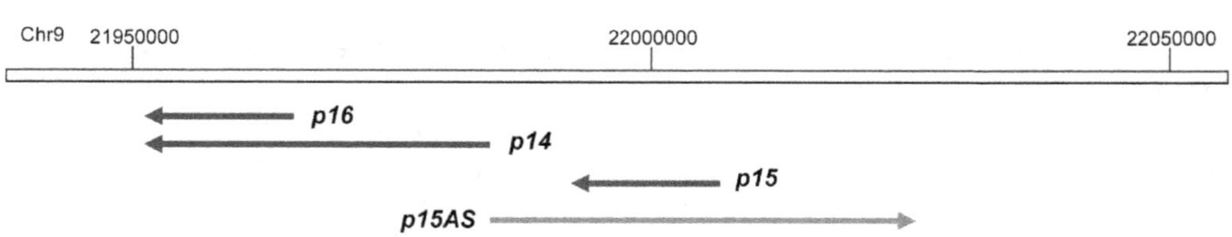

Figure 10.1 Identification of *P15* antisense RNA. The *p15* antisense transcript was identified by Rapid amplification of 5′ and 3′ cDNA ends (RACE) locating between chr9:21984790–22019563 (Yu *et al.*, 2008).

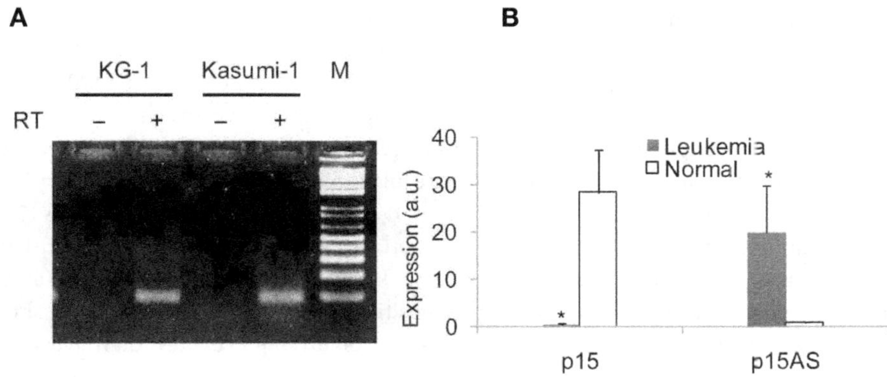

Figure 10.2 *p15* antisense expression in leukaemia cells. (A) Native *p15AS* transcripts were detected in leukaemia cell lines. *P15AS* (97 bp) bands were amplified from the cDNA reverse-transcribed from RNA by a strand-specific primer from leukaemia cell lines KG-1 and Kasumi-1. (B) *p15* and *p15AS* expression in leukaemic and normal lymphocytes, analysed by real-time RT-PCR. Leukaemia samples showed higher expression of *p15* antisense and lower expression of *p15* than normal lymphocytes. Both *p15* and its antisense expression levels were normalized with *GAPDH* (Yu *et al.*, 2008).

After one week of embryonic stem cell differentiation, DNA hypermethylation was surprisingly observed at the exogenous *p15* promoter region, suggesting that heterochromatin formation rather than DNA methylation is a prerequisite for epigenetic silencing. Together, the data suggest that even transient 'doses' of antisense RNA can induce long-lasting epigenetic silencing of tumour-suppressor genes. Antisense RNA, therefore, can be used as a potential diagnostic marker for cancer risk evaluation as well as a possible target for developing anti-tumour therapies.

p21 antisense

p21 is another important tumour-suppressor gene, encoding a potent cyclin-dependent kinase inhibitor 1A, also known as CDKN1A. This tumour suppressor binds to and inhibits the activity of cyclin–CDK2 or –CDK4 complexes, thus functioning as a regulator of cell cycle progression at G1 and a promoter of cell cycle arrest in response to stimuli. It is well positioned to function as both a sensor and effector of multiple anti-proliferative signals (Abbas and Dutta, 2009). Like many tumour-suppressor genes, *p21* has been observed to be epigenetically silenced in many tumour types (Fang and Lu, 2002). Recently, this tumour-suppressor gene has been experimentally shown to exhibit bidirectional transcription and its antisense RNA has been identified (Morris *et al.*, 2008). Suppression of *p21* antisense transcript with siRNA correlates with a distinct rise in the activity in the *p21* promoter, suggesting that silencing of the *p21* antisense transcript is sufficient for modulating a rise in *p21* transcription levels dependent of AGO2 (Morris *et al.*, 2008). The researchers also found that suppression of *p21* antisense transcription with siRNA results in a loss of the suppressive H3K27me3 epigenetic mark at the *p21* promoter, while suppression of *p21* sense expression with siRNA leads to the recruitment of AGO1 and H3K27me3 to the *p21* promoter. This finding suggests that bidirectional transcription of the sense and antisense genes is an endogenous gene regulatory mechanism whereby antisense RNA directs epigenetic regulatory complexes to a sense promoter. A model was further proposed for bidirectional gene regulation in human cells that exert a controlled equilibrium of endogenous *p21* expression where, at steady state, *p21* contains equal levels of both sense and antisense transcripts. Epigenetic silencing of *p21*, thus, may be the result of an imbalance in bidirectional transcription levels, which allows the unchecked antisense RNA to direct silent state epigenetic marks to the sense promoter, resulting in stable transcriptional gene silencing (Morris *et al.*, 2008). When depletion in *p21* antisense occurs, there is a reduction in the low-level antisense directed H3K27me3 suppressive mark and an increase in *p21* sense expression, implying that H3K27me3 is increased via the *p21* antisense transcript. Transcriptional activation of *p21* expression is the result of AGO2-dependent, post-transcriptional silencing of *p21* antisense transcript, functioning in AGO1-medicated transcriptional control of *p21* expression. Conversely, when *p21* sense expression is suppressed, regulatory pressure is exerted on the *p21* antisense transcript to allow for *p21* antisense directed AGO1 and H3K27me3 enrichment at the *p21* promoter. The evidence suggests that the balance between *p21* sense and antisense bidirectional transcription is sustained by the action of *p21* antisense RNA. Thus, this proposed molecular pathway, in which an imbalance in bidirectional transcription causes dominant antisense RNA expression, offers a plausible explanation for the epigenetic silencing of tumour-suppressor genes in human cancer cells (Fig. 10.3). These data not only support the notion that antisense RNA gives rise to epigenetic silencing of tumour-suppressor genes like *p15*, but also show that antisense RNAs of bidirectionally transcribed genes are operative in directing epigenetic marks to sense gene promoters.

Up-regulation of tumour-suppressor genes by antisense RNAs

While it was found that antisense RNAs are involved in the epigenetic down-regulation of TSGs, it was also reported that antisense RNAs can up-regulate the expression of certain TSGs through distinct mechanisms.

p53 antisense (*Wrap53*)

The p53 tumour suppressor, a transcription factor that triggers apoptosis in response to oncogene activation as well as other stress stimuli, plays a key role in the prevention of human cancer. Thus, *p53* mutations occur frequently in human tumours (Levine *et al.*, 1991). Recent research has revealed that *p53* antisense gene, also defined as 'WD Repeat containing Antisense to P53' (*Wrap53*), plays an important role in the regulation of *p53* mRNA and the induction of p53 protein by targeting the 5′ untranslated region of *p53* mRNA (Mahmoudi *et al.*, 2009). This new finding provides novel insight into *p53* regulation as well as mechanisms of antisense mediated gene regulation in human cells.

Wrap53 is located on chromosome 17p13 and overlaps the first exon of *p53* in an antisense fashion. *Wrap53* and *p53* are co-expressed and show a positive correlation in cells. Research shows that reduction in *Wrap53* expression triggers the decay of *p53* transcripts. Both siRNA knockdown of *Wrap53* and hindrance of the *Wrap53*/*p53* hybrid formation, using 2′-O-methyl oligonucleotides that bind to the complementary region, lower *p53* mRNA abundance and suppress *p53* induction upon DNA damage, suggesting *Wrap53*'s role in the regulation of *p53* via *Wrap53*/*p53* RNA interaction. On the other hand, overexpression of *Wrap53* increases *p53* mRNA and protein levels and sensitizes cells to p53-dependent apoptosis suggesting that *Wrap53*/*p53* RNA–RNA interaction is required to maintain normal cellular levels of *p53*. From these facts, a plausible mechanism can be presented, in which interaction between *Wrap53* and *p53* transcripts via their complementary region masks certain target sequences in *p53* mRNA, protecting it from degradation. Either removal of *Wrap53* RNA or blockage of *Wrap53*/*p53* interaction leads to *p53* degradation by exposing *p53* mRNA to regulatory factors (Mahmoudi *et al.*, 2009). This finding not only provides new insight into *p53* regulation but also proposes a novel mechanism for antisense-mediated gene regulation in human cells (Farnebo, 2009). Dysfunction of *Wrap53* may be a potent cause of cancer. Furthermore, the up- and down-regulations of *Wrap53* pose avenues of research in developing future therapeutic strategies for battling cancer.

WT1 antisense

The Wilms' tumour-suppressor gene (*WT1*), first discovered in 1990 (Call *et al.*, 1990; Gessler *et al.*, 1990; Haber *et al.*, 1990; Rose *et al.*, 1990), is mutated in about 10% of all Wilms' tumours (Rose *et al.*, 1990). In 1994, Campbell *et al.* reported the finding of an antisense mRNA overlapping the first exon of *WT1*, arising from a promoter within

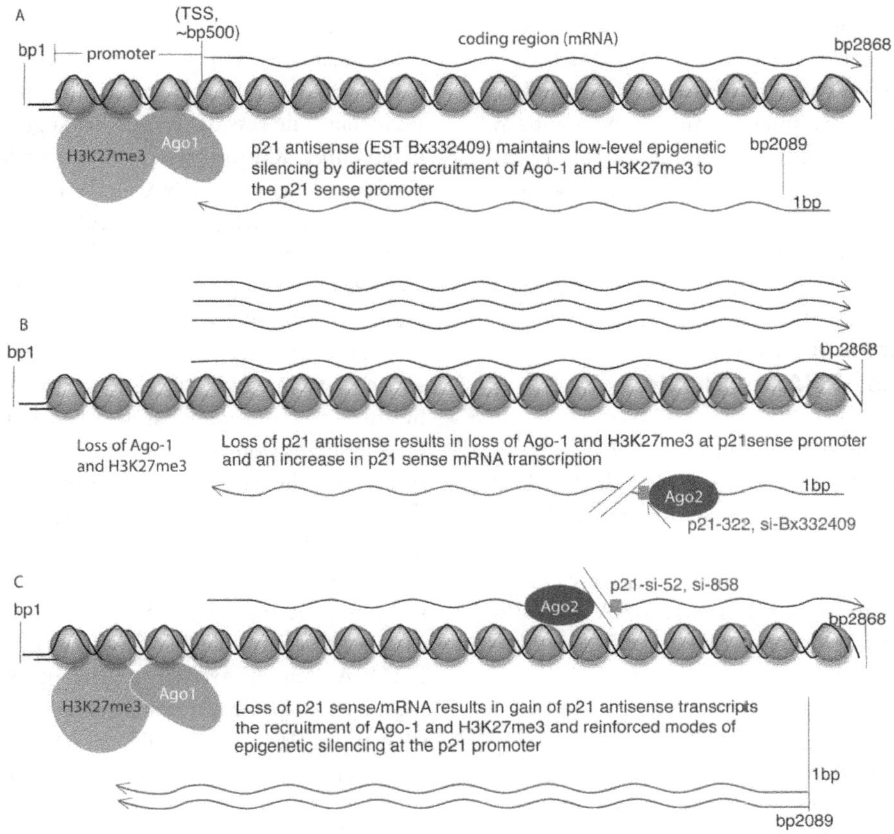

Figure 10.3 Model for bidirectional RNA-mediated transcriptional regulation of gene expression in human cells. (A) Bidirectional transcription is presumed to exert a controlled equilibrium of endogenous *p21* gene expression, whereby the antisense transcript directs AGO1 dependent low-level epigenetic regulation, H3K27me3, at the *p21* sense/mRNA promoter. (B) Upon suppression of the antisense transcript the directed H3K27me3 is diminished, allowing for enhanced *p21* sense/mRNA transcriptional gene activation. (C) When *p21* sense/mRNA expression is suppressed, a release of regulatory pressure is exerted on the *p21* antisense transcript allowing for *p21* antisense directed AGO1 and H3K27me3 enrichment at the *p21* sense/mRNA promoter. AGO1 has been observed to be required for the early stages of RNA directed transcriptional gene silencing, thus the early recruitment of AGO1 to the *p21* sense promoter may function as a re-enforcing loop to exert stable epigenetic silencing on p21 gene expression (Morris *et al.*, 2008).

intron 1 of *WT1* and is up-regulated by *WT1* itself (Campbell *et al.*, 1994). Antisense RNAs are generally linked to the down-regulation of their sense partners. However, Moorwood *et al.* (1998) presented evidence that, in the case of *WT1*, the antisense RNA transcript enhanced expression of *WT1*. In contrast to the expectation that low levels of antisense correlate with high levels of sense expression and vice versa, the researchers were surprised to observe that levels of expression were high for both sense and antisense transcripts localized at the same sites. The authors also showed that in a cultured cell line (T5A1), expression of *WT1* increased following the introduction of a construct expressing antisense RNA. They concluded that antisense RNA might induce *WT1* transcription, without the antisense transcript directly affecting the sense promoter, transcript stability, or translation rate. The authors observed that *WT1* expression was highest in cells with lower expression of the introduced antisense RNA and attributed the inverse correlation as owing to the antisense having a positive effect on one of the processes when expressed at lower levels and a negative effect when expressed at higher levels (Jinno *et al.*, 1994; Mitsuya *et al.*, 1997). They concluded that the antisense of *WT1*, functionally dissimilar to most known antisense transcripts, is perhaps a vestige of the imprinting mechanism that has attained a new role of sustaining appropriate expression levels of *WT1* (Moorwood *et al.*, 1998).

Mechanisms by which antisense RNAs modulate genes

Although numerous antisense RNAs and their regulatory function have been identified, the exact mechanisms whereby antisense RNAs exert their role in gene regulation have not yet been revealed. Here, we briefly summarize studies on the mechanisms of antisense RNA regulation, including transcriptional and posttranscriptional regulation.

Antisense RNA and DNA methylation

DNA methylation is an important epigenetic modification that affects gene regulation. Generally, DNA hypermethylation in the promoter leads to the down-regulation or silencing of a gene, whereas DNA hypomethylation gives rise to the activation of a gene, especially in tumour-suppressor genes (Jones and Baylin, 2002; Feinberg and Tycko, 2004). Since antisense RNA is known to have a repressive effect, it is potentially associated with the regulatory mechanisms of DNA methylation. As previously described, Yu *et al.* found that pP15–AS constructs induce *p15* silencing through heterochromatin formation. Bisulphite sequencing analysis in tumour cell lines failed to find any changes in DNA methylation of the *p15* promoter. However, when they transfected a pP15–AS construct into a mouse ES cell line, they surprisingly observed DNA hypermethylation in the *p15* promoter after differentiation of the embryonic stem cells, indicating that antisense RNA does indeed induce changes in DNA methylation (Yu *et al.*, 2008). But why could not any change in the DNA methylation of the *p15* promoter region be observed in cancer cells and un-differentiated stem cells? One possible explanation is that the DNA methylation mechanism was mutated in malignant cells and had not yet been established in un-differentiated stem cells. The researchers further suggested that heterochromatin formation is a precondition for gene silencing while DNA methylation remains a secondary effect (Yu *et al.*, 2008). This study provides convincing evidence that DNA hypermethylation of a tumour-suppressor gene can be induced by its antisense counterpart.

Further evidence of the relationship between antisense RNA and DNA methylation is provided by studies from Tufarelli *et al.* They showed that DNA methylation and silencing of *HBA2* correlate with the presence of antisense transcripts derived from truncated forms of the neighbouring gene *LUC7L* located on the opposite strand. A deletion removes the final three exons of *LUC7L*, including the transcription stop poly(A) signal, causing transcription of *LUC7L* to be carried into the HBA2 gene. The resulting aberrant copy of the *HBA2* antisense leads to hypermethylation and silencing of *HBA2* (Tufarelli *et al.*, 2003). Tufarelli *et al.* later turned their attention to antisense RNA, showing that, in both transgenic and embryonic stem cell models, transcription of antisense RNAs mediates the silencing and methylation of their associated CpG islands. A recent study by Ligtenberg *et al.* described a similar phenomenon. They reported that patients with Lynch syndrome carry germline deletions of the last exon of *TACSTD1*, a gene immediately upstream of MSH2, commonly linked to non-polyposis colorectal cancer. Owing to these deletions, transcription of *TACSTD1* extends into *MSH2*, generating antisense RNA against *MSH2* that initiates methylation and inactivation of the gene (Ligtenberg *et al.*, 2009).

Experiments involving siRNA pull-down have also produced indications of antisense RNA-induced DNA methylation. Using siRNA pull-down assays, Weinberg *et al.* (2006) discovered that in crude DNA extracts the antisense strand of *EF52* can bind to Flag-tagged mouse Dnmt3a, a *de novo* DNA methyltransferase. Dnmt3a had been reported to bind to double-stranded siRNAs (ds-siRNAs) (Jeffery and Nakielny, 2004). This finding suggests that antisense RNA may direct siRNA-mediated transcriptional gene silencing (TGS) through interactions with a complex containing Dnmt3a. The authors then tested whether the antisense transcript may function alone to direct TGS. By designing plasmids targeting the U3 region of the HIV-1 LTR/promoter, containing the promoter necessary to express antisense, sense, or combined dsRNA, they demonstrated that the antisense produces TGS in cells at levels comparable to that of dsRNA transfected samples. These results suggest a mechanism of action in which the antisense strand of siRNA is transported in the nucleus to a complex containing Dnmt3a. The Dnmt3a complex is guided by RNA homology to the promoter and direct histone methylation as the primary silencing event, leading to increased DNA methylation over time (Weinberg *et al.*, 2006). Together, these studies suggest that antisense RNA may silence sense gene via the DNA methylation mechanism.

Antisense RNA and histone modifications

Heterochromatin represents a transcriptional suppressive state of genes while heterochromatin formation is a result of histone modifications, including increased methylation in H3K9, H3K27, and decreased methylation in H3K4. A study observing siRNA-mediated transcriptional silencing and histone 3 lysine 9(H3K9) methylation in the absence of DNA methylation suggested that chromatin modifications may play a more important role in TSG than previously thought (Ting *et al.*, 2005). Natural antisense RNA may be a trigger for heterochromatin formation in TSG silencing in tumorigenesis. The work of Yu *et al.* showed, via chromatin immunoprecipitation (ChIP) analysis, that *p15AS* expression construct induced *p15* silencing through heterochromatin formation. In cells containing antisense transcripts, the *p15* promoter showed a marked increase in the dimethylation of H3K9 and a decrease in the dimethylation of H3K4 in cells. However, bisulphite sequencing analysis failed to show any changes in DNA methylation in the *p15* promoter, suggesting that heterochromatin is a prerequisite for gene silencing while DNA methylation remain a secondary effect (Yu *et al.*, 2008). Weinberg *et al.* also observed that H3K9 and H3K27 methylation resulting from antisense siRNA treatment corresponds with a transcriptionally silent chromatin state. As result, they suggested that antisense strands of siRNA alone can direct sequence-specific transcriptional gene silencing in human cells (Weinberg *et al.*, 2006). When Morris *et al.* studied the effects of bidirectional transcription on gene suppression, they found that suppression of *p21* sense expression by antisense transcripts causes the suppressive histone mark H3K27me3 to appear at the *p21* promoter (Morris *et al.*, 2008).

Although there has not yet been a study done in mammalian cells on the relationship between histone deacetylation and antisense-induced silencing, a detailed study in yeast reported that accumulation and stabilization of antisense RNA lead to targeted histone deacetylation and the silencing of sense RNA (Camblong *et al.*, 2007). A loss of function in Rrp6, an exonuclease component of the nuclear exosome, results in the stabilization of two *PHO84* antisense transcripts, leading to the repression of the PHO84 gene. Furthermore, knockdown of *PHO84* antisense production prevents *PHO84* repression even in the absence of Rrp6. Chromatin immunoprecipitation experiments showed that recruitment of Hda1 histone deacetylase exclusively to *PHO84* chromatin corresponds with the loss of Rrp6 function and the stabilization of *PHO84* antisense RNA, suggesting that Hda1 activity is dependent on antisense RNA. While the activators of Hda1 recruitment have not been identified, a plausible mechanism has been proposed in which *PHO84* antisense RNA triggers the initial binding of Hda1 to the PHO84 gene region by forming structures specifically recognized by histone-modifying enzymes, thus permitting its successive spreading to neighbouring regions. Additionally, Camblong *et al.* (2007) showed that H3K18 deacetylation is confined to *PHO84* antisense RNA. Based on these results, they proposed a new mechanism, by which the increased stabilization of antisense RNA combines with H3K18 deacetylation to affect sense gene silencing. Clearly, this proposed mechanism is not limited to yeast but is probably an alternative RNA-mediated TGS mechanism conserved in higher eukaryotes.

Antisense RNAs and polycomb-group (PcG) proteins

PcG proteins, first discovered in fruit flies, are a family of proteins able to remodel chromatin so that transcription factors cannot bind to promoter sequences, thus repressing gene expression (Schuettengruber *et al.*, 2007). Evidence of PcG protein-mediated antisense RNA silencing mainly arises from studies of the *Kcnq1* imprinted cluster. Recently, new light has been shed on the regulation of the *Kcnq1* imprinted cluster, offering insight into possible mechanisms of antisense RNA-induced silencing through PcG. Antisense *Kcnq1ot1* is an essential component of Kcnq1 gene imprinting, as it has been shown that its deletion results in the derepression of paternal imprints (Fitzpatrick *et al.*, 2007). However, precisely how *Kcnq1ot1* mediates silencing of the *Kcnq1* cluster remained unclear. To answer this question, Pandey *et al.* studied the lineage-specific role of *Kcnq1ot1* in chromatin regulation. They showed that *Kcnq1ot1* interacts with chromatin on levels several folds higher in placenta than in embryonic liver. ChIP-chip studies showed that Polycomb group proteins Ezh2 and G9a interact with *Kcnq1ot1* in the placenta, thus suggesting that *Kcnq1ot1* antisense RNA target and direct histone methyltransferase (HMTs) complexes to the *Kcnq1* cluster (Pandey *et al.*, 2008). Antisense ncRNA is an important element in the initiation of silencing and, along with various events, establish the repression of sense RNA and the nucleation of a repressive compartment.

Other studies on mechanisms of antisense RNA regulation

Morris *et al.* showed that both short and long antisense ncRNAs mediate control of transcription and require the presence of Argonaute 1 (AGO1) and Argonaute 2 (AGO2) (Morris *et al.*, 2008). Recently, Morris proposed a mechanism whereby antisense ncRNAs regulate transcription. Long and short antisense ncRNAs expressed at bidirectionally transcribed genes may fold into secondary structured ncRNAs that can interact with specific sites on the promoter of the sense strand, affecting the recruitment of AGO1, DNMT3a and HDAC1 to this target site. Thus, both types of antisense RNA modulate transcriptional silencing through epigenetic remodelling of the target RNA loci (Morris, 2009).

Yu *et al.* suggested that antisense-induced silencing is Dicer-independent. When they transfected a pP15–AS construct into Dicer-knockout mouse ES cells, the silencing induced by antisense was unaffected (Yu *et al.*, 2008). However, this effect remains to be confirmed in other genes and cell lines. Furthermore, tetracycline inducible vectors and *loxP/Cre* deletion experiments showed that stable silencing of *p15* by *p15AS* persisted even after *p15AS* expression was eliminated. The fact that *p15* promoter activity was not reactivated after tetracycline withdrawal or removal of the CMV *p15AS* promoter indicates that antisense RNA may function in the initiation of epigenetic silencing rather than its maintenance. Thus, transient bursts in antisense gene expression may induce long-lasting heritable effects. These results suggest that antisense-induced silencing may be characterized by a 'hit and run' mechanism (Yu *et al.*, 2008).

Future perspective

The mechanisms of antisense RNA-mediated TSG regulation are multilayered. In other words, modulation of TSGs by antisense RNA may be either up- or down-regulated; the regulation can occur transcriptionally, post-transcriptionally or both, depending on the nature of the sense- and antisense interactions. Although a general picture of the pathways involved in antisense RNA-induced gene regulation has emerged, many questions remain to be addressed. For instance, when regulating different TSGs, why do some antisense RNAs function as down-regulators while others act as up-regulators? Are the molecular mechanisms of antisense RNA regulation highly gene and species specific? Are they spatially and temporally restricted and, if so, how are these constraints controlled? Is there an antisense RNA-induced silencing complex (ARISC), similar to the RNA-induced initiation of transcriptional gene silencing (RITS) and RNA-induced silencing complex (RISC) found in RNAi? And should ARISC exist, it is unclear which components would be involved in the complex. The relationship between natural antisense RNA and siRNA has yet to be revealed, bringing speculation to whether sense–antisense RNA pairs are precursors of siRNA. Finally, we need to explore the possibility of using antisense RNA as a biomarker for the early diagnosis of tumour and developing new therapeutic strategies for tumour treatment by targeting and controlling these transcripts.

References

Abbas, T., and Dutta, A. (2009). p21 in cancer: intricate networks and multiple activities. Nat. Rev. *9*, 400–414.

Beiter, T., Reich, E., Williams, R.W., and Simon, P. (2009). Antisense transcription: a critical look in both directions. Cell. Mol. Life Sci. *66*, 94–112.

Call, K.M., Glaser, T., Ito, C.Y., Buckler, A.J., Pelletier, J., Haber, D.A., Rose, E.A., Kral, A., Yeger, H., Lewis, W.H., *et al.* (1990). Isolation and characterization of a zinc finger polypeptide gene at the human chromosome 11 Wilms' tumor locus. Cell *60*, 509–520.

Camblong, J., Iglesias, N., Fickentscher, C., Dieppois, G., and Stutz, F. (2007). Antisense RNA stabilization induces transcriptional gene silencing via histone deacetylation in *S. cerevisiae*. Cell *131*, 706–717.

Campbell, C.E., Huang, A., Gurney, A.L., Kessler, P.M., Hewitt, J.A., and Williams, B.R. (1994). Antisense transcripts and protein binding motifs within the Wilms tumour (WT1) locus. Oncogene *9*, 583–595.

Chen, J., Sun, M., Kent, W.J., Huang, X., Xie, H., Wang, W., Zhou, G., Shi, R.Z., and Rowley, J.D. (2004). Over 20% of human transcripts might form sense–antisense pairs. Nucleic Acids Res. *32*, 4812–4820.

Crowley, T.E., Nellen, W., Gomer, R.H., and Firtel, R.A. (1985). Phenocopy of discoidin I-minus mutants by antisense transformation in Dictyostelium. Cell *43*, 633–641.

Cui, I., and Cui, H. (2010). Antisense RNAs and epigenetic regulation. Epigenomics *2*, 139–150.

Fang, J.Y., and Lu, Y.Y. (2002). Effects of histone acetylation and DNA methylation on p21(WAF1) regulation. World J. Gastroenterol. *8*, 400–405.

Farnebo, M. (2009). Wrap53, a novel regulator of p53. Cell Cycle *8*, 2343–2346.

Feinberg, A.P., and Tycko, B. (2004). The history of cancer epigenetics. Nat. Rev. *4*, 143–153.

Fire, A., Xu, S., Montgomery, M.K., Kostas, S.A., Driver, S.E., and Mello, C.C. (1998). Potent and specific genetic interference by double-stranded RNA in *Caenorhabditis elegans*. Nature *391*, 806–811.

Fitzpatrick, G.V., Pugacheva, E.M., Shin, J.Y., Abdullaev, Z., Yang, Y., Khatod, K., Lobanenkov, V.V., and Higgins, M.J. (2007). Allele-specific binding of CTCF to the multipartite imprinting control region KvDMR1. Mol. Cell. Biol. *27*, 2636–2647.

Gessler, M., Poustka, A., Cavenee, W., Neve, R.L., Orkin, S.H., and Bruns, G.A. (1990). Homozygous deletion in Wilms tumours of a zinc-finger gene identified by chromosome jumping. Nature *343*, 774–778.

Haber, D.A., Buckler, A.J., Glaser, T., Call, K.M., Pelletier, J., Sohn, R.L., Douglass, E.C., and Housman, D.E. (1990). An internal deletion within an 11p13 zinc finger gene contributes to the development of Wilms' tumor. Cell *61*, 1257–1269.

Hastie, N.D., and Held, W.A. (1978). Analysis of mRNA populations by cDNA.mRNA hybrid-mediated inhibition of cell-free protein synthesis. Proc. Natl. Acad. Sci. U.S.A. *75*, 1217–1221.

He, Y., Vogelstein, B., Velculescu, V.E., Papadopoulos, N., and Kinzler, K.W. (2008). The antisense transcriptomes of human cells. Science *322*, 1855–1857.

Hollingsworth, R.E., and Lee, W.H. (1991). Tumor suppressor genes: new prospects for cancer research. J. Natl. Cancer Inst. *83*, 91–96.

Izant, J.G., and Weintraub, H. (1984). Inhibition of thymidine kinase gene expression by anti-sense RNA: a molecular approach to genetic analysis. Cell *36*, 1007–1015.

Izant, J.G., and Weintraub, H. (1985). Constitutive and conditional suppression of exogenous and endogenous genes by anti-sense RNA. Science *229*, 345–352.

Jeffery, L., and Nakielny, S. (2004). Components of the DNA methylation system of chromatin control are RNA-binding proteins. J. Biol. Chem. *279*, 49479–49487.

Jinno, Y., Yun, K., Nishiwaki, K., Kubota, T., Ogawa, O., Reeve, A.E., and Niikawa, N. (1994). Mosaic and polymorphic imprinting of the WT1 gene in humans. Nat. Gen. *6*, 305–309.

Jones, P.A., and Baylin, S.B. (2002). The fundamental role of epigenetic events in cancer. Nat. Rev. Genet. *3*, 415–428.

Katayama, S., Tomaru, Y., Kasukawa, T., Waki, K., Nakanishi, M., Nakamura, M., Nishida, H., Yap, C.C., Suzuki, M., Kawai, J., *et al.* (2005). Antisense transcription in the mammalian transcriptome. Science *309*, 1564–1566.

Kimelman, D., and Kirschner, M.W. (1989). An antisense mRNA directs the covalent modification of the transcript encoding fibroblast growth factor in Xenopus oocytes. Cell *59*, 687–696.

Kumar, M., and Carmichael, G.G. (1997). Nuclear antisense RNA induces extensive adenosine modifications and nuclear retention of target transcripts. Proc. Natl. Acad. Sci. U.S.A. *94*, 3542–3547.

Lacatena, R.M., and Cesareni, G. (1981). Base pairing of RNA I with its complementary sequence in the primer precursor inhibits ColE1 replication. Nature *294*, 623–626.

Levine, A.J. (1990). Tumor suppressor genes. Bioessays *12*, 60–66.

Levine, A.J., Momand, J., and Finlay, C.A. (1991). The p53 tumour-suppressor gene. Nature *351*, 453–456.

Ligtenberg, M.J., Kuiper, R.P., Chan, T.L., Goossens, M., Hebeda, K.M., Voorendt, M., Lee, T.Y., Bodmer, D., Hoenselaar, E., Hendriks-Cornelissen, S.J., *et al.* (2009). Heritable somatic methylation and inactivation of MSH2 in families with Lynch syndrome due to deletion of the 3′ exons of TACSTD1. Nat. Gen. *41*, 112–117.

Lubbert, M. (2003). Gene silencing of the p15/INK4B cell-cycle inhibitor by hypermethylation: an early or later epigenetic alteration in myelodysplastic syndromes? Leukemia *17*, 1762–1764.

Mahmoudi, S., Henriksson, S., Corcoran, M., Mendez-Vidal, C., Wiman, K.G., and Farnebo, M. (2009). Wrap53, a natural p53 antisense transcript required for p53 induction upon DNA damage. Mol. Cell *33*, 462–471.

Marshall, C.J. (1991). Tumor suppressor genes. Cell *64*, 313–326.

Mercer, T.R., Dinger, M.E., and Mattick, J.S. (2009). Long non-coding RNAs: insights into functions. Nat. Rev. Genet. *10*, 155–159.

Mitsuya, K., Sui, H., Meguro, M., Kugoh, H., Jinno, Y., Niikawa, N., and Oshimura, M. (1997). Paternal expression of WT1 in human fibroblasts and lymphocytes. Hum. Mol. Genet. *6*, 2243–2246.

Moorwood, K., Charles, A.K., Salpekar, A., Wallace, J.I., Brown, K.W., and Malik, K. (1998). Antisense WT1 transcription parallels sense mRNA and protein expression in fetal kidney and can elevate protein levels *in vitro*. J. Pathol. *185*, 352–359.

Morris, K.V. (2009). Long antisense non-coding RNAs function to direct epigenetic complexes that regulate transcription in human cells. Epigenetics *4*, 296–301.

Morris, K.V., Santoso, S., Turner, A.M., Pastori, C., and Hawkins, P.G. (2008). Bidirectional transcription directs both transcriptional gene activation and suppression in human cells. PLoS Genet. *4*, e1000258.

Osato, N., Yamada, H., Satoh, K., Ooka, H., Yamamoto, M., Suzuki, K., Kawai, J., Carninci, P., Ohtomo, Y., Murakami, K., *et al.* (2003). Antisense transcripts with rice full-length cDNAs. Genome Biol. *5*, R5.

Pandey, R.R., Mondal, T., Mohammad, F., Enroth, S., Redrup, L., Komorowski, J., Nagano, T., Mancini-Dinardo, D., and Kanduri, C. (2008). Kcnq1ot1 antisense noncoding RNA mediates lineage-specific transcriptional silencing through chromatin-level regulation. Mol. Cell *32*, 232–246.

Paterson, B.M., Roberts, B.E., and Kuff, E.L. (1977). Structural gene identification and mapping by DNA–mRNA hybrid-arrested cell-free translation. Proc. Natl. Acad. Sci. U.S.A. *74*, 4370–4374.

Rose, E.A., Glaser, T., Jones, C., Smith, C.L., Lewis, W.H., Call, K.M., Minden, M., Champagne, E., Bonetta, L., Yeger, H., *et al.* (1990). Complete physical map of the WAGR region of 11p13 localizes a candidate Wilms' tumor gene. Cell *60*, 495–508.

Rosenberg, U.B., Preiss, A., Seifert, E., Jackle, H., and Knipple, D.C. (1985). Production of phenocopies by Kruppel antisense RNA injection into Drosophila embryos. Nature *313*, 703–706.

Rosok, O., and Sioud, M. (2004). Systematic identification of sense–antisense transcripts in mammalian cells. Nature Biotechnol. *22*, 104–108.

Schuettengruber, B., Chourrout, D., Vervoort, M., Leblanc, B., and Cavalli, G. (2007). Genome regulation by polycomb and trithorax proteins. Cell *128*, 735–745.

Simons, R.W., and Kleckner, N. (1988). Biological regulation by antisense RNA in prokaryotes. Ann. Rev. Genet. *22*, 567–600.

Singer, M.F., Jones, O.W., and Nirenberg, M.W. (1963). The effect of secondary structure of the template activity of polyribonucleotides. Proc. Natl. Acad. Sci. U.S.A. *49*, 392–399.

Ting, A.H., Schuebel, K.E., Herman, J.G., and Baylin, S.B. (2005). Short double-stranded RNA induces transcriptional gene silencing in human cancer cells in the absence of DNA methylation. Nat. Gen. *37*, 906–910.

Tomizawa, J., Itoh, T., Selzer, G., and Som, T. (1981). Inhibition of ColE1 RNA primer formation by a plasmid-specified small RNA. Proc. Natl. Acad. Sci. U.S.A. *78*, 1421–1425.

Tufarelli, C., Stanley, J.A., Garrick, D., Sharpe, J.A., Ayyub, H., Wood, W.G., and Higgs, D.R. (2003). Transcription of antisense RNA leading to gene silencing and methylation as a novel cause of human genetic disease. Nat. Gen. *34*, 157–165.

Volk, R., Koster, M., Poting, A., Hartmann, L., and Knochel, W. (1989). An antisense transcript from the Xenopus laevis bFGF gene coding for an evolutionarily conserved 24 kd protein. EMBO J. *8*, 2983–2988.

Wang, X.J., Gaasterland, T., and Chua, N.H. (2005). Genome-wide prediction and identification of cis-natural antisense transcripts in Arabidopsis thaliana. Genome Biol. *6*, R30.

Weinberg, M.S., Villeneuve, L.M., Ehsani, A., Amarzguioui, M., Aagaard, L., Chen, Z.X., Riggs, A.D., Rossi, J.J., and Morris, K.V. (2006). The antisense strand of small interfering RNAs directs histone methylation and transcriptional gene silencing in human cells. RNA *12*, 256–262.

Weinberg, R.A. (1991). Tumor suppressor genes. Science *254*, 1138–1146.

Werner, A., and Berdal, A. (2005). Natural antisense transcripts: sound or silence? Physiol. Gen. *23*, 125–131.

Yelin, R., Dahary, D., Sorek, R., Levanon, E.Y., Goldstein, O., Shoshan, A., Diber, A., Biton, S., Tamir, Y., Khosravi, R., *et al.* (2003). Widespread occurrence of antisense transcription in the human genome. Nat. Biotechnol. *21*, 379–386.

Yu, W., Gius, D., Onyango, P., Muldoon-Jacobs, K., Karp, J., Feinberg, A.P., and Cui, H. (2008). Epigenetic silencing of tumour-suppressor gene p15 by its antisense RNA. Nature *451*, 202–206.

Zhou, X., Sunkar, R., Jin, H., Zhu, J.K., and Zhang, W. (2009). Genome-wide identification and analysis of small RNAs originated from natural antisense transcripts in Oryza sativa. Genome Res. *19*, 70–78.

ncRNAs in p53 Regulation

11

Salah Mahmoudi, Anna Vilborg and Marianne Farnebo

Abstract

The p53 tumour suppressor plays a key role in tumour prevention through its function as a sensor of cellular stress. Once activated, p53 triggers apoptosis or cell-cycle arrest allowing removal or repair of damaged cells. Reduced p53 activity leads to tumour susceptibility, while excessive p53 activity causes tissue degeneration. Hence, p53 levels and activity are subjected to tight control by a complex network of regulation at multiple levels. In this chapter we focus on the critical control of p53 mRNA and protein by non-coding RNAs (ncRNAs). WRAP53α is a natural antisense RNA that interferes with p53 mRNA and protects it from degradation. Loss of WRAP53α reduces p53 levels and abrogates p53 response to DNA damage. The mRNA of p53 is also targeted by a number of miRNAs that negatively regulates its stability Once p53 is translated into a protein, it is regulated by the ncRNA MEG3 that increases p53 protein levels and transcriptional activity. In addition, p53 activates ncRNA through its function as a transcriptional activator. p53 binds the promoters and controls the expression of miRNAs and long intergenic non-coding RNAs (lincRNAs) such as lincRNA-p21, which in turn mediates part of the p53 stress response. p53 can also interact with Drosha and thereby regulate miRNA biogenesis at a post-transcriptional level.

Introduction

The p53 tumour suppressor is arguably the most important player in preventing tumour formation and progression. p53 mutations are found with high frequency (about 50%) in human tumours, however virtually all tumours have inactivated the p53 pathway in some fashion (Soussi and Wiman, 2007) (www.iarc.fr/p53; p53.free.fr). The p53 protein functions as a transcription factor with a crucial role in orchestrating the cellular stress response. p53 can be activated by a large variety of stress factors, and reacts by triggering an appropriate response – the nature of which will vary with the triggering factor and the cellular background. The most classical outcomes of p53 activation are cell cycle arrest and apoptosis, and it is generally believed that these two, together with senescence – irreversible cell cycle arrest – are the p53 responses with greatest impact on preventing tumour formation (Vousden and Prives, 2009). However, it has lately become evident that p53 is involved in many other processes in addition to these. p53 can promote several forms of DNA repair (Helton and Chen, 2007), thus contributing to maintaining genomic integrity. Further, p53 may promote differentiation of stem- and progenitor cells into more specialized cell types, and can also prevent self-renewal of stem cells (Helton and Chen, 2007; Cicalese *et al.*, 2009; Zhao and Xu, 2010). Paradoxically, p53 can also induce genes involved in promoting survival. One of these genes is p21 – while being a classical cell cycle arrest-inducing p53 target, p21 antagonizes apoptosis. The list of pro-survival p53 targets also includes several antioxidant genes and genes involved in metabolism. The logic behind this surprising p53 function may be that the elimination of every cell ever exposed to any kind of stress is not desirable – while protecting us from getting cancer; it would lead to tissue degeneration (Janicke *et al.*, 2008; Vousden and Prives, 2009). In addition to its crucial role in cancer, p53 has been implicated in several other diseases, generally diseases connected to excessive cell death. These include diabetes, cell death after ischaemia, and various neurodegenerative diseases such as Huntington, Parkinson, and Alzheimer (Vousden and Prives, 2009; Vousden and Ryan, 2009). Since p53 is able to eliminate cells through apoptosis and senescence and to induce differentiation, thus reducing stem cell populations, p53 has also been suggested to promote ageing. Initial studies in mice expressing constitutively active p53 also seemed to confirm this hypothesis (Tyner *et al.*, 2002), However, subsequent mice models carrying an extra copy of p53 but under control of its normal regulatory elements demonstrated that properly controlled p53 did not induce ageing. Instead it actually promoted longevity, largely by preventing tumorigenesis (Garcia-Cao *et al.*, 2002; Matheu *et al.*, 2008). Owing to its critical function in deciding on life or death for the cell, strict regulation of p53 levels and activity is crucial. p53 protein levels are kept under tight control by multiple mechanisms (Xu, 2003; Braithwaite *et al.*, 2006). Further, p53 mRNA stability and translation is subject to regulation by a number of factors, including long non-coding RNA (Vilborg *et al.*, 2010). In this chapter, we will give an introduction to the role of non-coding RNA in the regulation of p53 function, with the focus on WRAP53α, an antisense regulator of p53.

ncRNA in the regulation of p53 mRNA

Overview of regulation of p53 mRNA

The p53 mRNA is post-transcriptionally regulated by a number of different mechanisms affecting translation or stability of the p53 mRNA. The regulation occurs at all parts of the p53 mRNA, i. e. the 5′ untranslated region (UTR), the coding region and the 3′UTR. Regulation via 5′UTR: (1) The p53 5′UTR is GC-rich and forms a stem-loop that hampers ribosome access during cap-dependent translation initiation, thus reducing the translation efficiency of p53 mRNA. This can be overcome by cap-independent translation initiation using an internal ribosomal entry site (IRES) located within the p53 mRNA (Grover *et al.*, 2009). Most probably cap-dependent and -independent translation of p53 occurs in parallel. A number of proteins bind to the 5′UTR of p53 and regulates the p53 mRNA translation (Takagi *et al.*, 2005). (2) Destabilizing elements within the p53 5′UTR affecting the stead-state level of p53 mRNA have been found in chicken and mouse p53 (Kim *et al.*, 2001). In addition, the p53 antisense transcript WRAP53α interacts with the p53 5′UTR and protects it from degradation (further discussed below) (Mahmoudi *et al.*, 2009). Regulation via the coding region: The coding sequence of p53 mRNA is subjected to regulation by Mdm2. The Mdm2 protein interacts with this region of p53 mRNA, leading to enhanced p53 translation and a block of Mdm2's capacity in degrading the p53 protein (Candeias *et al.*, 2008). This is somewhat surprising considering the well-established role of Mdm2 as a negative regulator of p53 protein, and emphasizes the complexity of the regulatory networks controlling p53 levels and activity. Regulation via 3′UTR: The 3′UTR is in general the hotspot-region for post-translational regulation and p53 contain many important sites for regulation. (1) p53 is targeted by a number of miRNAs (Le *et al.*, 2009; Zhang *et al.*, 2009; Li *et al.*, 2010; Tian *et al.*, 2010; Hu *et al.*, 2010). (2) the p53 3′UTR contains two AU-rich elements recognized by a number of RNA-binding proteins affecting mRNA degradation and translation efficiency (Fu *et al.*, 1999; Mazan-Mamczarz *et al.*, 2003; Galban *et al.*, 2003; Schumacher *et al.*, 2005; Zou *et al.*, 2006; Vilborg *et al.*, 2009). (3) the p53 3′UTR can fold back on itself and form a loop by base-pairing with a region in the 5′UTR. Formation of this loop enhances translation (Chen and Kastan, 2010). In Fig. 11.1A, we have summarized the factors regulating the p53 mRNA.

WRAP53α, an antisense regulator of p53 mRNA

Natural antisense transcripts (NATs) are a group of regulatory RNAs with sequence complementarity to other cellular RNAs referred to as sense RNAs. NATs have the ability to modulate expression of their sense RNA and several modes

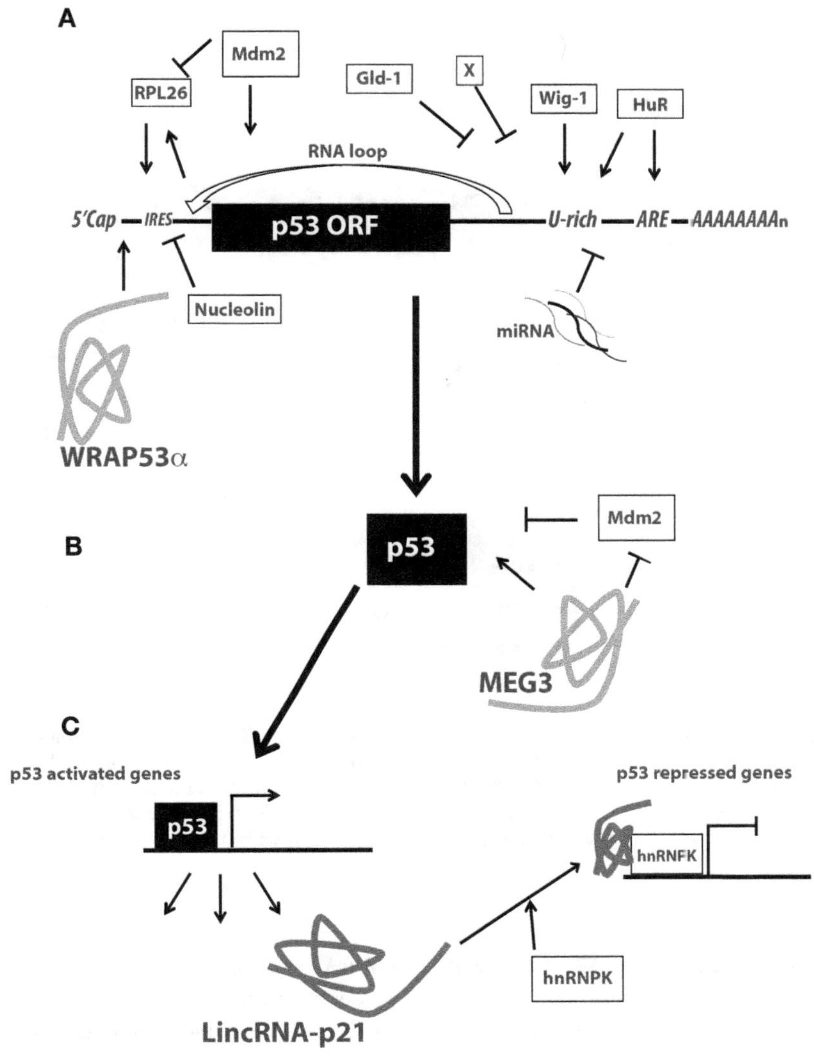

Figure 11.1 (A) Factors regulating mRNA stability and translation efficiency. X denotes a p53-regulating protein of unknown identity. (B) Regulation of p53 protein levels and transcriptional activity by the long ncRNA MEG3. (C) p53 can in turn activate the transcription of the long non-coding lincRNA-p21, which together with hnRNPK mediates transcriptional silencing of p53 transrepressed promoters.

of action have been proposed, including transcriptional modulation (caused by epigenetic alterations or steric hindrance of RNA polymerases), RNA masking and double-stranded RNA (dsRNA) mechanisms(Lapidot and Pilpel, 2006; Faghihi and Wahlestedt, 2009).

This class of RNAs includes both protein-coding and non-coding RNAs. They are either grouped as *trans*-NATs, which are transcribed from different location than their sense RNAs, or as *cis*-NATs, transcribed from the same genomic locus as the sense but from the opposite strand. Data from large-scale transcriptome projects indicate that overlapping genes are a common phenomenon in the human genome; in fact, more than 20% of all genes have antisense partners (Yelin *et al.*, 2003; Chen *et al.*, 2004; Osato *et al.*, 2007).

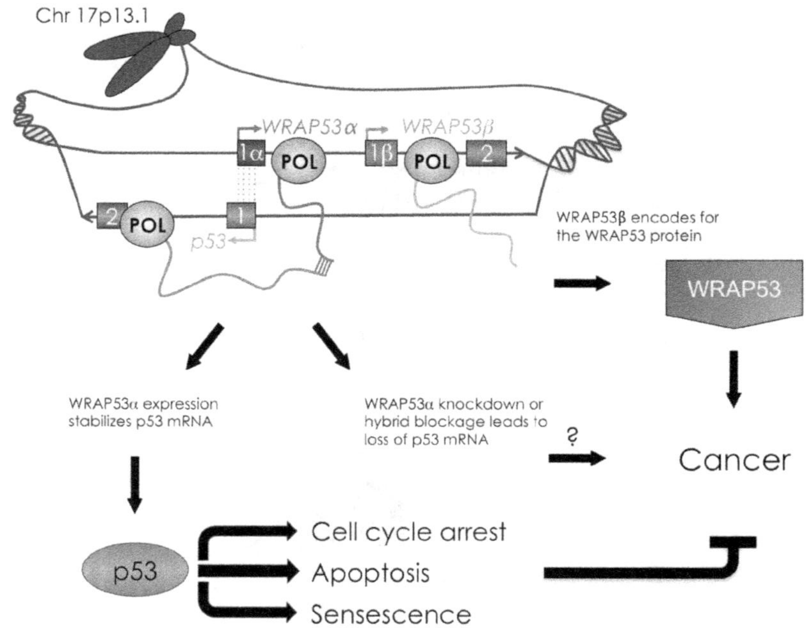

Figure 11.2 Model illustrating the *WRAP53* gene locus. WRAP53α transcripts originating from exon 1α overlaps p53 exon 1 and can via the complementarity region protect p53 mRNA from degradation. Knockdown of WRAP53α or blockage of WRAP53α/p53 hybrid formation reduces p53 levels. Conversely, overexpression of WRAP53α increases p53 levels and sensitizes cells to p53-induced apoptosis. Transcripts mainly originating from exon 1β gives rise to the WRAP53 protein with oncogenic abilities.

WRAP53α is a *cis*-NAT of p53 and regulates the action of p53 by stabilizing the p53 mRNA (Mahmoudi *et al.*, 2009). The *WRAP53* gene is located on chromosome 17p13 and overlaps the p53 gene on the opposite strand in a head-to-head fashion (Fig. 11.2). *WRAP53* has three alternative start exons (1α, 1β and 1γ) and exon 1α directly overlaps the first exon of *p53*. *WRAP53* transcripts containing exon 1α (referred to as WRAP53α throughout the text) interacts with p53 transcripts via their complementary region, resulting in increased stead-state levels of p53 mRNA. Knockdown of WRAP53α or inhibition of WRAP53α/p53 RNA interaction leads to rapid loss of p53 mRNA and subsequently p53 protein. Conversely, overexpression of exon 1α of WRAP53 leads to increased levels of p53 mRNA and protein, altogether suggesting that WRAP53α protects p53 mRNA from degradation (Mahmoudi *et al.*, 2009). Exactly how WRAP53α regulate p53 stability is not clear. Considering the perfect complementarity of the RNA sequences and the requirement of RNA interaction between WRAP53α and p53, it is conceivable that WRAP53α hybridizes with p53 to form RNA duplexes. This type of interaction could mask sequences involved in destabilizing p53 mRNA or enhance binding of stabilizing factors, thus protecting p53 from degradation. Upon binding WRAP53α could sterically block other factors from interfering with p53 mRNA. Alternatively, WRAP53α could change the conformation of p53 5′UTR to expose or mask binding sequences involved in regulating p53 mRNA stability. Existence of destabilizing elements within the p53 5′UTR have been reported in chicken and mouse (Kim *et al.*, 2001). Several functional microRNA binding sites have been identified in the 3′UTR of the p53 mRNA (Le *et al.*, 2009; Zhang *et al.*, 2009; Li *et al.*, 2010; Tian *et al.*, 2010; Hu *et al.*,

2010), but to date none have been described in the 5′UTR. However, it was recently shown that the 3′UTR of p53 mRNA can fold back on itself and interact with the 5′UTR (Chen and Kastan, 2010), which raises the possibility that WRAP53α could in fact affect regulatory regions in the p53 3′UTR. Future investigations will reveal which mechanism(s) is (are) the predominant one for WRAP53α regulation of p53 mRNA stability.

Data suggest that the mRNAs of WRAP53α and p53 interact directly. However, no such RNA-duplexes have been detected, possibly due to a transient and labile interaction. It is feasible that WRAP53α by a transient binding changes the secondary structure of the p53 mRNA, leading to a more stable conformation. Additionally, a short-lived RNA interaction may prevent activation of the cellular interferon-mediated pathway, which is triggered by any cytoplasmic double-stranded RNA more than 30 base pairs long and results in shutdown of protein synthesis, culminating in apoptosis (Wang and Carmichael, 2004). This pathway is generally triggered in responses to double-stranded RNA viruses (Wang and Carmichael, 2004), and stable endogenous sense/antisense RNA duplexes may cause its inappropriate activation. It is also plausible that the interaction between WRAP53α and p53 does not rely on base pairing, but is rather accomplished by the formation of secondary and tertiary structures. It is feasible that WRAP53α by a transient interaction changes the secondary structure of the p53 mRNA, leading to a more stable conformation.

Intriguingly, p53 was found expressed at approximately 100-fold higher levels than WRAP53α (Mahmoudi *et al.*, 2009). NATs are often found to be less abundant than their sense transcripts, and for this reason epigenetic alteration has been suggested as an attractive model of action for these NATs (Faghihi and Wahlestedt, 2009). However, WRAP53α seems to act on a post-transcriptional level to stabilize p53 mRNA. This is based on the following observations: (1) the regulation occurs both in the nucleus and in the cytoplasm, (2) luciferase reporter constructs under the control of the p53 promoter cDNA did not show any changes in luciferase activity upon altered levels of WRAP53α, (3) treatment with the histone deacetylase inhibitor, Trichostatin A (TSA), in WRAP53α depleted cells did not rescue the drop p53 levels (Mahmoudi and Farnebo *et al.*, unpublished), (4) silencing of WRAP53α only affects mature p53 mRNA and not its pre-mRNA, (5) reporter constructs (under the control of CMV promoter) expressing p53 exon 1 fused to luciferase cDNA showed an almost complete shutdown of luciferase activity in WRAP53α depleted cells and (6) using single-stranded 2′-*O*-methyl RNA oligos, targeting either p53 exon 1 or WRAP53 exon 1α, reduced p53 levels with no effect in WRAP53 expression (Mahmoudi *et al.*, 2009). These results support post-transcriptional regulation mediated via the 5′UTR of the p53 mRNA as the mode of action underlying WRAP53α-mediated regulation of p53.

The notion of a short-lived interaction between WRAP53α and p53 is further supported by the fact that a single WRAP53α transcript seems to be able to regulate 100 p53 transcripts through RNA–RNA interaction. A plausible model for the WRAP53α mediated regulation of p53 could entail a transient interaction between the two RNA molecules, resulting in a permanent modification of p53 mRNA and allowing WRAP53α to move on the next p53 mRNA molecule. Alternatively, the discrepancy in expression levels could be due to a labile WRAP53α transcript or the production of endogenous siRNAs derived from p53/WRAP53α RNA duplexes. In addition, it cannot be excluded that the reported low WRAP53α level is due to suboptimal detection of this mRNA. Nevertheless, silencing of WRAP53α using multiple siRNA oligos targeting different regions of the transcript leads to decreased mRNA and protein levels of p53, demonstrating that the less abundant WRAP53α transcript is capable of protecting p53 mRNA in excess. The finding that WRAP53α is required for both basal levels of p53 as well as p53 induction upon stress raises the possibility that loss/mutation of WRAP53α could contribute to cancer by reducing p53 levels and thereby leading to loss of p53 function (illustrated in Fig. 11.2). Thus, WRAP53α might act as a tumour suppressor; however, this assumption remains to be experimentally proven.

Apart from coding for the p53 antisense RNA,

the *WRAP53* gene also encodes a protein. This protein is mainly produced from transcripts initiated from exon 1β and these transcripts do not have the ability to regulate p53 since they lack exon 1α. However, some transcripts (arising from three different splice variants) contain the full-length open reading frame for the protein in addition to exon 1α (Mahmoudi *et al.*, 2009). It remains to be demonstrated if it is the protein-coding or non-coding variants (or both) of the WRAP53α transcripts that regulates p53. The WRAP53 protein (also denoted WDR79 and TCAB1) belongs to the WD40 family of proteins, and has recently been identified as a major player in the formation and function of specific nuclear structures called Cajal bodies (Venteicher *et al.*, 2009; Tycowski *et al.*, 2009; Mahmoudi *et al.*, 2010; Nizami *et al.*, 2010). The WRAP53 protein has RNA-binding capacity, raising the possibility that it could be involved in the regulation of p53 mRNA (Tycowski *et al.*, 2009). However, several findings clearly point against the involvement of the protein in the regulation of p53 (Mahmoudi *et al.*, 2009). Interestingly, the WRAP53 protein has been assigned oncogenic properties (Mahmoudi *et al.*, 2011), creating a very particular situation in which the *WRAP53* gene can act as either an oncogene and/or a tumour suppressor depending on the transcriptional start site. The concept of a protein acting both as an oncogene and a tumour suppressor depending on the context of its expression is not unheard of (Pierce *et al.*, 1999; Lamouille and Derynck, 2009). However, the fact that *WRAP53* may act as a tumour suppressor at the RNA level and as an oncogene at the protein level is exceptional (illustrated in Fig. 11.2).

Although some questions remain to be answered regarding the exact mechanism behind WRAP53α mediated regulation of p53 mRNA, the biological significance of this regulation is clear: WRAP53α is crucial for the p53 response to DNA damage. Depletion of WRAP53α or blockage of WRAP53α/p53 RNA hybrids prevents induction of p53 protein levels as well as transactivation of p53 target genes after treatment with DNA damaging agents. Supporting a key role for WRAP53α in stabilizing the p53 mRNA in response to DNA damage, both p53 and WRAP53α transcripts are induced by DNA damaging agents. Also, overexpression of WRAP53α sensitized cells for p53-induced apoptosis. Overall, these data clearly shows the importance of WRAP53α-mediated regulation of p53 expression both in unstressed cells and in response to DNA damage and further suggests a significant function for antisense transcripts in human physiology and disease.

ncRNA in the regulation of p53 protein levels and activity

Human maternally expressed gene 3 (MEG3) is an mRNA-like RNA of approximately 1.6 kb that act as a ncRNA (Schuster-Gossler *et al.*, 1998; Miyoshi *et al.*, 2000). MEG3 is suggested to be a tumour suppressor, since its expression is absent in many human cancer cell lines (Zhang *et al.*, 2003; Astuti *et al.*, 2005), and ectopic expression of MEG3 suppresses growth of several human cancer cell lines in culture (Zhang *et al.*, 2003). Further, ectopic expression of MEG3 leads to both increased p53 protein levels and increased p53 transcriptional activity on some p53 target genes. It is not entirely clear how MEG3 leads to increased p53 levels, but the ectopic expression of MEG3 is followed by a decrease in the p53-negative regulator Mdm2. MEG3 also enhances p53 binding to certain target promoters, thus specifically mediating enhanced expression of some p53 target genes (Zhou *et al.*, 2007). The MEG3 transcript has no complementary region to the promoters of either p53 or its target genes. Instead, maintained MEG3 secondary RNA structure is necessary for retained positive regulation of p53. It has been suggested to either act as a transcriptional co-activator to modulate p53 and/or Mdm2 transcription or to directly bind to Mdm2 and/or p53 to block Mdm2-mediated p53 degradation. Potentially the binding of MEG3 to p53 could be involved in selective p53 target gene activation. Alternatively, MEG3 could modulate promoters of certain p53 targets to favour their activation. For an illustration, see Fig. 11.1B.

Role of p53 in activation of ncRNAs

p53 itself can modulate post-transcriptional gene regulation in several ways. First of all, p53 regulates the transcription of several miRNAs, including transactivation of the tumour suppressive miR-34 family (Bommer *et al.*, 2007; Corney *et al.*, 2007; Raver-Shapira *et al.*, 2007; He *et al.*, 2007; Tarasov *et al.*, 2007). In further evidence of p53 involvement in the miRNA pathway, loss of miRNA biogenesis in Dicer knockout mice activates p53 by up-regulation of p19ARF, resulting in p53-dependent senescence (Mudhasani *et al.*, 2008). In addition, p53 can enhance the biogenesis of several growth-inhibitory miRNAs post-transcriptionally by interacting with Drosha (Suzuki *et al.*, 2009). p53 may itself engage in regulation of 5′UTRs – according to one study, p53 can inhibit its own translation by binding to its 5′UTR (Mosner *et al.*, 1995). Similarly, p53 has been shown to bind to the 5′UTRs of CDK4 (Miller *et al.*, 2000) and FGF-2 (Galy *et al.*, 2001; Gonzalez-Herrera *et al.*, 2006) and inhibit their IRES-dependent translation. p53 also transactivates a number of proteins involved in regulation of mRNA stability, such as Wig-1 (Zhu and Chen, 2000; Shu *et al.*, 2006; Vilborg *et al.*, 2009; Scoumanne *et al.*, 2010). In addition, p53 also transactivates a number of long intergenic noncoding RNAs (lincRNAs) (Huarte *et al.*, 2010). One of these, lincRNA-p21, acts as a transcriptional repressor in the p53 pathways (Huarte *et al.*, 2010). Many lincRNAs regulate gene expression by interacting with chromatin regulatory complexes (Khalil *et al.*, 2009). In accordance with such a mechanism, it was shown that lincRNA-p21 binds to the transcription factor hnRNPK, and is important for proper localization of hnRNPK to promoters repressed after p53 induction. Further, lincRNA-p21 expression is required for full induction of p53-dependent apoptosis in response to cellular stress (Huarte *et al.*, 2010). For an illustration, see Fig. 11.1C.

Speculative summary

The control of p53 is complex with tight regulation at several levels, including transcriptional, posttranscriptional and posttranslational control. The identification of multiple ncRNAs in the p53 pathway adds new insight into p53 regulation and highlights the biological relevance of ncRNAs in the mammalian genome. At first sight, additional mechanisms regulating the tightly controlled p53 may seem redundant. However, this ever-increasing complexity demonstrates the extreme importance of ensuring adequate p53 levels in order for the cells to achieve the correct response to a given signal. Further, ncRNAs in the p53 pathway are not limited to regulating the levels of p53, but may also enable the choice of appropriate p53 responses by contributing to the activation of different regulative circuits, such as in the case of MEG3 (Zhou *et al.*, 2007). The fact that WRAP53α levels are induced upon DNA damage suggests an important role of this pathway in response to DNA stress (Mahmoudi *et al.*, 2009). Although the initial increase of p53 levels upon DNA damage is mediated at a post-translational level, the cells ensure sustained *de novo* protein synthesis by also stabilizing the p53 mRNA. This allows an enhanced and prolonged p53 response that might for example be important for the shift between cell cycle arrest to apoptosis.

The important role of WRAP53α in p53 action also opens the possibility that dysfunction of WRAP53α itself may be a cause of cancer. Future studies will show if WRAP53α is deregulated in primary tumours. There is also a therapeutic aspect of this discovery. By increasing WRAP53α levels in p53 wild type harbouring tumours, p53 expression could be enhanced resulting in a stronger p53-dependent apoptosis of the tumour. On the other hand, reduction of WRAP53α in mutant p53 cancers could be a suitable approach to counteract a p53 gain-of-function phenotype (Zhou *et al.*, 2007). p53 is often mutated in the DNA binding domain, leaving the other parts of the mRNA intact. Thus, WRAP53α-mediated p53 regulation should be functional also in these tumours. In conclusion, modulation of WRAP53α has therapeutic potential in both p53 wild type and mutant tumours.

Strikingly, the overlapping region of WRAP53α and p53 shows higher conservation than any of the coding exons within the two genes. This together with the fact that the head-to-head organization is conserved in all mammals further strengthens the biological significance of WRAP53α in p53 regulation. In addition, the p53 family member p73 has also an antisense gene called *WDR8* that overlaps *p73* in a similar fashion as *WRAP53/p53* and encodes for a protein that belongs to the same family as WRAP53 (Mahmoudi *et al.*, 2009). Thus, it is also possible that p73 is regulated by WDR8 in a similar manner as WRAP53α/p53. Preliminary experiment from our group have, however, not shown such a regulation between this second pair of genes (Mahmoudi and Farnebo *et al.*, unpublished). This could be due to the fact that *WDR8/p73* overlap on an intron–intron basis, whereas *WRAP53/p53* overlap exon to exon.

Besides its role in p53's response to DNA damage, WRAP53α is also required for maintaining p53's steady-state levels (Mahmoudi *et al.*, 2009). Basal levels of p53 have functions that may differ markedly from that of stress-induced p53, such as regulation of metabolism, induction of anti-oxidant factors, and promotion of cell survival (Vousden and Prives, 2009). WRAP53α could by sustaining basal p53 levels also have a role in these p53 functions.

An interesting aspect of the *WRAP53* gene is its contradictory roles; as an antisense transcript originating from exon1α WRAP53 stabilizes p53 levels, and as a protein originating mainly from exon1β (referred to as WRAP53β transcripts) WRAP53 has oncogenic properties and promotes cancer cell survival (Mahmoudi S, 2011). WRAP53α and WRAP53β transcripts have different promoters and the induction of one seem to repress the expression of the other (Mahmoudi *et al.*, 2009). This raises an interesting scenario were the up-regulation of the WRAP53β transcripts could have double growth-promoting effects: by inducing the WRAP53 protein and simultaneously getting rid of p53 through blockage of WRAP53α expression. On the contrary, the expression of WRAP53α could act growth suppressive on a double level by inducing p53 expression and preventing WRAP53 protein expression. Future studies will certainly shed light on the apparently multifaceted genomic organization on WRAP53 and p53. However, this organization does raise the possibility that an overlap between some sense/antisense genes with protein-coding capacities might have evolved as a mechanism to balance the growth promoting and growth suppressive actions of these genes.

References

Astuti, D., Latif, F., Wagner, K., Gentle, D., Cooper, W.N., Catchpoole, D., Grundy, R., Ferguson-Smith, A.C., and Maher, E.R. (2005). Epigenetic alteration at the DLK1-GTL2 imprinted domain in human neoplasia: analysis of neuroblastoma, phaeochromocytoma and Wilms' tumour. Br. J. Cancer *92*, 1574–1580.

Bommer, G.T., Gerin, I., Feng, Y., Kaczorowski, A.J., Kuick, R., Love, R.E., Zhai, Y., Giordano, T.J., Qin, Z.S., Moore, B.B., *et al.* (2007). p53-mediated activation of miRNA34 candidate tumor-suppressor genes. Curr. Biol. *17*, 1298–1307.

Braithwaite, A.W., Del Sal, G., and Lu, X. (2006). Some p53-binding proteins that can function as arbiters of life and death. Cell Death Differ. *13*, 984–993.

Candeias, M.M., Malbert-Colas, L., Powell, D.J., Daskalogianni, C., Maslon, M.M., Naski, N., Bourougaa, K., Calvo, F., and Fahraeus, R. (2008). P53 mRNA controls p53 activity by managing Mdm2 functions. Nat. Cell. Biol. *10*, 1098–1105.

Chen, J., and Kastan, M.B. (2010). 5'–3'–UTR interactions regulate p53 mRNA translation and provide a target for modulating p53 induction after DNA damage. Genes Dev. *24*, 2146–2156.

Chen, J., Sun, M., Kent, W.J., Huang, X., Xie, H., Wang, W., Zhou, G., Shi, R.Z., and Rowley, J.D. (2004). Over 20% of human transcripts might form sense–antisense pairs. Nucleic Acids Res. *32*, 4812–4820.

Cicalese, A., Bonizzi, G., Pasi, C.E., Faretta, M., Ronzoni, S., Giulini, B., Brisken, C., Minucci, S., Di Fiore, P.P., and Pelicci, P.G. (2009). The tumor suppressor p53 regulates polarity of self-renewing divisions in mammary stem cells. Cell *138*, 1083–1095.

Corney, D.C., Flesken-Nikitin, A., Godwin, A.K., Wang, W., and Nikitin, A.Y. (2007). MicroRNA-34b and MicroRNA-34c are targets of p53 and cooperate in control of cell proliferation and adhesion-independent growth. Cancer Res. *67*, 8433–8438.

Faghihi, M.A., and Wahlestedt, C. (2009). Regulatory roles of natural antisense transcripts. Nat. Rev. Mol. Cell. Biol. *10*, 637–643.

Fu, L., Ma, W., and Benchimol, S. (1999). A translation repressor element resides in the 3' untranslated region of human p53 mRNA. Oncogene *18*, 6419–6424.

Galban, S., Martindale, J.L., Mazan-Mamczarz, K., Lopez de Silanes, I., Fan, J., Wang, W., Decker, J., and Gorospe, M. (2003). Influence of the RNA-binding protein HuR in pVHL-regulated p53 expression in renal carcinoma cells. Mol. Cell. Biol. *23*, 7083–7095.

Galy, B., Creancier, L., Prado-Lourenco, L., Prats, A.C., and Prats, H. (2001). p53 directs conformational change and translation initiation blockade of human fibroblast growth factor 2 mRNA. Oncogene *20*, 4613–4620.

Garcia-Cao, I., Garcia-Cao, M., Martin-Caballero, J., Criado, L.M., Klatt, P., Flores, J.M., Weill, J.C., Blasco, M.A., and Serrano, M. (2002). 'Super p53' mice exhibit enhanced DNA damage response, are tumor resistant and age normally. EMBO J. *21*, 6225–6235.

Gonzalez-Herrera, I.G., Prado-Lourenco, L., Teshima-Kondo, S., Kondo, K., Cabon, F., Arnal, J.F., Bayard, F., and Prats, A.C. (2006). IRES-dependent regulation of FGF-2 mRNA translation in pathophysiological conditions in the mouse. Biochem. Soc. Trans. *34*, 17–21.

Grover, R., Candeias, M.M., Fahraeus, R., and Das, S. (2009). p53 and little brother p53/47, linking IRES activities with protein functions. Oncogene *28*, 2766–2772.

He, L., He, X., Lim, L.P., de Stanchina, E., Xuan, Z., Liang, Y., Xue, W., Zender, L., Magnus, J., Ridzon, D., *et al.* (2007). A microRNA component of the p53 tumour suppressor network. Nature *447*, 1130–1134.

Helton, E.S., and Chen, X. (2007). p53 modulation of the DNA damage response. J. Cell. Biochem. *100*, 883–896.

Hu, W., Chan, C.S., Wu, R., Zhang, C., Sun, Y., Song, J.S., Tang, L.H., Levine, A.J., and Feng, Z. (2010). Negative regulation of tumor suppressor p53 by microRNA miR-504. Mol. Cell *38*, 689–699.

Huarte, M., Guttman, M., Feldser, D., Garber, M., Koziol, M.J., Kenzelmann-Broz, D., Khalil, A.M., Zuk, O., Amit, I., Rabani, M., *et al.* (2010). A large intergenic noncoding RNA induced by p53 mediates global gene repression in the p53 response. Cell *142*, 409–419.

Janicke, R.U., Sohn, D., and Schulze-Osthoff, K. (2008). The dark side of a tumor suppressor: anti-apoptotic p53. Cell Death Differ. *15*, 959–976.

Khalil, A.M., Guttman, M., Huarte, M., Garber, M., Raj, A., Rivea Morales, D., Thomas, K., Presser, A., Bernstein, B.E., van Oudenaarden, A., *et al.* (2009). Many human large intergenic noncoding RNAs associate with chromatin-modifying complexes and affect gene expression. Proc. Natl. Acad. Sci. U.S.A. *106*, 11667–11672.

Kim, H., You, S., Foster, L.K., Farris, J., and Foster, D.N. (2001). The rapid destabilization of p53 mRNA in immortal chicken embryo fibroblast cells. Oncogene *20*, 5118–5123.

Lamouille, S., and Derynck, R. (2009). Oncogene and tumour suppressor: the two faces of SnoN. EMBO J. *28*, 3459–3460.

Lapidot, M., and Pilpel, Y. (2006). Genome-wide natural antisense transcription: coupling its regulation to its different regulatory mechanisms. EMBO Rep. *7*, 1216–1222.

Le, M.T., Teh, C., Shyh-Chang, N., Xie, H., Zhou, B., Korzh, V., Lodish, H.F., and Lim, B. (2009). MicroRNA-125b is a novel negative regulator of p53. Genes Dev. *23*, 862–876.

Li, J., Donath, S., Li, Y., Qin, D., Prabhakar, B.S., and Li, P. (2010). miR-30 regulates mitochondrial fission through targeting p53 and the dynamin-related protein-1 pathway. PLoS Genet. *6*, e1000795.

Mahmoudi, S., Henriksson, S., Corcoran, M., Mendez-Vidal, C., Wiman, K.G., and Farnebo, M. (2009). Wrap53, a natural p53 antisense transcript required for p53 induction upon DNA damage. Mol. Cell *33*, 462–471.

Mahmoudi, S., Henriksson, S., Weibrecht, I., Smith, S., Soderberg, O., Stromblad, S., Wiman, K.G., and Farnebo, M. (2010). WRAP53 is essential for cajal body formation and for targeting the survival of motor neuron complex to cajal bodies. PLoS Biol. *8*, e1000521.

Mahmoudi S., Henriksson, S., Farnebo L., Roberg K., Farnebo M., (2011). WRAP53 promotes cancer cell survival and is a potential target for cancer therapy. Cell Death Dis. 2, e114.

Matheu, A., Maraver, A., and Serrano, M. (2008). The Arf/p53 pathway in cancer and aging. Cancer Res. *68*, 6031–6034.

Mazan-Mamczarz, K., Galban, S., Lopez de Silanes, I., Martindale, J.L., Atasoy, U., Keene, J.D., and Gorospe, M. (2003). RNA-binding protein HuR enhances p53 translation in response to ultraviolet light irradiation. Proc. Natl. Acad. Sci. U.S.A. *100*, 8354–8359.

Miller, S.J., Suthiphongchai, T., Zambetti, G.P., and Ewen, M.E. (2000). p53 binds selectively to the 5′ untranslated region of cdk4, an RNA element necessary and sufficient for transforming growth factor beta- and p53-mediated translational inhibition of cdk4. Mol. Cell. Biol. *20*, 8420–8431.

Miyoshi, N., Wagatsuma, H., Wakana, S., Shiroishi, T., Nomura, M., Aisaka, K., Kohda, T., Surani, M.A., Kaneko-Ishino, T., and Ishino, F. (2000). Identification of an imprinted gene, Meg3/Gtl2 and its human homologue MEG3, first mapped on mouse distal chromosome 12 and human chromosome 14q. Genes Cells *5*, 211–220.

Mosner, J., Mummenbrauer, T., Bauer, C., Sczakiel, G., Grosse, F., and Deppert, W. (1995). Negative feedback regulation of wild-type p53 biosynthesis. EMBO J. *14*, 4442–4449.

Mudhasani, R., Zhu, Z., Hutvagner, G., Eischen, C.M., Lyle, S., Hall, L.L., Lawrence, J.B., Imbalzano, A.N., and Jones, S.N. (2008). Loss of miRNA biogenesis induces p19Arf-p53 signaling and senescence in primary cells. J. Cell. Biol. *181*, 1055–1063.

Nizami, Z.F., Deryusheva, S., and Gall, J.G. (2010). Cajal bodies and histone locus bodies in Drosophila and Xenopus. Cold Spring Harb. Symp. Quant. Biol. *75*, 313–320.

Osato, N., Suzuki, Y., Ikeo, K., and Gojobori, T. (2007). Transcriptional interferences in cis natural antisense transcripts of humans and mice. Genetics *176*, 1299–1306.

Pierce, A.M., Schneider-Broussard, R., Gimenez-Conti, I.B., Russell, J.L., Conti, C.J., and Johnson, D.G. (1999). E2F1 has both oncogenic and tumor-suppressive properties in a transgenic model. Mol. Cell. Biol. *19*, 6408–6414.

Raver-Shapira, N., Marciano, E., Meiri, E., Spector, Y., Rosenfeld, N., Moskovits, N., Bentwich, Z., and Oren, M. (2007). Transcriptional activation of miR-34a contributes to p53-mediated apoptosis. Mol. Cell *26*, 731–743.

Schumacher, B., Hanazawa, M., Lee, M.H., Nayak, S., Volkmann, K., Hofmann, E.R., Hengartner, M., Schedl, T., and Gartner, A. (2005). Translational repression of C. elegans p53 by GLD-1 regulates DNA damage-induced apoptosis. Cell *120*, 357–368.

Schuster-Gossler, K., Bilinski, P., Sado, T., Ferguson-Smith, A., and Gossler, A. (1998). The mouse Gtl2 gene is differentially expressed during embryonic development, encodes multiple alternatively spliced transcripts, and may act as an RNA. Dev. Dyn. *212*, 214–228.

Scoumanne, A., Cho, S.J., Zhang, J., and Chen, X. (2010). The cyclin-dependent kinase inhibitor p21 is regulated by RNA-binding protein PCBP4 via mRNA stability. Nucleic Acids Res. *39*, 213–224.

Shu, L., Yan, W., and Chen, X. (2006). RNPC*1*, an RNA-binding protein and a target of the p53 family, is required for maintaining the stability of the basal and stress-induced p21 transcript. Genes Dev. *20*, 2961–2972.

Soussi, T., and Wiman, K.G. (2007). Shaping genetic alterations in human cancer: the p53 mutation paradigm. Cancer Cell *12*, 303–312.

Suzuki, H.I., Yamagata, K., Sugimoto, K., Iwamoto, T., Kato, S., and Miyazono, K. (2009). Modulation of microRNA processing by p53. Nature *460*, 529–533.

Takagi, M., Absalon, M.J., McLure, K.G., and Kastan, M.B. (2005). Regulation of p53 translation and induction after DNA damage by ribosomal protein L26 and nucleolin. Cell *123*, 49–63.

Tarasov, V., Jung, P., Verdoodt, B., Lodygin, D., Epanchintsev, A., Menssen, A., Meister, G., and Hermeking, H. (2007). Differential regulation of microRNAs by p53 revealed by massively parallel sequencing: miR-34a is a p53 target that induces apoptosis and G1-arrest. Cell Cycle *6*, 1586–1593.

Tian, S., Huang, S., Wu, S., Guo, W., Li, J., and He, X. (2010). MicroRNA-1285 inhibits the expression of p53 by directly targeting its 3′ untranslated region. Biochem. Biophys. Res. Commun. *396*, 435–439.

Tycowski, K.T., Shu, M.D., Kukoyi, A., and Steitz, J.A. (2009). A conserved WD40 protein binds the Cajal body localization signal of scaRNP particles. Mol. Cell *34*, 47–57.

Tyner, S.D., Venkatachalam, S., Choi, J., Jones, S., Ghebranious, N., Igelmann, H., Lu, X., Soron, G., Cooper, B., Brayton, C., *et al.* (2002). p53 mutant mice that display early ageing-associated phenotypes. Nature *415*, 45–53.

Venteicher, A.S., Abreu, E.B., Meng, Z., McCann, K.E., Terns, R.M., Veenstra, T.D., Terns, M.P., and Artandi, S.E. (2009). A human telomerase holoenzyme protein required for Cajal body localization and telomere synthesis. Science *323*, 644–648.

Vilborg, A., Glahder, J.A., Wilhelm, M.T., Bersani, C., Corcoran, M., Mahmoudi, S., Rosenstierne, M., Grander, D., Farnebo, M., Norrild, B., *et al.* (2009). The p53 target Wig-1 regulates p53 mRNA stability through an AU-rich element. Proc. Natl. Acad. Sci. U.S.A. *106*, 15756–15761.

Vilborg, A., Wilhelm, M.T., and Wiman, K.G. (2010). Regulation of tumor suppressor p53 at the RNA level. J. Mol. Med. *88*, 645–652.

Vousden, K.H., and Prives, C. (2009). Blinded by the light: the growing complexity of p53. Cell *137*, 413–431.

Vousden, K.H., and Ryan, K.M. (2009). p53 and metabolism. Nat. Rev. Cancer *9*, 691–700.

Wang, Q., and Carmichael, G.G. (2004). Effects of length and location on the cellular response to double-stranded RNA. Microbiol. Mol. Biol. Rev. *68*, 432–452.

Xu, Y. (2003). Regulation of p53 responses by post-translational modifications. Cell Death Differ. *10*, 400–403.

Yelin, R., Dahary, D., Sorek, R., Levanon, E.Y., Goldstein, O., Shoshan, A., Diber, A., Biton, S., Tamir, Y., Khosravi, R., *et al.* (2003). Widespread occurrence of antisense transcription in the human genome. Nat. Biotechnol. *21*, 379–386.

Zhang, X., Zhou, Y., Mehta, K.R., Danila, D.C., Scolavino, S., Johnson, S.R., and Klibanski, A. (2003). A pituitary-derived MEG3 isoform functions as a growth suppressor in tumor cells. J. Clin. Endocrinol. Metab. *88*, 5119–5126.

Zhang, Y., Gao, J.S., Tang, X., Tucker, L.D., Quesenberry, P., Rigoutsos, I., and Ramratnam, B. (2009). MicroRNA 125a and its regulation of the p53 tumor suppressor gene. FEBS Lett. *583*, 3725–3730.

Zhao, T., and Xu, Y. (2010). p53 and stem cells: new developments and new concerns. Trends Cell Biol. *20*, 170–175.

Zhou, Y., Zhong, Y., Wang, Y., Zhang, X., Batista, D.L., Gejman, R., Ansell, P.J., Zhao, J., Weng, C., and Klibanski, A. (2007). Activation of p53 by MEG3 non-coding RNA. J. Biol. Chem. *282*, 24731–24742.

Zhu, J., and Chen, X. (2000). MCG*10*, a novel p53 target gene that encodes a KH domain RNA-binding protein, is capable of inducing apoptosis and cell cycle arrest in G(2)-M. Mol. Cell. Biol. *20*, 5602–5618.

Zou, T., Mazan-Mamczarz, K., Rao, J.N., Liu, L., Marasa, B.S., Zhang, A.H., Xiao, L., Pullmann, R., Gorospe, M., and Wang, J.Y. (2006). Polyamine depletion increases cytoplasmic levels of RNA-binding protein HuR leading to stabilization of nucleophosmin and p53 mRNAs. J. Biol. Chem. *281*, 19387–19394.

Natural Antisense Transcripts Mediate Discordant and Concordant Regulation of Gene Expression

12

Mohammad Faghihi, Paul Halley, Zane Zeier and Claes Wahlestedt

Abstract

Recent characterization of functional non-protein-coding RNAs has revealed one prominent and complex class of regulatory RNAs called natural antisense transcripts. Natural antisense transcripts (also known as antisense RNAs or antisense transcripts) are RNA molecules that are transcribed from the opposite DNA strand from sense transcripts, and often overlap with mRNA of conventional sense genes. Despite pervasive transcription of natural antisense transcripts from mammalian genomes, their role in biology is not well understood. Functional validation studies indicate that antisense transcripts are not a uniform group of regulatory RNAs but instead belong to multiple categories with some common features. Interestingly, knockdown or blockade of endogenous antisense transcripts can have multiple outcomes depending upon whether they increase (concordant regulation) or decrease (discordant regulation) the concentration of the corresponding sense transcript. It has been proposed that discordant de-repression of sense transcript expression, resulting in upregulation of sense RNA expression, can be achieved by the inactivation of many but not all antisense transcripts. Collectively, recent evidence indicates that antisense transcripts are frequently functional and use diverse transcriptional and post-transcriptional gene regulatory mechanisms to carry out a wide variety of biological roles.

Natural antisense transcripts

Transcription of antisense RNA is tissue and cell line specific with a large fraction of antisense transcripts expressed in specific regions of the brain, suggesting their involvement in sophisticated brain functions and in gene dysregulation underlying complex neurological disorders. Recent large-scale expression-profiling studies have helped to define the nature of antisense transcripts in eukaryotic genomes and led to a consensus opinion that natural antisense transcripts, most of which are non-protein-coding RNAs, are transcribed abundantly throughout the mammalian genome. However, there are many unanswered questions concerning the biological functions of antisense RNAs and their heterogeneous modes of action in various cells have only begun to be elucidated. For instance, what fraction of antisense RNAs have biological significance, and through which regulatory mechanisms do they confer their activity? Evidence suggests that natural antisense transcripts can utilize various cellular pathways, but it is still unclear which intrinsic properties or extrinsic features, such as protein interactions, are decisive for the selection of a given pathway. Furthermore, how is the expression of non-protein-coding RNAs regulated in various cells types and during development? Further characterization of the structural diversity and complexity of expression patterns of anti-sense RNA species is needed in order to elucidate their various biological roles.

Natural antisense transcript definition

Natural antisense transcripts are RNA molecules that are transcribed from the opposite DNA strand from which sense transcripts are produced and in some cases overlap with the sense RNA, sense promoter or other sense regulatory elements. Transcription of natural antisense RNA can occur in cis or trans, with cis antisense RNA being transcribed from the same genomic locus as the sense RNA, therefore, the overlapping regions are complementary. In addition, antisense RNAs can encode proteins or be non-protein-coding transcripts, although the most prominent form of antisense transcription in the mammalian genome is a non-protein-coding antisense RNA partner of a protein-coding mRNA transcript (Fig. 12.1).

Antisense RNAs may also be transcribed in trans (from a distant loci), in which case they would only offer partial sequence complementarity to the sense transcript. It has now become clear that the genome of a mammalian organism often encodes two distinct RNAs from either strand of the same DNA (Nepveu and Marcu, 1986; Williams and Fried, 1986). Natural antisense transcripts have been suggested to regulate gene expression by causing transcriptional and post-transcriptional changes in sense mRNA levels (Kiyosawa *et al.*, 2003; Rosok and Sioud, 2005; Chan *et al.*, 2006; Li *et al.*, 2006)

Widespread expression of natural antisense transcripts

A large transcriptome profiling effort pursued by the FANTOM-3 consortium identified natural

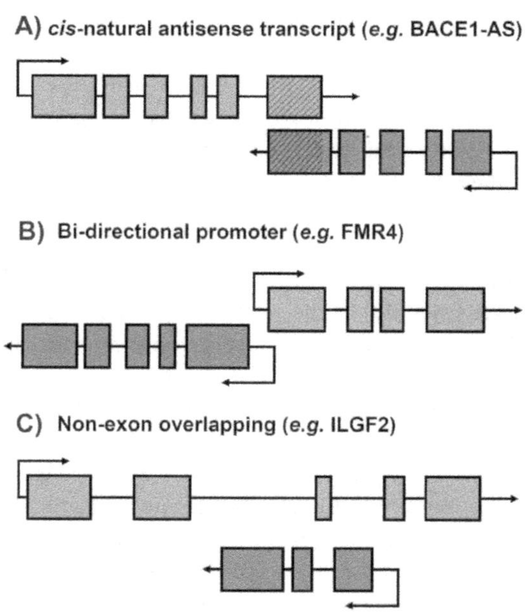

Figure 12.1 Illustration of human and mouse antisense transcription patterns at complex genomic loci.(A) *cis*-Natural antisense transcripts, like BACE1-AS (Faghihi *et al.*, 2008) in which two converging transcripts from opposite strands of DNA have exon overlapping sequence (depicted as hashed elements). Boxes in all three panels depict exons; lines illustrate introns and arrows represent the direction of transcription. Top rows indicate sense transcripts while bottom rows indicate antisense transcripts. (B) Bidirectional sense/antisense transcription from promoters such as the FMR1/FMR4 promoter (Khalil *et al.*, 2008) in which two diverging RNA transcripts share the same transcription start site (TSS), in some cases overlapping the 5′UTR. (C) Full or intronic transcripts in which antisense RNA resides inside the boundary of sense transcripts. Even if the processed RNA does not contain overlapping sequence, the possibility exists for RNA duplex formation between unprocessed transcripts. Alternatively, antisense RNA may implement its regulatory activity through interaction with corresponding DNA sequences.

antisense transcripts for more than 70% of transcription units (TU)[1] within mouse and human genomes (Katayama *et al.*, 2005). Over 20% of human and mouse mRNA had previously been predicted to form sense–antisense pairs (Lehner *et al.*, 2002; Shendure and Church, 2002; Yelin *et al.*, 2003; Chen *et al.*, 2004; Richards *et al.*, 2006; Zhang *et al.*, 2006).

In addition, well-documented experimental and evolutionary studies have ruled out the possibility that such genome-wide natural antisense production represents leakage of the RNA transcription machinery (Dahary *et al.*, 2005). Rather, this abundant antisense transcription is thought to have affected vertebrate genome evolution considerably (Dahary *et al.*, 2005), and antisense transcripts are suggested to play a pivotal role in the nervous system complexity found among higher organisms (St Laurent and Wahlestedt, 2007).

Large-scale detection of natural antisense transcripts

Various techniques have been used in order to identify and quantify natural antisense transcripts, including large-scale sequencing of cDNA clones (Okazaki *et al.*, 2002; Kiyosawa *et al.*, 2003; Carninci *et al.*, 2005; Katayama *et al.*, 2005), tiling arrays (Rinn *et al.*, 2003; Bertone *et al.*, 2004; Cawley *et al.*, 2004; Cheng *et al.*, 2005), bioinformatics analysis of RefSeq and EST databases (Fahey *et al.*, 2002; Lehner *et al.*, 2002; Shendure and Church, 2002; Yelin *et al.*, 2003; Chen *et al.*, 2004; Klimov *et al.*, 2006; Li *et al.*, 2006; Zhang *et al.*, 2006), hybridization techniques (Rosok and Sioud, 2004; Yu *et al.*, 2008), SAGE libraries (Quere *et al.*, 2004; Ge *et al.*, 2006), strand-specific microarrays (Gyorffy *et al.*, 2007; Vallon-Christersson *et al.*, 2007; Ge *et al.*, 2008), and most recently, asymmetrical strand-specific analysis of gene expression (ASSAGE) (He *et al.*, 2008) and global run-on sequencing (GRO-seq) (Core *et al.*, 2008). These studies have demonstrated widespread antisense transcription from mammalian genomes, giving rise to an abundant and heterogeneous antisense transcriptome, however, additional sequencing experimentation will be required in order to adequately characterize low copy number antisense transcripts. Nevertheless, the majority of transcriptional units (TUs) appear to contain antisense transcripts (Katayama *et al.*, 2005), and antisense transcripts have been documented for many active promoters (Core *et al.*, 2008; Preker *et al.*, 2008; Seila *et al.*, 2008). Therefore, the presence of antisense transcripts should no longer be considered a curiosity but rather a pervasive feature of mammalian genomes.

[1]A transcription unit (TU) is a group of expressed sequence tags or mRNAs, usually with alternative splice patterns, in which they share exonic overlap of at least one nucleotide and exist in the same chromosomal orientation

Characteristics of antisense RNAs

Antisense transcripts are not evenly distributed across the genome; they are more abundant at the ends of protein-coding genes (Finocchiaro *et al.*, 2007; He *et al.*, 2008). Specifically, antisense transcription is enriched 250 nucleotides upstream of transcription start sites (TSS) (Core *et al.*, 2008; Seila *et al.*, 2008) and 1.5 kb downstream of sense genes (Sun *et al.*, 2005; Core *et al.*, 2008).

The relative basal expression levels of sense and antisense transcripts in different tissues and cell lines may be either positively or negatively correlated (Katayama *et al.*, 2005; Okada *et al.*, 2008). Moreover, levels of antisense transcription in different human cell lines is not always coupled to the expression of its sense gene partner, which suggests the use of alternative transcriptional regulatory elements and promoters (He *et al.*, 2008). Antisense RNAs have a tendency to undergo fewer splicing events and are typically less abundant compared with sense transcripts (He *et al.*, 2008).

It has been proposed that discordant derepression of sense transcription, resulting in the up regulation of sense RNA expression, can be achieved by removal of the antisense-mediated steric blockade of many but not all antisense transcripts (Katayama *et al.*, 2005; Wahlestedt, 2006).

These and other variable intrinsic properties or regulation of gene expression must operate through a variety of mechanisms.

Natural antisense transcripts must therefore be

assumed to exert positive or negative regulation at different levels of gene expression, invoking diverse and largely uncharacterized co-regulatory factors. Hence, potential involvement of natural antisense transcripts should be considered in a wide range of future studies on the regulation of gene expression.

There are broadly two types of regulation between sense and antisense transcripts, namely concordant and discordant regulation.

Examples of concordant regulation

Antisense transcripts that concordantly regulate gene expression include BACE1-AS (Faghihi *et al.*, 2008), aHIF1 (Neckers, 1999), asEPO-R (Zhang *et al.*, 2008) and Zeb2-AS (Beltran *et al.*, 2008), which augment the level of the sense mRNA, or corresponding protein levels.

The BACE1-AS transcript

We have recently identified a conserved non-protein-coding antisense transcript for β-secretase-1 (BACE1), a critical enzyme involved in Alzheimer's disease pathophysiology. The BACE1-antisense transcript (BACE1-AS) concordantly up-regulates BACE1 mRNA and subsequently BACE1 protein expression *in vitro* and *in vivo* (Faghihi *et al.*, 2008).

The Zeb2-AS transcript

Concordant regulation is also reported for Zeb2-AS, a natural antisense transcripts to the transcriptional repressor of E-cadherin called zinc finger homeoBox 1b (zeb2). Zeb2-AS inhibits 5′UTR splicing, which in turn makes an internal ribosome entry site (IRES) accessible, causing increased Zeb2 protein levels without changing Zeb2 mRNA levels. This natural antisense transcript is an example of concordant regulation in which antisense RNA induces Zeb2 protein up-regulation and subsequently induces epithelial-mesenchymal transition (Beltran *et al.*, 2008).

The asEPO-R transcript

Erythropoietin receptor (EPO-R) mRNA and protein levels are shown to be positively controlled (concordantly regulated) by a cis-antisense RNA called asEPO-R (Zhang *et al.*, 2008). Both sense–antisense transcripts as well as EPO-R protein are increased in canine lung after pneumonectomy, suggesting a role for antisense regulation during lung growth. Although asEPO-R has two potential open reading frames (ORFa), evidence suggests that concordant upregulation of EPO-R is protein independent (Zhang *et al.*, 2008).

Examples of discordant regulation

In discordant regulation, e.g. naPINK1 (Scheele *et al.*, 2007), RevΔ5-desaturase (Dreesen *et al.*, 2006), zfh-5AS (Komine *et al.*, 2006) or antisense-frq (Kramer *et al.*, 2003), the antisense transcripts have negative (opposing) effects on sense transcripts.

RevΔ5-desaturase transcript

The non-protein-coding antisense RNA for Δ5-desaturase regulates fatty acid metabolism, during the transition between fasting and re-feeding, by altering the expression of the sense gene. This regulation was shown to be discordant i.e. a diet enriched in fish oil produced a reciprocal increase in antisense and decreased levels of the sense transcript (Dreesen *et al.*, 2006).

zfh-5AS transcript

Consistent with discordant regulation, gene-targeted knock-out of the zfh-5 antisense transcript, a long spliced and polyadenylated RNA, caused a corresponding up-regulation of zfh-5 mRNA in the brain of mice (Komine *et al.*, 2006).

frq-AS transcript

Another documented discordant regulatory mechanism is reported for an antisense transcripts of frequency (frq) mRNA, related to Neurospora crassa circadian clock function. This non-protein-coding RNA transcript is important in synchronizing internal and external time by reducing frq mRNA transcription, as well as FRQ protein (Kramer *et al.*, 2003).

Features of concordant versus discordant regulation

Intriguingly, both concordant and discordant regulatory mechanisms are primarily unidirectional such that the antisense transcript regulates the sense RNA (Wahlestedt, 2006).

In most reported cases of concordant regulation, natural antisense transcripts show a low degree of RNA processing. In these cases, antisense RNAs are not spliced, or display short introns (Chen *et al.*, 2005), and have a much shorter half-life than their coding sense partner (Faghihi *et al.*, 2008). Therefore, it might take much less time for antisense transcription and processing to occur relative to sense mRNA transcripts. Such rapid transcription and processing of antisense transcripts is similar to what has been suggested for ('nimble' genes) (Chen *et al.*, 2005), and implies that they may be involved in the acute stress responses.

In contrast, discordant natural antisense transcripts are typically processed RNA transcripts possessing multiple exons and are more prominent in the cytoplasm. This cytoplasmic abundance suggests a long lasting regulatory role through cytoplasmic RNA duplex formation (see proposed cytoplasmic mechanisms).

Regulatory roles of antisense transcripts

Natural antisense transcripts can regulate gene expression at various levels including chromatin architecture/epigenetic memory, transcription, transcript localization, translation and turnover (Kiyosawa *et al.*, 2003; Rosok and Sioud, 2005; Chan *et al.*, 2006; Li *et al.*, 2006). Natural antisense transcripts have also demonstrated involvment in methylation, demethylation (Imamura *et al.*, 2004), parental gene imprinting (Sleutels *et al.*, 2000), chromosome X inactivation (Lee *et al.*, 1999), RNA splicing (Krystal *et al.*, 1990; Munroe and Lazar, 1991; Enerly *et al.*, 2005), transport (Khochbin *et al.*, 1992), polyadenylation (Volk *et al.*, 1989; Kiyosawa *et al.*, 2005; Mihola *et al.*, 2007), editing and stabilization (Kumar and Carmichael, 1997; Rossignol *et al.*, 2002). Proposed mechanisms for antisense-mediated regulation of sense mRNA can therefore be categorized into four main groups: mechanisms related to transcription, RNA–DNA interactions, nuclear RNA–RNA interactions and cytoplasmic RNA–RNA interactions (Faghihi and Wahlestedt, 2009).

Transcription-related mechanism

According to this model, the antisense RNA does not possess inherent regulatory activity; instead it is the act of transcription in the antisense direction itself that modulates sense transcription. One proposed mechanisms includes transcriptional collision and may be involved in genomic rearrangements.

RNA–DNA interactions

This gene regulatory model is based on direct or indirect RNA–DNA and RNA–chromatin interactions (Bernstein and Allis, 2005) where transcripts may bind to the corresponding DNA strand, resulting in DNA methylation directly (Fig. 12.2), or provide a scaffold for histone-modifying enzyme (HME) recruitment and subsequent alterations of the chromatin status (Fig. 12.3). Such RNA–DNA interactions allow natural antisense transcripts to participate in various epigenetic regulation mechanisms including DNA methylation (Tufarelli *et al.*, 2003), DNA demethylation (Imamura *et al.*, 2004), chromatin modification (Morris *et al.*, 2008; Yu *et al.*, 2008), and monoallelic expression (for example, genomic imprinting (Kanduri, 2008), X-chromosome inactivation (Ohhata *et al.*, 2008) and random monoallelic exclusion of the autosomal loci).

Alteration of DNA and chromatin in autosomal loci

Chromatin modification induced by natural antisense transcripts

Recent studies provide evidence for direct RNA–protein interaction between Ezh2, a subunit of Polycomb repressive complex 2 (PRC2), and many ncRNA transcripts (Zhao *et al.*, 2010). Other studies of X inactivation (Zhao *et al.*, 2008) and the HOX gene cluster (Rinn *et al.*,

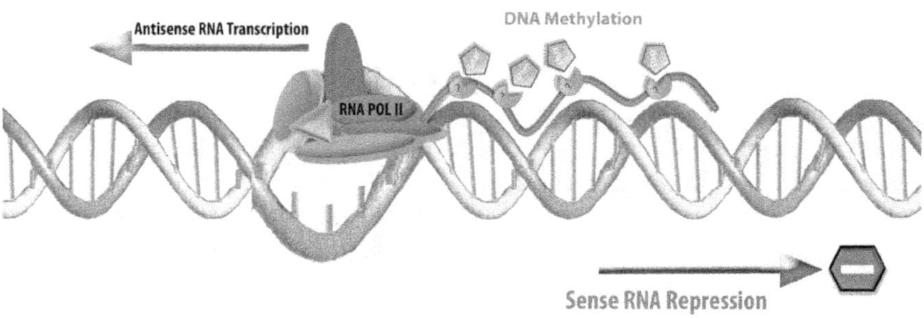

Figure 12.2 Induction of DNA methylation by natural antisense transcripts. RNA–DNA interaction model in which the newly formed antisense RNA transcript can interact, directly or indirectly, with DNA methyltransferases (DMT) and guide methylation of DNA, resulting in repression of sense RNA transcription.

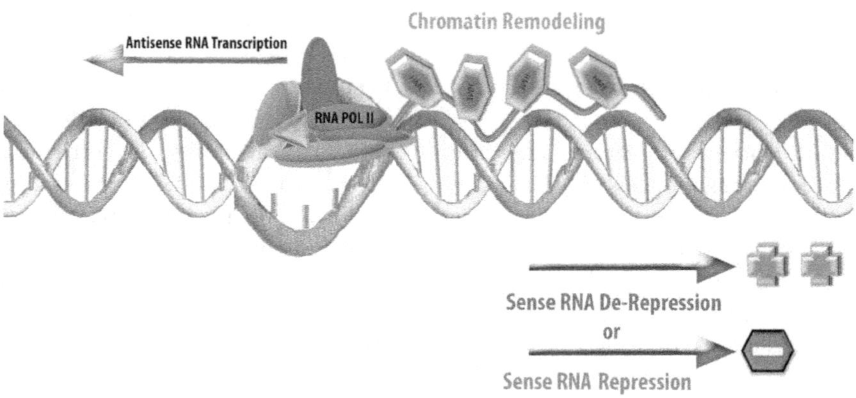

Figure 12.3 Induction of chromatin modifications by natural antisense transcripts. Alternatively, the newly formed antisense RNA transcript can recruit histone-modifying enzymes (HMEs) and modify chromatin, thereby modulating chromatin architecture and epigenetic memory. Activation of sense RNA transcription, by removing the repressive effects of promoter-directed RNA transcripts, has been observed for several gene loci (Janowski *et al.*, 2007; Morris *et al.*, 2008; Schwartz *et al.*, 2008). This de-repression effect may be particularly prominent when an antisense RNA is located in or around the transcription start site of the sense gene. Argonaute2 (AGO2) protein is thought to be involved in this largely uncharacterized process (not shown). The resultant chromatin remodelling would repress or de-repress sense RNA transcription. Similar models have been proposed for imprinted genes as well as for mammalian X-chromosome inactivation.

2007) showed involvement of RNA transcript in PRC2-mediated induction of H3K27met3, repressive chromatin marks. For example, PRC2 transcriptome profiling has identified over 9000 PRC2-interacting RNAs in embryonic stem cells, many of them categorized as antisense RNAs transcripts (Zhao *et al.,* 2010). PCR2 is a protein complex consisting of four core subunits Eed, Suz12, RbAp48 and Ezh2, which catalyses trimethylation of histone H3-lysine 27 (H3K27met3) (Schwartz and Pirrotta, 2008). Other evidence of RNA cofactors in PRC2 includes the human HOX-D locus (Rinn *et al.,* 2007), and X chromosome inactivation (Zhao *et al.,* 2008). We believe that antisense RNA could be manipulated in order to obtain a locus-specific alteration of chromatin modification. Therefore, removal or blockade of antisense transcription could result in epigenetic changes, and in turn alter sense mRNA expression.

Recent studies showed that instead of traditional binary divisions of chromatin into hetero- or eu-chromatin, there are five principal chromatin types (Filion *et al.*, 2011), which argue that chromatin states are more flexible and dynamic than previously thought. In addition, these chromatin marks could be induced or modified by non-protein-coding RNA expression. Interestingly, antisense RNA-mediated regulation of gene expression has a high level of specificity. Considering the low abundance and high potency of many antisense RNA transcripts, we believe that antisense RNA-mediated regulation, at least for a subset of antisense transcripts, occurs in the nucleus and at the level of transcription. Local accumulation of antisense RNA or formation of sense–antisense RNA duplexes, which can then be processed into small RNA intermediates, may trigger a cascade of events that result in chromatin modification, and ultimately activation or deactivation of sense transcription. This modification is local and expanded only to the promoter of sense mRNA, but not to neighbouring genes. Characterization of these events is currently under investigation.

Typically, small RNAs are not involved

Some antisense-mediated epigenetic changes are independent of Dicer, arguing against a role for small RNA mediators (Yu *et al.*, 2008). Therefore, antisense RNA processing into small RNA (mediated by Dicer in the cytoplasm) is not necessarily required for epigenetic modifications. Instead, antisense transcripts accumulate locally and trigger DNA or chromatin modifications (Pandey *et al.*, 2008). These modifications then expand to neighbouring regions, even though these adjacent regions do not exhibit complementarity to the original antisense transcript (Mohammad *et al.*, 2008). The secondary expansion of the modifications may be restricted to the promoter or enhancer of a single gene (random monoallelic exclusion), or may include a cluster of genes, such as in the case of genomic imprinting of the Kcnq1 locus (Kanduri, 2008). Finally, this expansion process occasionally involves an entire chromosome, such as X chromosome inactivation in females.

Examples of ncRNA-mediated DNA and chromatin modification in autosomal loci

Natural antisense transcripts have been proposed to cause DNA methylation (Tufarelli *et al.*, 2003), DNA demethylation (Imamura *et al.*, 2004) and chromatin modifications of non-imprinted autosomal loci (Morris *et al.*, 2008; Yu *et al.*, 2008). Modifications of chromatin structure induced by non-coding RNA are suggested by the observation that many histone methyl transferase complexes lack DNA-binding domains but possess RNA-binding motifs (Bernstein and Allis, 2005). Suppression of transcription is usually caused by DNA and chromatin modification at the promoter region of the sense strand (Text Box 1). For example, an antisense RNA for the α-globulin 2 gene (HBA2) can induce DNA methylation, leading to silencing of the α-globulin 2 gene (Tufarelli *et al.*, 2003). Both trimethylated K4 on Histone H3 (a mark of active transcription) and trimethylated K27 on histone H3 (a mark of repressed chromatin) were reported to associate with non-protein-coding RNAs, suggesting a scaffolding model where the RNA guides chromatin modification (Woo and Kingston, 2007).

Chromatin alterations at active promoters

Overlapping transcription of small (<50 nucleotides) RNAs (Kapranov *et al.*, 2007; Borel *et al.*, 2008) as well as promoter-associated small RNAs (PASRs) (Kapranov *et al.*, 2007; Borel *et al.*, 2008), termini-associated small RNAs (TASRs) (Kapranov *et al.*, 2007; Borel *et al.*, 2008) and promoter upstream transcripts (PROMPTs) (Preker *et al.*, 2008) has been documented using strand-specific genomic tilling arrays in the ENCODE region of the human genome (Kapranov *et al.*, 2007; Borel *et al.*, 2008). Transcription start site-associated RNAs (TSSa-RNAs) (20–90 nucleotides) have also been identified using deep sequencing techniques (Core *et al.*, 2008; Seila *et al.*, 2008). The enrichment of antisense transcripts in both promoter and terminal regions has been confirmed by the unbiased ASSAGE technique (He *et al.*, 2008). TSSa-RNAs flank active promoters in both the sense and antisense directions with respect to downstream protein-coding genes

Box 12.1 RNA transcripts proposed to induce DNA or chromatin modifications

RNA transcripts proposed to induce DNA or chromatin modifications (in order of discovery) are:

- *LUC7L gene.* An antisense RNA for the α-globulin gene (*HBA2*) can induce promoter DNA methylation, leading to transcriptional silencing of the α-globulin gene (Tufarelli *et al.*, 2003). Expression of HBA2 is silenced and its CpG island becomes completely methylated during early development. Antisense RNA mediates silencing and methylation of the associated CpG island in the affected individual as well as in a mouse transgenic model.
- EF1α-promoter-associated RNA. RNA-mediated modification of the promoter region is shown for elongation factor 1α (EF1α) (Han *et al.*, 2007). Low-copy promoter-associated RNAs are recognized by siRNA and function as a recognition motif to direct epigenetic silencing complexes to the corresponding sense gene promoter and, consequently, to mediate transcriptional silencing in human cells.
- P15AS transcript. Transcriptional silencing of the tumour-suppressor gene p15 via DNA methylation and heterochromatin formation induced by the antisense transcript, p15AS (Yu *et al.*, 2008).
- PR-AT transcript. Promoter-directed antigen RNA (agRNA) is a class of synthetic RNA that can bind to the TSS of genes and can activate or block transcription of the target gene dependent upon the cellular context (Janowski *et al.*, 2005). Natural antisense transcript for progesterone receptor, PR-AT1 and PR-AT2, is required for agRNA-induced gene activation or suppression. PR-AT1 and PR-AT2 bind to the Argonaut (Ago) protein and provide a scaffold for suppressor or activator proteins to assemble in the progesterone receptor promoter region (Schwartz *et al.*, 2008).
- P21-AS transcript. Low-copy promoter-directed antisense transcript of the tumour-suppressor gene p21 mediates epigenetic modification of the sense promoter region (Morris *et al.*, 2008). Suppression of sense mRNA is directed by antisense-mediated induction of trimethylation at histone H3 lysine 27 (H3K27me3) at the *p21* sense promoter region (Morris *et al.*, 2008).

(Core *et al.*, 2008; Seila *et al.*, 2008). Divergent promoter activity, which is believed to produce this small RNA group, is documented for more than half of mouse and human genes (Core *et al.*, 2008; Seila *et al.*, 2008).

Evidence for local chromatin modifications induced by antisense RNAs

The transcribed regions of PROMPTs and TSSa-RNAs are known to overlap with paused RNA polymerase II (RNAPII) and active chromatin marks, such as trimethylation of histone H3 lysine 4 (H3K4me3), suggesting a role for local RNA accumulation in maintaining the dynamic chromatin state required for promoter activity (Preker *et al.*, 2008; Seila *et al.*, 2008). The abundance of short RNA transcripts at the transcription initiation and termination sites (Core *et al.*, 2008; Seila *et al.*, 2008), as well as transcription of unstable RNAs upstream of genes that are usually co-localized with particular chromatin marks (Preker *et al.*, 2008), imply that these local and transient RNA transcripts are involved in the regulation of sense gene expression, and could possibly participate in chromatin remodelling. We hypothesize that the frequent presence of antisense RNAs in or around the promoter region of actively transcribed genes, in very low concentrations, suggests the possibility of local RNA-induced chromatin modifications.

Involvement of antisense RNAs in genomic imprinting

Antisense RNAs have also been heavily implicated in the process of genomic imprinting. Imprinted genes are those for which only one allele, either maternal or paternal, is actively transcribed, and these genes are often associated with natural antisense transcripts. In 2001, it was reported that 15% of imprinted genes had paired antisense transcripts (Reik and Walter, 2001), however in the following years, this figure significantly increased to 24–47% of human and mouse imprinted loci (Zhang *et al.*, 2006), and up to 81% of all imprinted transcriptional units of the mouse genome, when antisense transcripts complimentary to introns were included (Katayama *et al.*, 2005). There have been several studies suggesting pivotal roles for antisense transcripts in the regulation of imprinted genes, such as IGF2R (Wutz *et al.*, 1997), KCNQ1 (Smilinich *et al.*, 1999), UBE3A (Rougeulle *et al.*, 1998), ATP10C, MKRN3, MAGEL2, NDN (Runte *et al.*, 2001), Slc22a2, Slc22a3 (Sleutels *et al.*, 2002), GNAS (Hayward and Bonthron, 2000) and Gnas (Wroe *et al.*, 2000). Some of these imprinted genes, including the insulin-like growth-factor type-2 receptor (IGFER) (Sleutels *et al.*, 2002) and the KQT-like potassium voltage-gated channel (KCNQ1) imprinting control region (Pandey *et al.*, 2008), exhibit guided chromatin and DNA modification by antisense RNA that expands to include neighbouring genes. These effects are not mediated through the RNAi pathway (Tomari and Zamore, 2005), but rather antisense RNAs appear to recruit repressor complexes that modify chromatin into an inactive state. Suppressive chromatin modifications spread in both directions to neighbouring genes, similar to X chromosome inactivation, but with limited penetrance (Pandey *et al.*, 2008).

Antisense RNAs are required for the balanced expression of genes on the X chromosome in female mammals, due to their involvement in X chromosome inactivation. This inactivation and dosage compensation occurs through heterochromatin formation along the X chromosome. Indeed, two long non-protein-coding genes, known as the X-inactive specific transcript (XIST) and the X-inactive-specific antisense transcript (TSIX), are transcribed from the X chromosome inactivation centre, and control the silencing of the X chromosome (Ohhata *et al.*, 2008). Expression of XIST, a 17kb RNA transcript that triggers X inactivation (Brown *et al.*, 1992), is negatively regulated in cis by the antisense TSIX, which is transcribed along the entire XIST gene (Lee *et al.*, 1999; Ohhata *et al.*, 2008). Elements of the mechanism involved have been demonstrated, whereby TSIX silences XIST through modification of the chromatin structure in the XIST promoter region. Premature termination of TSIX transcription, by the introduction of a polyadenylation signal, abolishes the repressive chromatin configuration at the XIST promoter on the mutated X-chromosome (Ohhata *et al.*, 2008). The X chromosome inactivation centre is both necessary and sufficient for X chromosome inactivation.

All of these well-documented reports concerning X chromosome inactivation, genomic imprinting and methylation or chromatin modifications of autosomal loci suggest that a large proportion of natural antisense transcripts are involved in RNA-DNA binding, and control transcription through RNA-directed epigenetic modifications.

Nuclear RNA duplex formation

Another suggested mechanism for antisense transcript-mediated gene regulation is based on the predicted formation of a nuclear sense–antisense RNA duplex (Fig. 12.4). This may result in several outcomes, all of which serve to modulate sense mRNA expression. Nuclear RNA duplexes may contribute to alternative splicing, or hybrid nuclear RNA could act as a substrate for editing enzymes that could alter the localization, transport and stability of the sense mRNA transcript. Antisense transcripts could also potentially form a triplex with DNA strands and cause alternative initiation and termination of the sense RNA (Enerly *et al.*, 2005).

Alternative splicing and termination

Antisense RNAs may also bind to sense RNA transcripts, masking the splice sites and thereby changing the balance between splice variants (Fig. 12.5). In the case of N-MYC, the antisense transcript N-CYM, forms a RNA duplex with the donor site of the first exon of the N-MYC gene and modulates its splice variants (Krystal *et al.*, 1990). The thyroid hormone receptor alpha gene (TRα) is another example of this whereby the antisense transcript RevErbAα influences splicing of both TRα1 and TRα2 mRNAs (Miyajima *et al.*, 1989; Lazar *et al.*, 1990; Hastings *et al.*, 1997). Using a similar arrangement, antisense transcripts

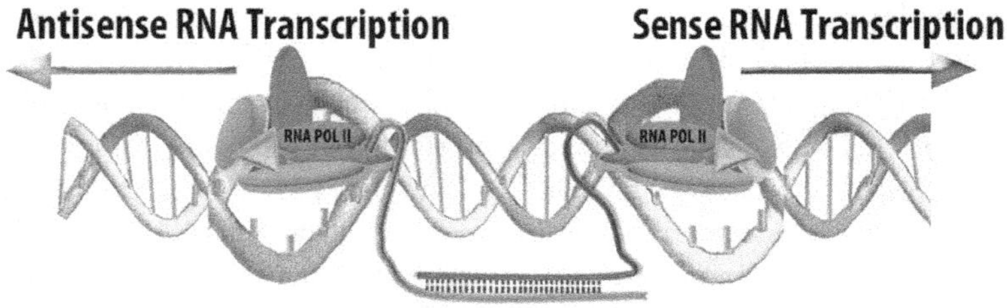

Figure 12.4 Nuclear sense–antisense RNA pairing. Nuclear RNA duplex formation might occur locally after transcription, thereby inhibiting sense RNA processing. The RNA processing includes capping, polyadenylation, nuclear localization and transport, which all might be affected by nuclear sense–antisense RNA duplex formation.

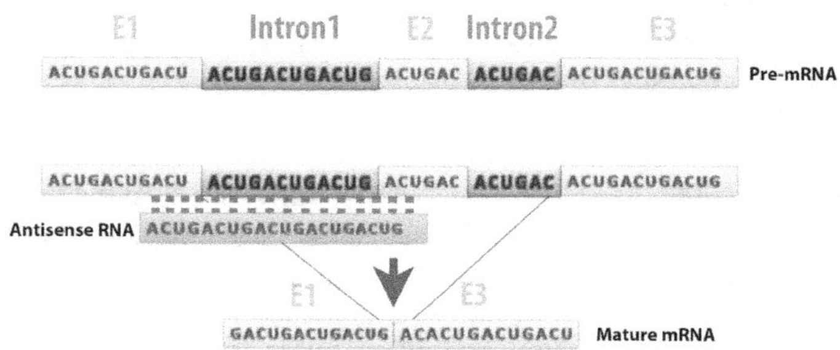

Figure 12.5 Nuclear and cytoplasmic sense–antisense RNA pairing. Natural antisense transcripts can cover donor and acceptor splice sites in the sense pre-mRNA transcript to change alternative splicing patterns.

could also cause alternative polyadenylation and termination of transcription.

Transport, nuclear retention and editing

Natural antisense transcripts may also modulate nuclear mRNA transport, by forming duplexes between sense and antisense RNAs. For instance, the antisense RNA for the non-spliced p53 RNA binds to intron-1, and prevents the transport of this p53 RNA to the cytoplasm (Khochbin *et al.*, 1992). This nuclear retention of antisense transcripts is commonly observed and may account for some antisense RNA-mediated regulation (Kumar and Carmichael, 1997; Prasanth *et al.*, 2005). Various cellular stressors, such as hypoxia, serum starvation and hydrogen peroxide, have been shown to change the nuclear retention pattern of antisense transcripts, thereby altering the levels of their sense partners (Faghihi *et al.*, 2008). It is likely that nuclear retention is mediated through direct interactions with nuclear proteins or other nuclear RNAs, although the precise mechanism for each case has not yet been discovered.

Natural antisense transcripts have also been linked to mRNA editing (Hatzoglou *et al.*, 2002; Peters *et al.*, 2003) (Fig. 12.6). Interactions between the Drosophila melanogaster 4f-rnp gene and its antisense transcript, sas-10, are reported to induce adenosine-to-inosine (A-to-I) editing in the overlapping region of 4f-rnp mRNA (Peters *et al.*, 2003). This type of A-to-I RNA editing is induced by double-stranded RNA formation, resulting in the recruitment of the enzyme ADAR (adenosine deaminase that acts on RNA), which deaminates the targeted adenosine to inosine (Ohman, 2007). There are few documented reports on the involvement of natural antisense transcripts in A-to-I RNA editing

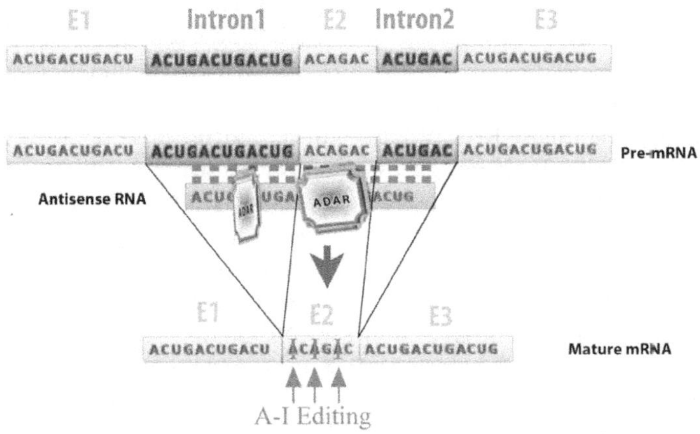

Figure 12.6 Nuclear and cytoplasmic sense–antisense RNA pairing. RNA editing is another possible consequence of nuclear RNA duplex formation, which causes A-to-I editing by the ADAR (adenosine deaminase that act on RNA) enzyme. RNA editing can induce nuclear retention and degradation of hyper-edited transcript by inosine-specific nucleases or alternatively, editing can result in alteration of the amino acid sequence.

(Kimelman and Kirschner, 1989; Dolnick, 1993; Peters *et al.*, 2003) and the overlapping regions of many sense–antisense pairs contain A–G mismatches (Hatzoglou *et al.*, 2002; Faghihi *et al.*, 2008), suggesting possible RNA editing. Although bioinformatic analyses argue against extensive editing of overlapping sense–antisense regions it does not exclude the possibility of alternative forms or degrees of editing. Nevertheless, editing is sometimes followed by the degradation of sense RNA via inosine-specific RNAse activity (Kumar and Carmichael, 1997; Scadden and Smith, 1997).

Alternative termination generates sense–antisense pairs

If two oppositely oriented neighbouring genes undergo alternative polyadenylation and termination, they may form an overlapping sense–antisense pattern. For instance, alternative polyadenylation of two oppositely oriented genes, the programmed cell death-2 (PDCD2) and TATA box binding protein (TBP), induces formation of a sense–antisense pair (Mihola *et al.*, 2007). Alternative termination of transcripts originating from the tumour suppressor p53-binding protein (TP53Bp1)/tubulin, gamma associated complex-4 (TUBGCP4) and the cyclin E2 (CCNE2)/FLJ20530 loci have also been reported to produce similar sense–antisense pairs (Yelin *et al.*, 2003). Therefore, it has been suggested that antisense RNAs may also play a role in controlling the balance of transcripts differing in their 3′ terminus (Dahary *et al.*, 2005).

Overlapping sense–antisense pairs are not the result of transcriptional 'leakage'

Given that at least half of all human genes encode transcripts with alternative 3′ termini (Iseli *et al.*, 2002), it was initially hypothesized that antisense transcription was a 'leakage' of RNA transcription machinery originating from un-terminated transcription of downstream genes. However, evolutionary studies concerning genomic organization of neighbouring genes conflicted with this transcriptional leakage hypothesis (Dahary *et al.*, 2005). Moreover, unambiguous sequencing of human cell transcriptomes (He *et al.*, 2008) did not support this idea. Indeed, there were poor correlations between the density of antisense-originated sequence tags and the density of sequence tags originated from the closest downstream genes (He *et al.*, 2008).

Cytoplasmic RNA duplex formation

The fourth proposed mechanism of sense–antisense interference, involves the formation of a duplex between sense and antisense RNA in the cytoplasm (Fig. 12.7A–C). Cytoplasmic RNA hairpins may affect sense mRNA stability or translation. This sense–antisense RNA duplex may also, in theory, cover microRNA (miRNA)-binding sites or serve as a hairpin template for generating endogenous small interfering RNA (siRNA).

Changes in mRNA stability

Cytoplasmic sense–antisense duplex formation can alter sense mRNA stability. We previously showed that cytoplasmic sense–antisense duplexes are not normally subject to Dicer cleavage in mammalian cells, and thus do not generally form a basis for endogenous RNA interference (Faghihi and Wahlestedt, 2006). However, duplex formation is still biologically relevant should antisense transcripts modulate the stability of their sense partners.

The overlapping region between sense and antisense transcripts may affect mRNA stability by reducing mRNA decay (the process by which mRNA undergoes endo- or exonucleolytic degradation by various RNases). Indeed, we have recently demonstrated that a BACE1 antisense transcript (BACE1-AS) increases the stability of BACE1 mRNA through a mechanism involving RNA duplex formation. We hypothesize that transient RNA duplex formation may alter the secondary or tertiary structure of BACE1 and thereby increase its stability (Faghihi *et al.*, 2008).

Antisense transcripts for inducible nitric oxide synthase (iNOS), which is an important gene in inflammatory diseases, also increase the stability of iNOS mRNA (Matsui *et al.*, 2008) through their interactions with the AU-rich element (ARE)-binding HuR protein. The HuR protein, in turn, may suppress RNA degradation by inhibiting deadenylase or exonuclease enzymes (Matsui *et al.*, 2008). In addition, a natural antisense transcript for basic fibroblast growth factor (bFGF) has been reported to alter the stability of bFGF mRNA (Volk *et al.*, 1989; Zuniga Mejia Borja *et al.*, 1993), possibly through a similar mechanism. In contrast, the antisense transcript for hypoxia inducible factor, HIF-1α (aHIF) is an example of a decrease in RNA stability induced by antisense RNA. The aHIF non-protein-coding RNA destabilizes one isoform of HIF-α mRNA and shifts the balance in favour of the other variant (Cayre *et al.*, 2003; Uchida *et al.*, 2004). Two isoforms of HIF, HIF-1α and HIF-2α, are both responsive to hypoxia but HIF-1α has a more ubiquitous expression pattern than HIF-2α. Prolonged hypoxia causes the induction of the aHIF antisense transcript, which in turn binds to the sense mRNA,

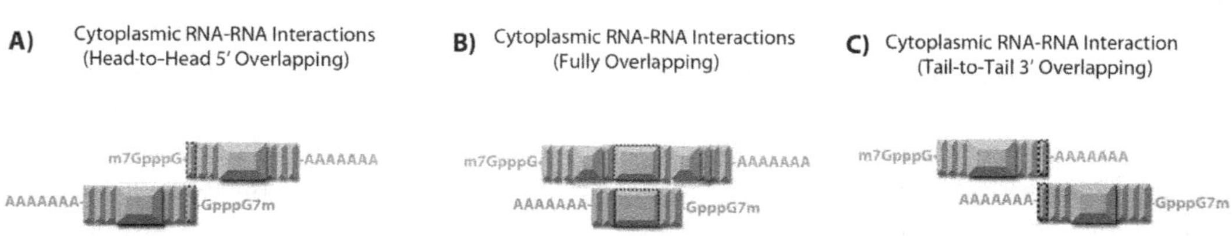

Figure 12.7 Cytoplasmic sense–antisense duplex formation. Cytoplasmic sense–antisense duplex formation can occur in three different formats; (A) head-to-head 5′ overlapping, (B) fully overlapping in which antisense is embedded in the sense transcript or (C) tail-to-tail 3′ overlapping configuration. Exons are depicted as vertical boxes and the overlapping region is shown with a border. Cytoplasmic sense–antisense RNA duplex formation can possibly exert its own effects on sense mRNA. Hiding or exposing AU-rich elements in sense transcript can affect RNA stability. Changes in the RNA secondary structure, after binding to antisense RNA, can alter translation, sub cellular localization and accessibility of the RNA degradation machinery. According to this model, natural antisense transcripts can potentially 'mask' miRNA-binding sites and release the miRNA-induced block of translation. Finally, Dicer can also process sense–antisense RNA duplexes to endogenous siRNAs.

leading to a reduction in HIF-1α transcript levels. Importantly, destabilization takes place by exposing the AU rich elements (ARE) in HIF-1α mRNA following antisense binding to its 3′UTR (Rossignol *et al.*, 2002, 2004; Uchida *et al.*, 2004).

These AU rich elements (ARE) are regions in RNA transcripts with frequent 'A' and 'U' nucleotides (such as 'AUUUA') that target the RNA for degradation. Alterations in HIF-1α secondary structure expose the ARE and make this RNA prone for degradation. Stabilization of mRNA by the opposite mechanism, i.e. covering the AU rich element, has been suggested for an antisense transcript of the Bcl–2/IgH hybrid gene (Capaccioli *et al.*, 1996).

Translation block by antisense RNAs

Translational inhibition is yet another proposed function for some naturally occurring antisense molecules, such as the B cell maturation antigen (BCMA) transcript, whereby over-expression of an antisense RNA has been reported to reduce sense protein, but not sense mRNA levels (Laabi *et al.*, 1994; Hatzoglou *et al.*, 2002). BCMA belongs to the tumour necrosis factor receptor (TNF-R) family and is controlled at the translational level by an antisense RNA transcript. While this BCMA antisense transcript is a fully processed RNA with an ORF, its discordant regulation of the sense protein appears to take place independent of antisense coding potential (Hatzoglou *et al.*, 2002).

Another documented example of translational inhibition is the transcription factor PU.1, an important regulator of haematopoiesis and a suppressor of leukaemia transformation. The translation of PU.1 is inhibited by a non-protein-coding antisense transcript (Ebralidze *et al.*, 2008), which is a polyadenylated RNA with a lower concentration but a longer half- life time than the sense PU.1 transcript. This form of translational control may be due to processed antisense RNAs in the cytoplasm binding to the sense transcript and stalling translation between the initiation and elongation steps (Ebralidze *et al.*, 2008). It is also possible that formation of duplexes may alter ribosome entry, and consequently the protein output of a target gene.

Cytoplasmic RNA duplex formation and 'masking' of microRNA binding sites

A number of natural antisense transcripts may have the ability to cover microRNA (miRNA)-binding sites through the formation of cytosolic RNA duplexes. This appears to be the case for the already described BACE1 antisense transcript (BACE1-AS), in addition to its ability to increase the stability of BACE1 mRNA (Faghihi *et al.*, 2008). The 3′UTR of mRNAs are frequently shown to contain target sites for miRNAs, and at least 34% of natural antisense transcripts in the FANTOM3 data set showed a tail-to-tail configuration with a 3′UTR overlapping region. In fact, in contrast to plants, where miRNA binding sites are predominantly found in coding regions, most animal miRNAs are predicted to have their binding site in the 3′ UTR of target mRNAs (Arteaga-Vazquez *et al.*, 2006). It must be mentioned however that, while most web tools for identifying miRNA binding sites are designed to screen 3′ UTR regions of transcripts, binding of miRNAs to the coding region, or even 5′ UTR of target mRNAs, has been shown, not only in plants but also recently in animals (Rhoades *et al.*, 2002; Lytle *et al.*, 2007; Duursma *et al.*, 2008). We propose that one of the regulatory functions of antisense transcripts may be to 'mask' miRNA-binding sites on the sense mRNA.

Natural antisense RNA and the formation of endogenous siRNA

RNA interference

RNA interference (RNAi) is a surveillance mechanism, through which a cell responds to exogenous double-stranded RNA (dsRNA) molecules by destroying mRNAs containing sequences homologous to the dsRNA (Fire *et al.*, 1998). Duplex RNAs are first cleaved by an RNase III enzyme, known as Dicer, to generate short interfering RNAs (siRNAs). SiRNAs include two strands, the guide and passenger strands, and become incorporated into a multi-protein RNA-induced silencing complex (RISC) that unwinds the helical structure of the siRNA duplex. The guide strand is retained in the RISC and guides the entire complex to a target mRNA. Based on the complementarity between the target mRNA sequence and the guide strand sequence, RISC initiates the endonucleolytic cleavage or translational arrest of the target mRNA (Saxena *et al.*, 2003).

Processing of short RNA (>200 nucleotides in length) from long and largely un-annotated nuclear RNAs was suggested from high resolution tilling array studies of human cell lines (Kapranov *et al.*, 2007). This raised the possibility that long non-protein-coding RNAs may control mRNA levels, through the generation of siRNAs. Indeed, gene regulation by endogenous siRNAs has been frequently observed in organisms possessing RNA-dependent RNA polymerase (RdRP) (Ruby *et al.*, 2006; Pak and Fire, 2007; Sijen *et al.*, 2007).

Endogenous siRNA

The presence of endogenous processing machinery for exogenous siRNAs, which mediate sequence-specific knockdown of targeted genes, implies that endogenous siRNAs should also exist. Endogenous plant siRNAs, derived from natural antisense transcripts, have been observed in Arabidopsis thaliana, where they regulate salt tolerance (Borsani *et al.*, 2005). Two types of these siRNAs were shown to be generated from the overlapping region of pyrroline-5-carboxylate dehydrogenase (P5CDH), a stress-related gene, and SRO5, a gene of unknown function (Borsani *et al.*, 2005). It appears that endogenous plant siRNAs are derived from the sense–antisense RNA duplex formation of several genes: such as the Sho gene antisense transcript (Zubko and Meyer, 2007), and the small RNA-generating RLK (SRRLK)/AtRAP antisense transcript (Katiyar-Agarwal *et al.*, 2007). Furthermore 64% of protein-coding natural antisense transcripts in A. thaliana have been reported to generate endogenous siRNAs (Jin *et al.*, 2008).

In mammals, no RNA-dependent RNA polymerase (RdRP) activity has yet been found, and, as a result, the biogenesis and functions of endogenous siRNAs remain largely unknown (Okamura and Lai, 2008). However, in recent years, endogenous siRNAs originating from mRNAs, and their corresponding antisense transcripts, have been identified in both mouse oocytes (Watanabe *et al.*, 2008) and human HepG2 liver carcinoma cells (Kawaji *et al.*, 2008). In fact, both Piwi-interacting RNAs[2] (piRNAs, Dicer independent) (Aravin *et al.*, 2007; Klattenhoff and Theurkauf, 2008) and (approximately) 21-nucleotide siRNAs, originating from mRNAs, were found in mouse oocytes (Watanabe *et al.*, 2008). Transposable elements, inverted repeat structures, sense–antisense genes (cis-natural antisense transcript) and antisense transcripts from remote loci (trans-natural antisense transcripts) have also been recognized as sources of dsRNAs and subsequent, Dicer-dependent, endogenous siRNA production (Kawaji *et al.*, 2008; Watanabe *et al.*, 2008). It is therefore conceivable that endogenous siRNAs are produced from naturally occurring antisense transcripts in mammalian cells.

Intramolecular versus intermolecular double-stranded RNA formation

Inverted repeats with an intra-molecular double-stranded RNA (dsRNA) structure are more likely to act as precursors of endogenous siRNAs as they can form more stable dsRNA structures required for their cleavage by Dicer. However, endogenous siRNA has also been reported from intermolecular sense–antisense double-stranded RNA formation. One example of this is the overlapping transcripts in the kinesin family member 4A (KIF4A) and PDZ domain-containing 11 (Pdzd11) locus that generate an endogenous siRNA derived from its cis-antisense RNA. KIF4A and Pdzd11 are two genes located on opposite strands of the X chromosome, where they are orientated in a head-to-head manner. The overlapping transcripts in the Pdzd11/Kif4A locus are shown to generate endogenous siRNA. Importantly, almost all of the endogenous siRNAs (117 unique sequences) are derived from the overlapping region of the sense and antisense RNA, suggesting that these

[2]Piwi-interacting RNA: A small (25–35 nucleotides long) RNA species that is processed from single-stranded precursor RNA, independent of Dicer, and forms a complex with the Piwi protein. piRNAs are though to be involved in transposon silencing and stem-cell function.

endogenous siRNAs are produced from an intermolecular dsRNA formed between oppositely oriented transcripts. In Dicer mutants, levels of the siRNAs derived from the Pdzd11/Kif4 locus are decreased, and Pdzd11 and Kif4 mRNA levels are increased, suggesting that endogenous siRNA production is cytoplasmic and Dicer dependent, and that expression of both sense and antisense RNAs is regulated by the RNAi pathway (Watanabe *et al.*, 2008). Therefore, endogenous siRNAs may regulate both sense and antisense transcript levels.

Endogenous siRNA is not a prime consequence of sense–antisense duplex RNA formation

Although antisense RNAs are abundant in mammalian cells, high-throughput small RNA sequencing has not yielded many endogenous siRNA mapping to antisense transcripts, suggesting that RNAi is not their predominant mode of action. The co-expression of natural antisense transcripts with their sense counterparts (Faghihi and Wahlestedt, 2006), as well as the frequently observed concordant regulation of sense and antisense RNAs in many tissues and cell lines, also argues against endogenous siRNA being the sole mechanism of antisense-mediated regulation of gene expression. Additionally, many co-expressed sense–antisense transcripts in *D. melanogaster* S2 cells do not generate endogenous siRNAs (Czech *et al.*, 2008; Okamura *et al.*, 2008).

It is unclear how the majority of co-expressed sense and antisense RNAs escape the endogenous siRNA formation pathway (Faghihi and Wahlestedt, 2006). It is also not clear whether there is active selection for entry into the RNAi pathway and endogenous siRNA formation. Subcellular compartmentalization and coating of transcripts by proteins might act as potential barriers for the formation of dsRNA and subsequent endogenous siRNAs. Nevertheless, biogenesis of endogenous siRNAs from some antisense transcripts has been documented. The reports mentioned above have suggested a biological role for endogenous siRNAs in mammals and show that organisms lacking RNA-dependent RNA polymerase activity can produce functional endogenous siRNAs from naturally occurring dsRNAs (Kawaji *et al.*, 2008; Tam *et al.*, 2008; Watanabe *et al.*, 2008).

Four models of antisense transcripts regulatory function

In the mechanism sections of this chapter, I presented examples of functional natural antisense transcripts in order to show the multilayered involvement of these molecules in regulating gene expression. It can be concluded that antisense transcripts are not a uniform group of regulatory elements, but that they display some common features, including sequence complementarity to conventional sense genes. In summary, the proposed mechanisms of antisense transcript actions can be categorized into four models, each corresponding to putative regulatory mechanism for certain antisense RNA.

Although there are examples for each model, it is important to note that not all four models have equally strong experimental support. For instance, the frequently observed convergent promoter activity, and engagement of multiple RNA Polymerase II complexes in the opposite orientation of active promoters, would argue against the transcriptional collision model as a widely used mode of action. On the other hand, nuclear and cytoplasmic RNA–RNA interactions, which result in various outcomes, such as RNA editing, splicing and dicing, have been observed for several functionally validated antisense transcripts, but this is not a predominant basis of natural antisense regulation. Furthermore, antisense RNA-induced alteration in sense promoter DNA methylation has only a few documented examples and, considering the static nature of DNA methylation, it might be operational only in certain early developmental stages. Conversely, antisense RNA-induced chromatin remodelling may be a more feasible and dynamic mode of action for many low-copy natural antisense transcripts. In the latter case, antisense RNA might predominantly act locally to maintain or modify chromatin structure, and ultimately activate or suppress sense gene expression. Natural antisense transcripts are involved in different gene regulatory pathways, but it is still not clear which intrinsic properties of natural antisense RNA molecules, or extrinsic features such as protein interactions, cellular and developmental context, are decisive for any given pathway.

It is becoming clear that the expression of natural antisense transcripts and other non-protein-coding RNAs are pervasive in the human genome, although the mechanisms of their production and preferential sites of action are poorly understood. Particularly, the exact molecular machineries behind the antisense RNA-induced chromatin modifications are not currently known. It will be important to learn which histone-modifying enzymes might be involved, how they interact with antisense RNAs and which histone modifications they induce over time. It will also be of interest to define the exact contexts in which perturbation of the same antisense transcript (Schwartz *et al.*, 2008) results in either activation or repression of sense RNA expression. It is remarkable that these fundamental and powerful regulatory mechanisms relating to natural antisense transcription remain to be elucidated in large part. It can be safely assumed that examination of natural antisense phenomena will remain a rich field of investigation for the foreseeable future.

Conclusion

We know that many natural antisense transcripts are functional RNA molecules that control sense mRNA expression at different levels. Most natural antisense transcripts are non-coding RNAs, which help to mediate sense gene expression in response to a variety of environmental stimuli and to keep tight regulation of sense protein expression (allowing proteins to perform their physiological functions while avoiding the serious disease-related consequences of over or under expression).

Although protein synthesis was the first assigned function for RNA molecules, it is becoming increasingly evident that the more pervasive function of RNA molecules is a regulatory one. This hypothesis is considerably supported by the content of this current chapter in which we summarize the proposed regulatory functions of naturally occurring antisense transcripts. Considering the other reported functional long non-protein-coding RNAs (macroRNAs), and small regulatory RNA (such as miRNA, piRNA, rasiRNA), and their enormous physiological impact, it is now feasible to claim that RNA molecules primarily act in a regulatory manner, and that this could far outweigh their traditionally assigned functions in protein synthesis as messenger, transfer and ribosomal RNAs.

References

Aravin, A.A., Hannon, G.J., and Brennecke, J. (2007). The Piwi-piRNA pathway provides an adaptive defense in the transposon arms race. Science *318*, 761–764.

Arteaga-Vazquez, M., Caballero-Perez, J., and Vielle-Calzada, J.P. (2006). A family of microRNAs present in plants and animals. Plant Cell *18*, 3355–3369.

Beltran, M., Puig, I., Pena, C., Garcia, J.M., Alvarez, A.B., Pena, R., Bonilla, F., and de Herreros, A.G. (2008). A natural antisense transcript regulates Zeb2/Sip1 gene expression during Snail1-induced epithelial-mesenchymal transition. Genes Dev. *22*, 756–769.

Bernstein, E., and Allis, C.D. (2005). RNA meets chromatin. Genes Dev. *19*, 1635–1655.

Bertone, P., Stolc, V., Royce, T.E., Rozowsky, J.S., Urban, A.E., Zhu, X., Rinn, J.L., Tongprasit, W., Samanta, M., Weissman, S., *et al.* (2004). Global identification of human transcribed sequences with genome tiling arrays. Science *306*, 2242–2246.

Borel, C., Gagnebin, M., Gehrig, C., Kriventseva, E.V., Zdobnov, E.M., and Antonarakis, S.E. (2008). Mapping of small RNAs in the human ENCODE regions. Am. J. Hum. Genet. *82*, 971–981.

Borsani, O., Zhu, J., Verslues, P.E., Sunkar, R., and Zhu, J.K. (2005). Endogenous siRNAs derived from a pair of natural cis-antisense transcripts regulate salt tolerance in Arabidopsis. Cell *123*, 1279–1291.

Brown, C.J., Hendrich, B.D., Rupert, J.L., Lafreniere, R.G., Xing, Y., Lawrence, J., and Willard, H.F. (1992). The human XIST gene: analysis of a 17kb inactive X-specific RNA that contains conserved repeats and is highly localized within the nucleus. Cell *71*, 527–542.

Capaccioli, S., Quattrone, A., Schiavone, N., Calastretti, A., Copreni, E., Bevilacqua, A., Canti, G., Gong, L., Morelli, S., and Nicolin, A. (1996). A bcl-2/IgH antisense transcript deregulates bcl-2 gene expression in human follicular lymphoma t(14;18) cell lines. Oncogene *13*, 105–115.

Carninci, P., Kasukawa, T., Katayama, S., Gough, J., Frith, M.C., Maeda, N., Oyama, R., Ravasi, T., Lenhard, B., Wells, C., *et al.* (2005). The transcriptional landscape of the mammalian genome. Science *309*, 1559–1563.

Cawley, S., Bekiranov, S., Ng, H.H., Kapranov, P., Sekinger, E.A., Kampa, D., Piccolboni, A., Sementchenko, V., Cheng, J., Williams, A.J., *et al.* (2004). Unbiased mapping of transcription factor binding sites along human chromosomes 21 and 22 points to widespread regulation of noncoding RNAs. Cell *116*, 499–509.

Cayre, A., Rossignol, F., Clottes, E., and Penault-Llorca, F. (2003). aHIF but not HIF-1alpha transcript is a poor prognostic marker in human breast cancer. Breast Cancer Res. *5*, R223–230.

Chan, W.Y., Wu, S.M., Ruszczyk, L., Law, E., Lee, T.L., Baxendale, V., Lap-Yin Pang, A., and Rennert, O.M. (2006). The complexity of antisense transcription revealed by the study of developing male germ cells. Genomics 87(6), 681–92.

Chen, J., Sun, M., Kent, W.J., Huang, X., Xie, H., Wang, W., Zhou, G., Shi, R.Z., and Rowley, J.D. (2004). Over 20% of human transcripts might form sense–antisense pairs. Nucleic Acids Res. *32*, 4812–4820.

Chen, J., Sun, M., Hurst, L.D., Carmichael, G.G., and Rowley, J.D. (2005). Human antisense genes have unusually short introns: evidence for selection for rapid transcription. Trends Genet. *21*, 203–207.

Cheng, J., Kapranov, P., Drenkow, J., Dike, S., Brubaker, S., Patel, S., Long, J., Stern, D., Tammana, H., Helt, G., *et al.* (2005). Transcriptional maps of 10 human chromosomes at 5-nucleotide resolution. Science *308*, 1149–1154.

Core, L.J., Waterfall, J.J., and Lis, J.T. (2008). Nascent RNA sequencing reveals widespread pausing and divergent initiation at human promoters. Science *322*, 1845–1848.

Czech, B., Malone, C.D., Zhou, R., Stark, A., Schlingeheyde, C., Dus, M., Perrimon, N., Kellis, M., Wohlschlegel, J.A., Sachidanandam, R., *et al.* (2008). An endogenous small interfering RNA pathway in Drosophila. Nature *453*, 798–802.

Dahary, D., Elroy-Stein, O., and Sorek, R. (2005). Naturally occurring antisense: transcriptional leakage or real overlap? Genome Res. *15*, 364–368.

Dolnick, B.J. (1993). Cloning and characterization of a naturally occurring antisense RNA to human thymidylate synthase mRNA. Nucleic Acids Res. *21*, 1747–1752.

Dreesen, T.D., Adamson, A.W., Tekle, M., Tang, C., Cho, H.P., Clarke, S.D., and Gettys, T.W. (2006). A newly discovered member of the fatty acid desaturase gene family: a non-coding, antisense RNA gene to delta5-desaturase. Prostaglandins Leukot. Essent. Fatty Acids *75*, 97–106.

Duursma, A.M., Kedde, M., Schrier, M., le Sage, C., and Agami, R. (2008). miR-148 targets human DNMT3b protein coding region. RNA 14(5), 872–7.

Ebralidze, A.K., Guibal, F.C., Steidl, U., Zhang, P., Lee, S., Bartholdy, B., Jorda, M.A., Petkova, V., Rosenbauer, F., Huang, G., *et al.* (2008). PU.1 expression is modulated by the balance of functional sense and antisense RNAs regulated by a shared cis-regulatory element. Genes Dev. *22*, 2085–2092.

Enerly, E., Sheng, Z., and Li, K.B. (2005). Natural antisense as potential regulator of alternative initiation, splicing and termination. *In Silico* Biol. *5*, 367–377.

Faghihi, M.A., and Wahlestedt, C. (2006). RNA interference is not involved in natural antisense mediated regulation of gene expression in mammals. Genome Biol. *7*, R38.

Faghihi, M.A., and Wahlestedt, C. (2009). Regulatory roles of natural antisense transcripts. Nat. Rev. Mol. Cell. Biol. *10*, 637–643.

Faghihi, M.A., Modarresi, F., Khalil, A.M., Wood, D.E., Sahagan, B.G., Morgan, T.E., Finch, C.E., St Laurent, G., 3rd, Kenny, P.J., and Wahlestedt, C. (2008). Expression of a noncoding RNA is elevated in Alzheimer's disease and drives rapid feed-forward regulation of beta-secretase. Nat. Med. *14*, 723–730.

Fahey, M.E., Moore, T.F., and Higgins, D.G. (2002). Overlapping antisense transcription in the human genome. Comp. Funct. Genomics 3, 244–253.

Filion, G.J., van Bemmel, J.G., Braunschweig, U., Talhout, W., Kind, J., Ward, L.D., Brugman, W., de Castro, I.J., Kerkhoven, R.M., Bussemaker, H.J., *et al.* Systematic protein location mapping reveals five principal chromatin types in Drosophila cells. Cell *143*, 212–224.

Finocchiaro, G., Carro, M.S., Francois, S., Parise, P., DiNinni, V., and Muller, H. (2007). Localizing hotspots of antisense transcription. Nucleic Acids Res. *35*, 1488–1500.

Fire, A., Xu, S., Montgomery, M.K., Kostas, S.A., Driver, S.E., and Mello, C.C. (1998). Potent and specific genetic interference by double-stranded RNA in *Caenorhabditis elegans*. Nature *391*, 806–811.

Ge, X., Wu, Q., Jung, Y.C., Chen, J., and Wang, S.M. (2006). A large quantity of novel human antisense transcripts detected by LongSAGE. Bioinformatics *22*, 2475–2479.

Ge, X., Rubinstein, W.S., Jung, Y.C., and Wu, Q. (2008). Genome-wide analysis of antisense transcription with Affymetrix exon array. BMC Genomics *9*, 27.

Gyorffy, A., Surowiak, P., Tulassay, Z., and Gyorffy, B. (2007). Highly expressed genes are associated with inverse antisense transcription in mouse. J. Genet. *86*, 103–109.

Han, J., Kim, D., and Morris, K.V. (2007). Promoter-associated RNA is required for RNA-directed transcriptional gene silencing in human cells. Proc. Natl. Acad. Sci. U.S.A. *104*, 12422–12427.

Hastings, M.L., Milcarek, C., Martincic, K., Peterson, M.L., and Munroe, S.H. (1997). Expression of the thyroid hormone receptor gene, erbAalpha, in B lymphocytes: alternative mRNA processing is independent of differentiation but correlates with antisense RNA levels. Nucleic Acids Res. *25*, 4296–4300.

Hatzoglou, A., Deshayes, F., Madry, C., Lapree, G., Castanas, E., and Tsapis, A. (2002). Natural antisense RNA inhibits the expression of BCMA, a tumour necrosis factor receptor homologue. BMC Mol. Biol. *3*, 4.

Hayward, B.E., and Bonthron, D.T. (2000). An imprinted antisense transcript at the human GNAS1 locus. Hum. Mol. Genet. *9*, 835–841.

He, Y., Vogelstein, B., Velculescu, V.E., Papadopoulos, N., and Kinzler, K.W. (2008). The antisense transcriptomes of human cells. Science 322(5909), 1855–7.

Imamura, T., Yamamoto, S. Ohgane, J., Hattori, N., Tanaka, S., and Shiota, K. (2004). Non-coding RNA directed DNA demethylation of Sphk1 CpG island. Biochem. Biophys. Res. Commun. *322*, 593–600.

Iseli, C., Stevenson, B.J., de Souza, S.J., Samaia, H.B., Camargo, A.A., Buetow, K.H., Strausberg, R.L.,

Simpson, A.J., Bucher, P., and Jongeneel, C.V. (2002). Long-range heterogeneity at the 3′ ends of human mRNAs. Genome Res. *12*, 1068–1074.

Janowski, B.A., Huffman, K.E., Schwartz, J.C., Ram, R., Hardy, D., Shames, D.S., Minna, J.D., and Corey, D.R. (2005). Inhibiting gene expression at transcription start sites in chromosomal DNA with antigene RNAs. Nat. Chem. Biol. *1*, 216–222.

Janowski, B.A., Younger, S.T., Hardy, D.B., Ram, R., Huffman, K.E., and Corey, D.R. (2007). Activating gene expression in mammalian cells with promoter-targeted duplex RNAs. Nat. Chem. Biol. *3*, 166–173.

Jin, H., Vacic, V., Girke, T., Lonardi, S., and Zhu, J.K. (2008). Small RNAs and the regulation of cis-natural antisense transcripts in Arabidopsis. BMC Mol. Biol. *9*, 6.

Kanduri, C. (2008). Functional insights into long antisense noncoding RNA Kcnq1ot1 mediated bidirectional silencing. RNA Biol. *5(4)*, 208–11.

Kapranov, P., Cheng, J., Dike, S., Nix, D.A., Duttagupta, R., Willingham, A.T., Stadler, P.F., Hertel, J., Hackermuller, J., Hofacker, I.L., *et al.* (2007). RNA maps reveal new RNA classes and a possible function for pervasive transcription. Science *316*, 1484–1488.

Katayama, S., Tomaru, Y., Kasukawa, T., Waki, K., Nakanishi, M., Nakamura, M., Nishida, H., Yap, C.C., Suzuki, M., Kawai, J., *et al.* (2005). Antisense transcription in the mammalian transcriptome. Science *309*, 1564–1566.

Katiyar-Agarwal, S., Gao, S., Vivian-Smith, A., and Jin, H. (2007). A novel class of bacteria-induced small RNAs in Arabidopsis. Genes Dev. *21*, 3123–3134.

Kawaji, H., Nakamura, M., Takahashi, Y., Sandelin, A., Katayama, S., Fukuda, S., Daub, C.O., Kai, C., Kawai, J., Yasuda, J., *et al.* (2008). Hidden layers of human small RNAs. BMC Genomics 9, 157.

Khalil, A.M., Faghihi, M.A., Modarresi, F., Brothers, S.P., and Wahlestedt, C. (2008). A novel RNA transcript with antiapoptotic function is silenced in fragile x syndrome. PLoS ONE 3, e1486.

Khochbin, S., Brocard, M.P., Grunwald, D., and Lawrence, J.J. (1992). Antisense RNA and p53 regulation in induced murine cell differentiation. Ann. N.Y. Acad. Sci. *660*, 77–87.

Kimelman, D., and Kirschner, M.W. (1989). An antisense mRNA directs the covalent modification of the transcript encoding fibroblast growth factor in Xenopus oocytes. Cell *59*, 687–696.

Kiyosawa, H., Yamanaka, I., Osato, N., Kondo, S., and Hayashizaki, Y. (2003). Antisense transcripts with FANTOM2 clone set and their implications for gene regulation. Genome Res. *13*, 1324–1334.

Kiyosawa, H., Mise, N., Iwase, S., Hayashizaki, Y., and Abe, K. (2005). Disclosing hidden transcripts: mouse natural sense–antisense transcripts tend to be poly(A) negative and nuclear localized. Genome Res. *15*, 463–474.

Klattenhoff, C., and Theurkauf, W. (2008). Biogenesis and germline functions of piRNAs. Development *135*, 3–9. Klimov, D., Skoblov, M., Ryazantzev, A., Tyazhelova, T., and Baranova, A. (2006). *In silico* search for natural antisense transcripts reveals their differential expression in human tumors. J. Bioinform. Comput. Biol. *4*, 515–521.

Komine, Y., Nakamura, K., Katsuki, M., and Yamamori, T. (2006). Novel transcription factor zfh-5 is negatively regulated by its own antisense RNA in mouse brain. Mol. Cell Neurosci. *31*, 273–283.

Kramer, C., Loros, J.J., Dunlap, J.C., and Crosthwaite, S.K. (2003). Role for antisense RNA in regulating circadian clock function in Neurospora crassa. Nature *421*, 948–952.

Krystal, G.W., Armstrong, B.C., and Battey, J.F. (1990). N-myc mRNA forms an RNA-RNA duplex with endogenous antisense transcripts. Mol. Cell. Biol. *10*, 4180–4191.

Kumar, M., and Carmichael, G.G. (1997). Nuclear antisense RNA induces extensive adenosine modifications and nuclear retention of target transcripts. Proc. Natl. Acad. Sci. U.S.A. *94*, 3542–3547.

Laabi, Y., Gras, M.P., Brouet, J.C., Berger, R., Larsen, C.J., and Tsapis, A. (1994). The BCMA gene, preferentially expressed during B lymphoid maturation, is bidirectionally transcribed. Nucleic Acids Res. *22*, 1147–1154.

Lazar, M.A., Hodin, R.A., Cardona, G., and Chin, W.W. (1990). Gene expression from the c-erbA alpha/Rev-ErbA alpha genomic locus. Potential regulation of alternative splicing by opposite strand transcription. J. Biol. Chem. *265*, 12859–12863.

Lee, J.T., Davidow, L.S., and Warshawsky, D. (1999). Tsix, a gene antisense to Xist at the X-inactivation centre. Nat. Genet. *21*, 400–404.

Lehner, B., Williams, G., Campbell, R.D., and Sanderson, C.M. (2002). Antisense transcripts in the human genome. Trends Genet. *18*, 63–65.

Li, Y.Y., Qin, L., Guo, Z.M., Liu, L., Xu, H., Hao, P., Su, J., Shi, Y., He, W.Z., and Li, Y.X. (2006). *In silico* discovery of human natural antisense transcripts. BMC Bioinformatics 7, 18.

Lytle, J.R., Yario, T.A., and Steitz, J.A. (2007). Target mRNAs are repressed as efficiently by microRNA-binding sites in the 5′ UTR as in the 3′ UTR. Proc. Natl. Acad. Sci. U.S.A. *104*, 9667–9672.

Matsui, K., Nishizawa, M., Ozaki, T., Kimura, T., Hashimoto, I., Yamada, M., Kaibori, M., Kamiyama, Y., Ito, S., and Okumura, T. (2008). Natural antisense transcript stabilizes inducible nitric oxide synthase messenger RNA in rat hepatocytes. Hepatology *47*, 686–697.

Mihola, O., Forejt, J., and Trachtulec, Z. (2007). Conserved alternative and antisense transcripts at the programmed cell death 2 locus. BMC Genomics *8*, 20.

Miyajima, N., Horiuchi, R., Shibuya, Y., Fukushige, S., Matsubara, K., Toyoshima, K., and Yamamoto, T. (1989). Two erbA homologs encoding proteins with different T3 binding capacities are transcribed from opposite DNA strands of the same genetic locus. Cell *57*, 31–39.

Mohammad, F., Pandey, R.R., Nagano, T., Chakalova, L., Mondal, T., Fraser, P., and Kanduri, C. (2008). Kcnq1ot1/Lit1 noncoding RNA mediates transcriptional silencing by targeting to the perinucleolar region. Mol. Cell. Biol. *28*, 3713–3728.

Morris, K.V., Santoso, S., Turner, A.M., Pastori, C., and Hawkins, P.G. (2008). Bidirectional transcription directs both transcriptional gene activation and suppression in human cells. PLoS Genet. *4*, e1000258.

Munroe, S.H., and Lazar, M.A. (1991). Inhibition of c-erbA mRNA splicing by a naturally occurring antisense RNA. J. Biol. Chem. *266*, 22083–22086.

Neckers, L.M. (1999). aHIF: the missing link between HIF-1 and VHL? J. Natl. Cancer Inst. *91*, 106–107. Neeman, Y., Dahary, D., Levanon, E.Y., Sorek, R., and Eisenberg, E. (2005). Is there any sense in antisense editing? Trends Genet. *21*, 544–547.

Nepveu, A., and Marcu, K.B. (1986). Intragenic pausing and anti-sense transcription within the murine c-myc locus. EMBO J. *5*, 2859–2865. Ohhata, T., Hoki, Y., Sasaki, H., and Sado, T. (2008).

Crucial role of antisense transcription across the Xist promoter in Tsix-mediated Xist chromatin modification. Development *135*, 227–235.

Ohman, M. (2007). A-to-I editing challenger or ally to the microRNA process. Biochimie *89*, 1171–1176.

Okada, Y., Tashiro, C., Numata, K., Watanabe, K., Nakaoka, H., Yamamoto, N., Okubo, K., Ikeda, R., Saito, R., Kanai, A., *et al.* (2008). Comparative expression analysis uncovers novel features of endogenous antisense transcription. Hum. Mol. Genet. *17*, 1631–1640.

Okamura, K., and Lai, E.C. (2008). Endogenous small interfering RNAs in animals. Nat. Rev. Mol. Cell. Biol. *9*, 673–678.

Okamura, K., Balla, S., Martin, R., Liu, N., and Lai, E.C. (2008). Two distinct mechanisms generate endogenous siRNAs from bidirectional transcription in Drosophila melanogaster. Nat. Struct. Mol. Biol. *15*, 998.

Okazaki, Y., Furuno, M., Kasukawa, T., Adachi, J., Bono, H., Kondo, S., Nikaido, I., Osato, N., Saito, R., Suzuki, H., *et al.* (2002). Analysis of the mouse transcriptome based on functional annotation of 60,770 full-length cDNAs. Nature *420*, 563–573.

Pak, J., and Fire, A. (2007). Distinct populations of primary and secondary effectors during RNAi in *C. elegans*. Science *315*, 241–244.

Pandey, R.R., Mondal, T., Mohammad, F., Enroth, S., Redrup, L., Komorowski, J., Nagano, T., Mancini-Dinardo, D., and Kanduri, C. (2008). Kcnq1ot1 antisense noncoding RNA mediates lineage-specific transcriptional silencing through chromatin-level regulation. Mol. Cell *32*, 232–246.

Peters, N.T., Rohrbach, J.A., Zalewski, B.A., Byrkett, C.M., and Vaughn, J.C. (2003). RNA editing and regulation of Drosophila 4f-rnp expression by sas-10 antisense readthrough mRNA transcripts. RNA *9*, 698–710.

Prasanth, K.V., Prasanth, S.G., Xuan, Z., Hearn, S., Freier, S.M., Bennett, C.F., Zhang, M.Q., and Spector, D.L. (2005). Regulating gene expression through RNA nuclear retention. Cell *123*, 249–263.

Preker, P., Nielsen, J., Kammler, S., Lykke-Andersen, S.,

Christensen, M.S., Mapendano, C.K., Schierup, M.H., and Jensen, T.H. (2008). RNA exosome depletion reveals transcription upstream of active human promoters. Science *322*, 1851–1854.

Quere, R., Manchon, L., Lejeune, M., Clement, O., Pierrat, F., Bonafoux, B., Commes, T., Piquemal, D., and Marti, J. (2004). Mining SAGE data allows large-scale, sensitive screening of antisense transcript expression. Nucleic Acids Res. *32*, e163.

Reik, W., and Walter, J. (2001). Genomic imprinting: parental influence on the genome. Nat. Rev. Genet. *2*, 21–32.

Rhoades, M.W., Reinhart, E.J., Lim, L.P., Burge, C.B., Bartel, B., and Bartel, D.P. (2002). Prediction of plant microRNA targets. Cell *110*, 513–520.

Richards, M., Tan, S.P., Chan, W.K., and Bongso, A. (2006). Reverse Serial Analysis of Gene Expression (SAGE) Characterization of orphan SAGE Tags from human embryonic stem cells identifies the presence of novel transcripts and antisense transcription of key pluripotency genes. Stem Cells *24*, 1162–1173.

Rinn, J.L., Euskirchen, G., Bertone, P., Martone, R., Luscombe, N.M., Hartman, S., Harrison, P.M., Nelson, F.K., Miller, P., Gerstein, M., *et al.* (2003). The transcriptional activity of human chromosome 22. Genes Dev. *17*, 529–540.

Rinn, J.L., Kertesz, M., Wang, J.K., Squazzo, S.L., Xu, X., Brugmann, S.A., Goodnough, L.H., Helms, J.A., Farnham, P.J., Segal, E., *et al.* (2007). Functional demarcation of active and silent chromatin domains in human HOX loci by noncoding RNAs. Cell *129*, 1311–1323.

Rosok, O., and Sioud, M. (2004). Systematic identification of sense–antisense transcripts in mammalian cells. Nat. Biotechnol. *22*, 104–108.

Rosok, O., and Sioud, M. (2005). Systematic search for natural antisense transcripts in eukaryotes (review). Int. J. Mol. Med. *15*, 197–203.

Rossignol, F., Vache, C., and Clottes, E. (2002). Natural antisense transcripts of hypoxia-inducible factor 1alpha are detected in different normal and tumour human tissues. Gene *299*, 135–140.

Rossignol, F., de Laplanche, E., Mounier, R., Bonnefont, J., Cayre, A., Godinot, C., Simonnet, H., and Clottes, E. (2004). Natural antisense transcripts of HIF-1alpha are conserved in rodents. Gene *339*, 121–130.

Rougeulle, C., Cardoso, C., Fontes, M., Colleaux, L., and Lalande, M. (1998). An imprinted antisense RNA overlaps UBE3A and a second maternally expressed transcript. Nat. Genet. *19*, 15–16.

Rougeulle, C., and Heard, E. (2002). Antisense RNA in imprinting: spreading silence through Air. Trends Genet. *18*, 434–437.

Ruby, J.G., Jan, C., Player, C., Axtell, M.J., Lee, W., Nusbaum, C., Ge, H., and Bartel, D.P. (2006). Large-scale sequencing reveals 21U-RNAs and additional microRNAs and endogenous siRNAs in *C. elegans*. Cell *127*, 1193–1207.

Runte, M., Huttenhofer, A., Gross, S., Kiefmann, M., Horsthemke, B., and Buiting, K. (2001). The IC-SNURF-SNRPN transcript serves as a host for multiple small nucleolar RNA species and as an antisense RNA for UBE3A. Hum. Mol. Genet. *10*, 2687–2700.

Saxena, S., Jonsson, Z.O., and Dutta, A. (2003). Small RNAs with imperfect match to endogenous mRNA repress translation. Implications for off-target activity of small inhibitory RNA in mammalian cells. J. Biol. Chem. *278*, 44312–44319.

Scadden, A.D., and Smith, C.W. (1997). A ribonuclease specific for inosine-containing RNA: a potential role in antiviral defence? EMBO J. *16*, 2140–2149.

Scheele, C., Petrovic, N., Faghihi, M.A., Lassmann, T., Fredriksson, K., Rooyackers, O., Wahlestedt, C., Good, L., and Timmons, J.A. (2007). The human PINK1 locus is regulated *in vivo* by a non-coding natural antisense RNA during modulation of mitochondrial function. BMC Genomics *8*, 74.

Schwartz, J.C., Younger, S.T., Nguyen, N.B., Hardy, D.B., Monia, B.P., Corey, D.R., and Janowski, B.A. (2008). Antisense transcripts are targets for activating small RNAs. Nat. Struct. Mol. Biol. *15*, 842–848.

Schwartz, Y.B., and Pirrotta, V. (2008). Polycomb complexes and epigenetic states. Curr. Opin. Cell. Biol. *20*, 266–273.

Seila, A.C., Calabrese, J.M., Levine, S.S., Yeo, G.W., Rahl, P.B., Flynn, R.A., Young, R.A., and Sharp, P.A. (2008). Divergent transcription from active promoters. Science *322*, 1849–1851.

Shendure, J., and Church, G.M. (2002). Computational discovery of sense–antisense transcription in the human and mouse genomes. Genome Biol. 3, RESEARCH0044.

Sijen, T., Steiner, F.A., Thijssen, K.L., and Plasterk, R.H. (2007). Secondary siRNAs result from unprimed RNA synthesis and form a distinct class. Science *315*, 244–247.

Sleutels, F., Barlow, D.P., and Lyle, R. (2000). The uniqueness of the imprinting mechanism. Curr. Opin. Genet. Dev. *10*, 229–233.

Sleutels, F., Tjon, G., Ludwig, T., and Barlow, D.P. (2003). Imprinted silencing of Slc22a2 and Slc22a3 does not need transcriptional overlap between Igf2r and Air. EMBO J. *22*, 3696–3704.

Sleutels, F., Zwart, R., and Barlow, D.P. (2002). The non-coding Air RNA is required for silencing autosomal imprinted genes. Nature *415*, 810–813.

Smilinich, N.J., Day, C.D., Fitzpatrick, G.V., Caldwell, G.M., Lossie, A.C., Cooper, P.R., Smallwood, A.C., Joyce, J.A., Schofield, P.N., Reik, W., *et al.* (1999). A maternally methylated CpG island in KvLQT1 is associated with an antisense paternal transcript and loss of imprinting in Beckwith-Wiedemann syndrome. Proc. Natl. Acad. Sci. U.S.A. *96*, 8064–8069.

St Laurent, G., 3rd, and Wahlestedt, C. (2007). Noncoding RNAs: couplers of analog and digital information in nervous system function? Trends Neurosci. *30*, 612–621.

Sun, M., Hurst, L.D., Carmichael, G.G., and Chen, J. (2005). Evidence for a preferential targeting of 3′-UTRs by cis-encoded natural antisense transcripts. Nucleic Acids Res. *33*, 5533–5543.

Tam, O.H., Aravin, A.A., Stein, P., Girard, A., Murchison, E.P., Cheloufi, S., Hodges, E., Anger, M., Sachidanandam, R., Schultz, R.M., *et al.* (2008). Pseudogene-derived small interfering RNAs regulate gene expression in mouse oocytes. Nature *453*, 534–538.

Thakur, N., Tiwari, V.K., Thomassin, H., Pandey, R.R., Kanduri, M., Gondor, A., Grange, T., Ohlsson, R., and Kanduri, C. (2004). An antisense RNA regulates the bidirectional silencing property of the Kcnq1 imprinting control region. Mol. Cell. Biol. *24*, 7855–7862.

Tomari, Y., and Zamore, P.D. (2005). Perspective: machines for RNAi. Genes Dev. *19*, 517–529.

Tufarelli, C., Stanley, J.A., Garrick, D., Sharpe, J.A., Ayyub, H., Wood, W.G., and Higgs, D.R. (2003). Transcription of antisense RNA leading to gene silencing and methylation as a novel cause of human genetic disease. Nat. Genet. *34*, 157–165.

Uchida, T., Rossignol, F., Matthay, M.A., Mounier, R., Couette, S., Clottes, E., and Clerici, C. (2004). Prolonged hypoxia differentially regulates hypoxia-inducible factor (HIF)–1alpha and HIF-2alpha expression in lung epithelial cells: implication of natural antisense HIF-1alpha. J. Biol. Chem. *279*, 14871–14878.

Vallon-Christersson, J., Staaf, J., Kvist, A., Medstrand, P., Borg, A., and Rovira, C. (2007). Non-coding antisense transcription detected by conventional and single-stranded cDNA microarray. BMC Genomics *8*, 295.

Volk, R., Koster, M., Poting, A., Hartmann, L., and Knochel, W. (1989). An antisense transcript from the *Xenopus laevis* bFGF gene coding for an evolutionarily conserved 24 kd protein. EMBO J. *8*, 2983–2988.

Wahlestedt, C. (2006). Natural antisense and noncoding RNA transcripts as potential drug targets. Drug Discov. Today *11*, 503–508.

Watanabe, T., Takeda, A., Tsukiyama, T., Mise, K., Okuno, T., Sasaki, H., Minami, N., and Imai, H. (2006). Identification and characterization of two novel classes of small RNAs in the mouse germline: retrotransposon-derived siRNAs in oocytes and germline small RNAs in testes. Genes Dev. *20*, 1732–1743.

Watanabe, T., Totoki, Y., Toyoda, A., Kaneda, M., Kuramochi-Miyagawa, S., Obata, Y., Chiba, H., Kohara, Y., Kono, T., Nakano, T., *et al.* (2008). Endogenous siRNAs from naturally formed dsRNAs regulate transcripts in mouse oocytes. Nature *453*, 539–543.

Williams, T., and Fried, M. (1986). A mouse locus at which transcription from both DNA strands produces mRNAs complementary at their 3′ ends. Nature *322*, 275–279.

Woo, C.J., and Kingston, R.E. (2007). HOTAIR lifts noncoding RNAs to new levels. Cell *129*, 1257–1259.

Wroe, S.F., Kelsey, G., Skinner, J.A., Bodle, D., Ball, S.T., Beechey, C.V., Peters, J., and Williamson, C.M. (2000). An imprinted transcript, antisense to Nesp, adds complexity to the cluster of imprinted genes at the mouse Gnas locus. Proc. Natl. Acad. Sci. U.S.A. *97*, 3342–3346.

Wutz, A., Smrzka, O.W., Schweifer, N., Schellander, K., Wagner, E.F., and Barlow, D.P. (1997). Imprinted expression of the Igf2r gene depends on an intronic CpG island. Nature *389*, 745–749.

Yang, N., and Kazazian, H.H., Jr. (2006). L1 retrotransposition is suppressed by endogenously encoded small interfering RNAs in human cultured cells. Nat. Struct. Mol. Biol. *13*, 763–771.

Yelin, R., Dahary, D., Sorek, R., Levanon, E.Y., Goldstein, O., Shoshan, A., Diber, A., Biton, S., Tamir, Y., Khosravi, R., *et al.* (2003). Widespread occurrence of antisense transcription in the human genome. Nat. Biotechnol. *21*, 379–386.

Yu, W., Gius, D., Onyango, P., Muldoon-Jacobs, K., Karp, J., Feinberg, A.P., and Cui, H. (2008). Epigenetic silencing of tumour-suppressor gene p15 by its antisense RNA. Nature *451*, 202–206.

Zhang, Q., Zhang, J., Moe, O.W., and Hsia, C.C. (2008). Synergistic up-regulation of erythropoietin receptor (EPO-R) expression by sense and antisense EPO-R transcripts in the canine lung. Proc. Natl. Acad. Sci. U.S.A. *105*, 7612–7617.

Zhang, Y., Liu, X.S., Liu, Q.R., and Wei, L. (2006). Genome-wide *in silico* identification and analysis of cis natural antisense transcripts (cis-NATs) in ten species. Nucleic Acids Res. *34*, 3465–3475.

Zhao, J., Ohsumi, T.K., Kung, J.T., Ogawa, Y., Grau, D.J., Sarma, K., Song, J.J., Kingston, R.E., Borowsky, M., and Lee, J.T. Genome-wide identification of polycomb-associated RNAs by RIP-seq. Mol. Cell *40*, 939–953.

Zhao, J., Sun, B.K., Erwin, J.A., Song, J.J., and Lee, J.T. (2008). Polycomb proteins targeted by a short repeat RNA to the mouse X chromosome. Science *322*, 750–756.

Zubko, E., and Meyer, P. (2007). A natural antisense transcript of the Petunia hybrida Sho gene suggests a role for an antisense mechanism in cytokinin regulation. Plant J. *52*, 1131–1139.

Zuniga Mejia Borja, A., Meijers, C., and Zeller, R. (1993). Expression of alternatively spliced bFGF first coding exons and antisense mRNAs during chicken embryogenesis. Dev. Biol. *157*, 110–118.

Exploring the Genomic Dark Matter: Non-coding RNAs and Epigenetic Regulation of Transcription as a New Therapeutic Platform

13

Kevin V. Morris

Abstract

A growing body of evidence is beginning to emerge in human cells suggesting that particular species of non-coding RNAs can regulate gene transcription by modulating loci specific epigenetic states. While such observations were in the past relegated to imprinted genes, it is now becoming apparent that several genes in differentiated cells may be under some form of non-coding RNA based transcriptional and epigenetic control. Importantly, this form of regulation may be highly influenced by selective pressures and function in the governance and adaptability of the cell. Many studies have been carried out to date, which have begun to discern the mechanism of action whereby non-coding RNAs modulate gene transcription. Some evidence points to a role of long non-coding RNAs in controlling gene transcription by actively recruiting epigenetic silencing complexes to homology containing loci in the genome. While other studies point to a role for long intergenic non-coding RNAs in scaffold like features that are most probably equally implicit in the regulation of gene expression. The results of these studies will be considered in detail as well as the implications that a vast array of non-coding RNA based regulatory networks may be operative in human cells. Knowledge of this emerging RNA based epigenetic regulatory network has implications in cellular evolution as well as an entirely new area of pharmacopoeia, namely RNA mediated epigenetic regulation of gene expression.

Introduction

An enigma in modern day molecular biology has been to determine what exactly the role is for the non-coding regions of the human genome. These regions, dubbed the genomic dark matter consist of vast tracts of non-coding DNAs which do not appear to code for proteins. Many have wondered why the cell would waste energy faithfully replicating these non-coding DNAs if they aren't actually involved in making proteins. Some have argued that these regions, also described as junk DNA, function as book marks to partition the genome in such a manner as to allow gene compartmentalization (Alberts 1994). In eukaryotes a large percentage, estimates in humans range from 70% to 90% of the genome is composed of non-coding DNA (reviewed in Taft *et al.*, 2007; Morris, 2009). While these regions were thought to be relatively inert, evidence through the use of genome-wide analytical tools such as deep sequencing have begun to suggest that these non-coding DNAs are in fact transcribed into RNA (Mattick and Makunin 2006). As these RNAs are transcribed from non-coding DNAs they are referred to as non-coding RNAs (ncRNAs). But what, if anything, are these ncRNAs doing? And why would the cell transcribe vast tracks of DNA to never produce a protein?

Evidence addressing this central question, i.e. what is the role for ncRNAs in the human cell, has begun to emerge with the notion that some of ncRNAs are functionally involved in the regulation of gene transcription by playing a role in actively targeting epigenetic regulatory complexes to particular loci in the cell. Such a notion is counter to the central dogma in molecular biology that DNA generates RNA, which functions to make protein. The notion that ncRNAs can in essence control the transcription of DNA into RNA and that these regulatory states can be passed on to daughter cells, i.e. retained through cell division, is only now becoming appreciated. Interestingly, recent observations suggesting that ncRNAs are involved in the epigenetic regulation of gene transcription, stem from earlier studies utilizing the new tool of RNA interference (RNAi).

Small interfering RNAs and the regulation of gene transcription

RNAi is best known from experiments which demonstrated that small interfering RNA (siRNAs) could be used to functionally target and direct the degradation of a particular mRNA (reviewed in Hannon, 2002; Hannon and Rossi, 2004). This form of RNA based regulation was termed; RNAi mediated post-transcriptional gene silencing (PTGS). In PTGS, siRNAs are targeted to messenger RNA by implicit sequence homology. This targeting results in the recruitment of the native cellular DICER machinery to the targeted mRNA where degradation the homology containing transcript is initiated by the action of Argonaute 2 (Ago-2) ultimately resulting in gene suppression. While PTGS is a potent mechanism for the degradation of siRNA targeted mRNAs, its gene-suppressive effect is transient, lasting only until new transcripts are generated or the siRNAs are degraded. In some organisms, lower eukaryotes, RNAi based PTGS can be processive and the signal amplified (reviewed in Matzke *et al.*, 2001; Hannon, 2002; Hannon and Rossi, 2004; Matzke and Birchler, 2005). This, however, has not been shown to be the case with human cells.

Over the last decade much emphasis and excitement has centred on siRNA directed PTGS, as this methodology of gene suppression is amendable to quick knockouts of genes for experimental studies. While several studies have been carried out using RNAi to suppress gene expression in a PTGS based manner one underappreciated study has proven insightful. In this study, carried out in human cells, siRNAs were designed to target a genes promoter instead of the mRNA/coding region. The result of this siRNA targeting of a gene promoter was the suppression of gene expression at the transcriptional level, upstream of the well documented PTGS pathway (Morris *et al.*, 2004). This mode of RNA based transcriptional suppression involved an epigenetic component, as trichostatin A and 5′-azacytidine suppressed the siRNA mediated transcriptional gene silencing (TGS). Notably, the observed siRNA mediated TGS was similar to previous work carried out in plants and yeast (reviewed in Matzke and Birchler, 2005).

During the course of investigation as to how these promoter targeted siRNAs were capable of directing TGS in human cells one striking observation was generated. In this body of work Weinberg and colleagues discovered, quite serendipitously, that only the antisense strand of the siRNA was required for TGS in human cells (Weinberg *et al.*, 2006). Other studies have validated this requirement (Turner *et al.*, 2009), suggesting that an endogenous mechanism in human cells whereby ncRNAs regulate gene transcription may involve forms of RNA that are not double-stranded. Later work by several groups have collectively led to insights suggesting a model for how small antisense RNAs and/or siRNAs can function in controlling gene transcription in human cells (reviewed in Morris 2009; Fig. 13.1). While much has been gleamed from these bodies of work it had remained until recently unknown whether or not human cells generated endogenous ncRNAs to control transcription or if the observed effects to date (summarized in Fig. 13.1) were the result of small ncRNAs triggering some vestigial pathway in human cells? In essence, these observations might be telling us far more about endogenous modes of ncRNA regulation in human cells than have previously been appreciated, but this eventuality had until recently remained to be determined.

Uncovering an endogenous mechanism of ncRNA-based epigenetic regulation in human cells

Despite earlier observations indicating that siRNAs, derived *de novo* could affect gene

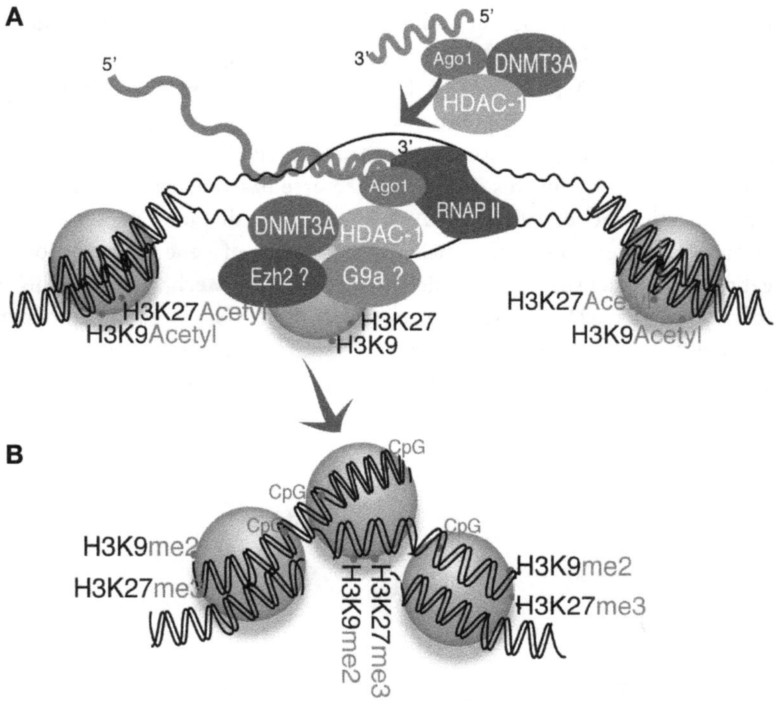

Figure 13.1 Small antisense ncRNA mediated regulation of gene transcription. (A) *de novo* derived small antisense ncNAs, designed to target a promoter of choice, can interact with promoter-associated RNAs at the transcribed promoter (Han *et al.*, 2007; Hawkins *et al.*, 2009) and facilitate the recruitment of epigenetic remodelling complexes (Suzuki 2005; Kim *et al.*, 2006; Weinberg *et al.*, 2006; Han *et al.*, 2007; Suzuki *et al.*, 2008; Hawkins *et al.*, 2009; Turner *et al.*, 2009). (B) The result of small antisense ncRNA targeting is the epigenetic remodelling and silencing of transcription at the targeted promoter loci.

transcription, little was known whether or not there were active ncRNAs modulating gene transcription in human cells. Some studies provided evidence that micro-RNAs (miRNAs) could function as endogenous regulators of gene transcription (Omoto *et al.*, 2004; Omoto and Fujii, 2005; Klase *et al.*, 2007; Kim *et al.*, 2008; Tan *et al.*, 2009). Hints to a deeper layer of RNA based transcriptional regulation arose from one study which set out to target *de novo* derived siRNAs to the p21 and E-cadherin promoters, specifically targeting A,T rich regions in these promoters. Notably, A,T rich regions are generally found in the coding bodies of genes while promoters tend to exhibit CpG islands and palindromic sequences with high GC content, often times used for transcription factor or protein binding (Alberts, 1994). In this study it was determined that these p21 and E-cadherin promoter targeted siRNAs were able to activate gene expression, though the activation of transcription was not demonstrated by nuclear run-on analysis or RNA polymerase II localization at the p21 or E-cadherin promoters (Li *et al.*, 2006; Place *et al.*, 2008). In these studies it was also determined that Ago-2 was required for the observed activation (Li *et al.*, 2006; Place *et al.*, 2008). This observation is noteworthy as siRNAs shown to modulate PTGS tend to require Ago-2 while several observations with siRNAs directing TGS where observed to require the action of Argonaute 1 (Ago-1), DNA methyltransferase 3a (DNMT3a) and histone deacetylase 1 (HDAC-1) (reviewed in detail in (Morris 2009; Fig. 13.1). Though notably some investigators find a role for Ago-2 in siRNA directed TGS, these studies however specifically targeted siRNAs to the TATAA start site in the candidate promoter (Janowski *et al.*, 2006; Napoli *et al.*, 2009), possibly suggesting a different mechanism of action, i.e. the obstruction of RNAPII activity by direct binding and maybe even a DNA/RNA interaction at the TATAA (discussed in Morris 2005, 2009).

The question as to how siRNAs, targeted to A,T rich regions in the p21 and E-cadherin promoters were utilizing Ago-2 to activate gene expression was addressed in a recent study. In this body of work it was determined that the activating siRNAs targeted to A,T rich promoter sites were actually capable of targeting long antisense ncRNAs which were found to overlap the p21 or E-cadherin coding region and were actively involved in regulating the transcriptional expression of p21 and E-cadherin (Morris *et al.*, 2008). Other tumour-suppressor genes, such as p15, also appeared to be under this newly discovered regulatory mechanism of long antisense ncRNA-based epigenetic regulation (Yu *et al.*, 2008). A model has begun to emerge based on several different bodies of work suggesting that long antisense ncRNAs are active regulators of gene transcription in human cells (Fig. 13.2). Mechanistically, this model suggests that the siRNA targeting and degradation of the gene specific long antisense ncRNAs, affects the ability of these RNAs to interact with and guide

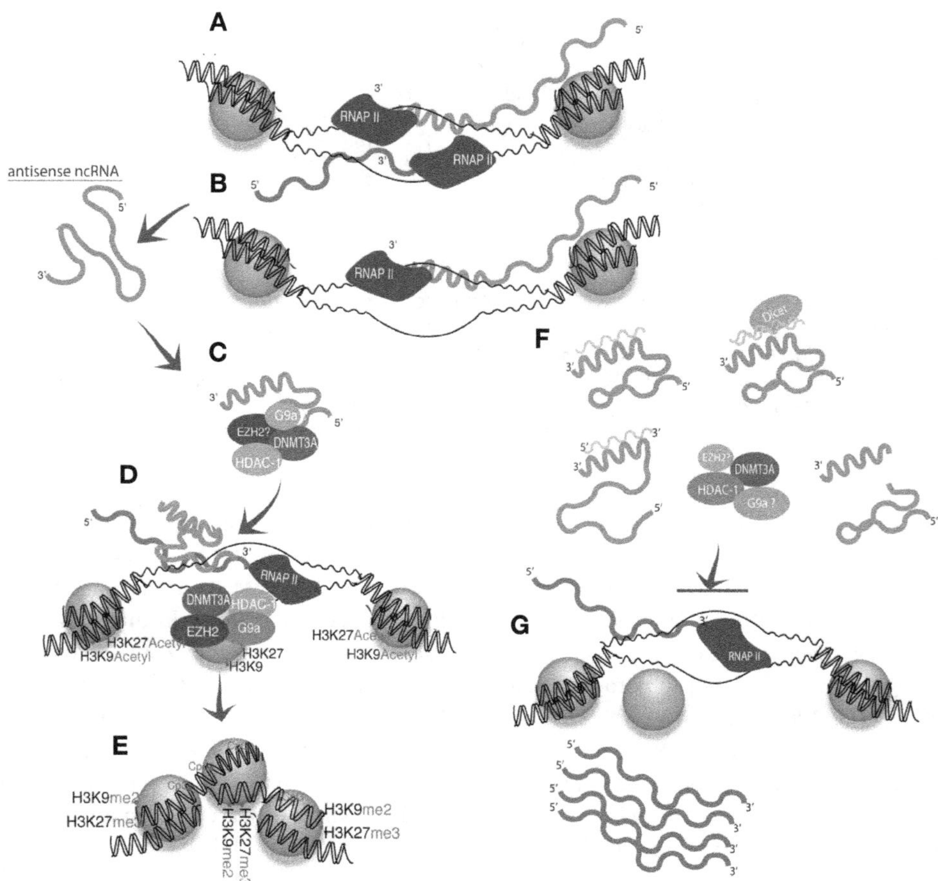

Figure 13.2 Antisense ncRNA-mediated transcriptional silencing and activation. (A) Bidirectionally transcribed genes produce long antisense non-coding RNAs which can (B) fold into secondary structures and interact with various proteins as well as target these proteins to homology containing loci in the genome. (C) The antisense RNA protein complexes can then target the corresponding homology containing loci in the genome where (D) epigenetic remodelling of the local chromatin via directed histone and DNA methylation occurs. (E) When the antisense non-coding RNAs are targeted by either single stranded oligonucleotides or siRNAs the secondary structure is changed and (F) interactions with the epigenetic remodelling protein complexes is blocked. (G) The result is a loss of antisense non-coding RNA directed epigenetic regulation of gene expression and subsequent activation of gene transcription.

epigenetic silencing complexes to their particular targeted loci in the promoter of the sense/mRNA loci for the gene. The result of this targeting is the activation of gene expression (Fig. 13.2; Knee and Murphy, 1997; Tufarelli *et al.*, 2003; Cho *et al.*, 2005; Katayama *et al.*, 2005; Wahlestedt, 2006; Camblong *et al.*, 2007; Ebralidze *et al.*, 2008; Morris *et al.*, 2008; Yu *et al.*, 2008; Li *et al.*, 2009). Thus siRNA mediated transcriptional gene activation is in essence suppressing the endogenous ncRNA suppressor, the result of which is activation or a loss of endogenous epigenetic brakes being instilled in the particular targeted gene by the action of the long antisense ncRNA (Fig. 13.2 and reviewed in Morris, 2009a,b,c).

Taken together these new findings strongly implicate long antisense ncRNAs as epigenetic regulators of gene expression in human cells. Notably, the regulation instilled by this form of ncRNA-based control has the potential if unchecked to functionally alter gene expression in a manner that is both directly targeted as well as long lasting. The direct targeting is due to ncRNA sequence homology in the particular target loci while the long-lasting effects are the result of epigenetic changes to the local chromatin architecture. Indeed evidence with small ncRNA targeted TGS has demonstrated that stable long-term epigenetic silencing can be established after a relatively short duration of small ncRNA targeting to a gene promoter (Hawkins *et al.*, 2009; Turner *et al.*, 2009). In this study a 20-fold excess of promoter targeted small ncRNA for ~3–4 days resulted in the stable suppression of the targeted gene for 28 days (Hawkins *et al.*, 2009). The impact of a set of molecules that govern this form of control cannot be overstated as long lasting effects can, by extension, have an impact on the evolution of the cell. Interestingly, to date the majority of those genes exhibiting bidirectional transcription, e.g. antisense ncRNA expression, appear to be involved in the cell-cycle functioning as tumour-suppressor genes or oncogenes (reviewed in Morris and Vogt, 2010). These observations might speak to a role for long antisense ncRNA-based regulation in the evolution of cancer (Morris and Vogt, 2010). For example the disregulation, or overt expression of a tumour-suppressor gene specific long antisense ncRNA could result in stable epigenetic silencing of that particular tumour-suppressor gene. Such could be the case with p21 or p15, both tumour-suppressor genes found to express long antisense ncRNAs, which are active in their regulation of transcription (Morris *et al.*, 2008; Yu *et al.*, 2008).

The observations that antisense ncRNAs are involved in regulating gene expression in differentiated cells (reviewed in Morris 2009) adds to the growing complexity of known ncRNAs involved in gene and cellular regulation. Examples include the Xist antisense ncRNA TSIX, which is involved in dosage compensation and X-inactivation in undifferentiated primordial cells (reviewed Lee 2009). Other examples of ncRNAs involved in epigenetic regulation can be found in the plethora of imprinting genes (reviewed in Latos and Barlow 2009). Some of these long ncRNAs function by targeting the recruitment of different epigenetic regulatory complexes to their intended targets in *trans* (Brideau *et al.*, 2010). One recent study on the Hox cluster of imprinted genes demonstrated that a long ncRNA Hotair functions as a bifunctional switch, able to recruit protein complexes that function in both activation and suppression of gene expression by differentially modulating the local chromatin architecture (Gupta *et al.*, 2010; Wang *et al.*, 2011; Fig. 13.3).

More recent studies have demonstrated a role long intergenic ncRNAs (lincRNAs). The lincRNAs, first discovered by high-throughput deep sequencing approaches, were found to be expressed within the middle of partitioned genes on vast open tracks of non-coding DNA (Guttman *et al.*, 2009). The lincRNAs have been found to be active modulators of stem cell differentiation, with some functional in reprogramming the chromatin state, via the recruitment of p53 to particular loci, ultimately resulting in cancer metastasis (Huarte *et al.*, 2010; Loewer *et al.*, 2010; Rinn *et al.* 2011; Fig. 13.4). Clearly, the emerging evidence suggests in human cells, that many more genes than previously envisaged, beyond X-inactivation and imprinting related genes, might actually be regulated by long ncRNAs. In fact ncRNAs might be actively switching on and off genes in an orchestral regulation that governs the fidelity of the cell and functions in cellular adaption (discussed in greater detail in Morris 2009). Furthermore, a possible underappreciated function for miRNAs might

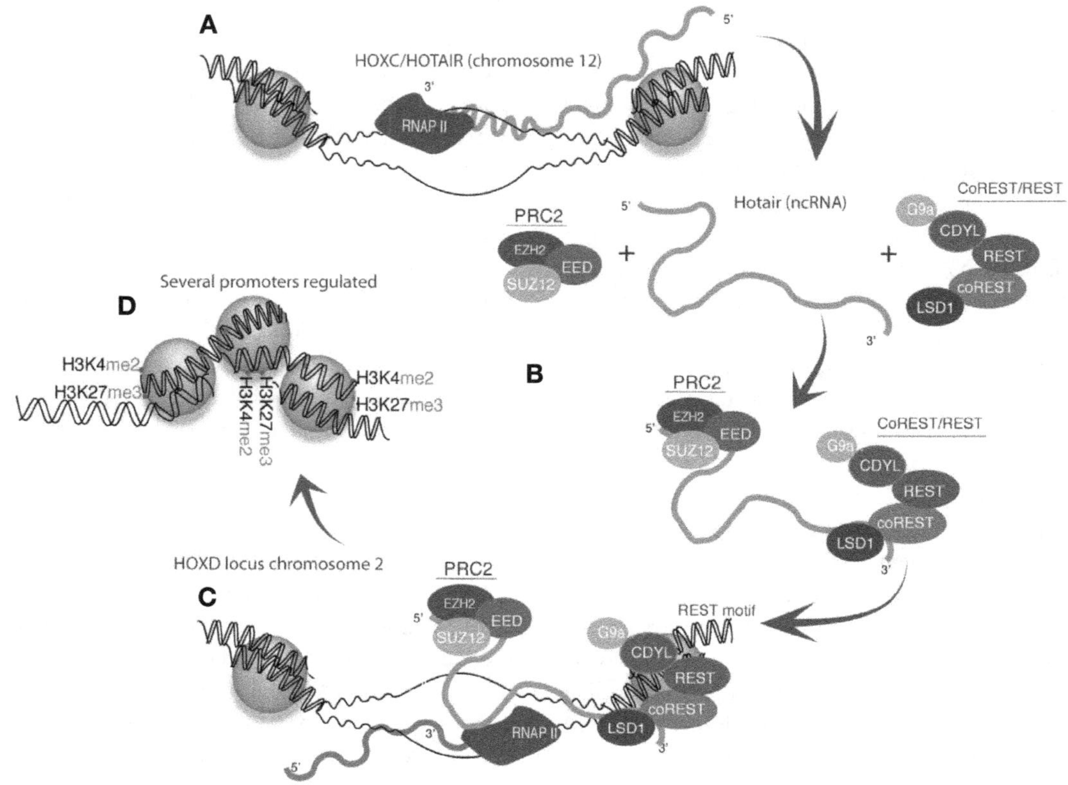

Figure 13.3 Hot air dual functional long ncRNA. (A) The HOXC long ncRNA, HotAir is expressed from chromosome 12 which can then (B) interact with PRC2 (histone repressive complex) and CoREST/REST (histone activating complex). (C) The entire protein/Hotair ncRNA complex can then target the HOXD cluster on chromosome 2 leading to (D) epigenetic regulation of several promoters associated with the HOXD cluster. This model is depicted based on the work from Howard Chang's laboratory (Gupta *et al.*, 2010).

also be that they function as active regulators of long antisense ncRNAs by sensing the ratios of sense to antisense compliments and adjusting the efficacy of antisense ncRNA targeting by interacting with the antisense ncRNAs directly in a manner that is suppressive of their targeting function (as depicted in Fig. 13.2E–G with shRNA or asRNA targeting of the endogenous antisense ncRNA, and discussed in greater detail in Morris, 2009).

New insights, a role for pseudogenes as modulators of endogenous TGS

Studies with the tumour-suppressor genes p21 and p15 indicated that bidirectional transcription is operative in human cells and that the long antisense ncRNAs, from these particular bidirectional transcribed genes, is functional in epigenetically regulating the sense/mRNA expression (reviewed in Morris, 2009). The majority of those genes that have been found to exhibit bidirectional transcription, in that they express both sense/mRNAs as well as antisense ncRNAs, appear to be involved in cell cycle regulation (reviewed in Morris and Vogt, 2010). More interesting is the observation that those genes exhibiting bidirectional characteristics are often times found to be epigenetically silenced in human cancers (Morris and Vogt, 2010). These studies, while informative with regards to a role for long antisense ncRNAs in epigenetic regulation of gene expression were unable to decipher whether or not the endogenous antisense ncRNAs

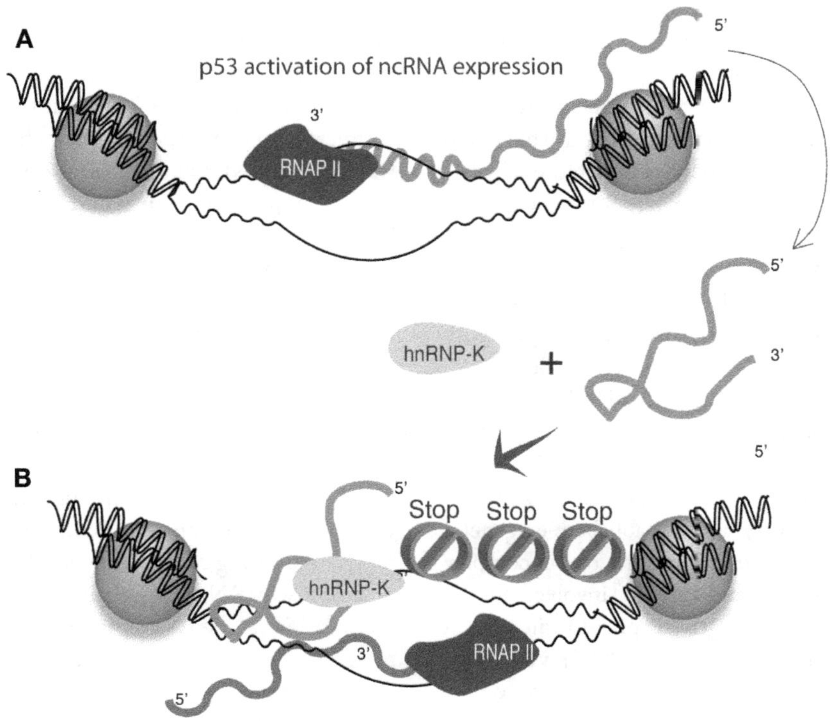

Figure 13.4 p53 modulated Linc-RNAs involved in gene regulation. Activation of p53 results in (A) concordant activation of particular lincRNAs which (B) can interact with hnRNP-K and modulate the expression of several p53 regulated genes in a network based manner. This model is based on work by John Rinn's lab (Huarte *et al.*, 2010).

emanated *in cis* or *trans*. One would assume that cis based bidirectional transcription would prove difficult for the cell. Despite the notion that each RNAPII would be reading the other compliment strand, RNAPII collulsion would be expected to occu (Greger *et al.*, 2000).

A recent study has shed some light on the issue of where the TGS-inducing antisense ncRNAs emanate. In this study an Oct-4 regulatory long-antisense ncRNA was observed to function in *trans* (Hawkins 2010). The Oct-4 TGS inducing antisense ncRNA appeared to be expressed from an Oct-4 pseudogene, Oct-4 pseudogene 5 (Oct-4pg5), which is found on an entirely different chromosome than Oct-4 (Hawkins 2010). Notably, this particular antisense ncRNA did not appear to be poly adenylated, contained Oct-4 intronic sequences, and was found to contain sequence homology spanning the Oct-4 promoter. Notably, the suppression of this Oct-4pg5 antisense ncRNA, using RNAi, resulted in transcriptional activation of Oct-4 with a concomitant loss of H3K27 tri-methylation and H3K9 di-methylation at the Oct-4 promoter (Hawkins 2010). Interestingly, the suppression of Ezh2 and G9a using RNAi also had an impact on Oct-4 transcription, the loss of these factors resulted in increased transcription of Oct-4, suggesting that Oct-4 is indeed under epigenetic regulation via the action of Ezh2, G9a, and Oct-4pg5 antisense ncRNA (Hawkins 2010). This observation is, however, at odds with previous findings that siRNAs or small antisense ncRNA mediated TGS which required Ago-1, DNMT3a and HDAC-1 to function (Fig. 13.1). As the suppression of Ago-1, DNMT3a, and HDAC-1 had only a modest effect on Oct-4 expression compared with effects of Ezh2 and G9a suppression. The one common connection between small and long antisense ncRNA mediated TGS appears to be the observation of increased H3K27 trimethylation specifically at the targeted loci (Figs. 13.1 and 13.2). H3K27 trimethylation is the result of Ezh2 activity. Interestingly, DNMT3a, which has been shown previously to bind small ncRNAs (Jeffery and Nakielny 2004) and promoter targeted small antisense ncRNAs, has also been found to be directly localized at small antisense ncRNA target loci with H3K27 tri-methylation (Weinberg *et al.*, 2006; Hawkins *et al.*, 2009; Turner *et al.*, 2009). Notably, DNMT3a has also been shown to complex with both Ezh2 and HDAC-1 (Fuks *et al.*, 2001, 2003; Vire *et al.*, 2005). HDAC-1, like DNMT3a is also required for small antisense ncRNA directed TGS in human cells (Suzuki *et al.*, 2008; Fig. 13.1). In fact these three proteins HDAC-1, Ezh2, and DNMT3a are very well established epigenetic remodelling proteins known to be in complex with one another and directly involved in the epigenetic control of gene expression in human cells (Fuks *et al.*, 2000, 2001, 2003; Vire *et al.*, 2005). While it had remained unknown as to how this complex is targeted to particular loci in the genome, the recent studies carried out with small ncRNA directed TGS, indicate a role for ncRNAs in targeting this epigenetic remodelling complex to specific loci in human cells.

Collectively, these observations suggest that there is some overlap in the molecular pathway by which antisense ncRNAs functionally modulate transcription. The endogenous pathway appears to be utilizing longer forms of antisense ncRNAs and maybe more reliant on Ezh2 and G9a (Hawkins 2010; Fig. 13.2). While the exogenously introduction of small antisense ncRNAs or siRNAs appears to depend more on DNMT3a, HDAC-1 and Ago-1.

In an effort to more comprehensively characterize the endogenous mechanism by which Oct-4pg5 antisense ncRNA mediated TGS of Oct-4, the Oct-4pg5 antisense ncRNA was directly immunoprecipitated and the associated protein complex assessed by mass spectrophotometry. Curiously, the Oct-4pg5 antisense ncRNA was found to associate with several known RNA binding proteins as well as several previously uncharacterized proteins (Hawkins 2010). Some of these proteins were characterized further for their requirement in Oct-4pg5 antisense ncRNA mediated TGS of Oct-4. Two of the sixteen proteins, PurA and Nucleolin, were selected based on a priori knowledge that they functionally bind RNA/DNA and are involved in microtubule transport. When these proteins were suppressed using RNAi, the Oct-4 mRNA was also appreciably suppressed (Hawkins 2010). These observations suggested a role for PurA and Nucleolin in the trafficking and degradation of the Oct4-pg5 ncRNA. The suppression of PurA and Nucleolin inhibited nucleolar trafficking of Oct4-pg5 and thus an increased accumulation of Oct4-pg5 antisense ncRNA was present in the cell leading to increased TGS of Oct-4 (Hawkins 2010).

In another body of work utilizing genome-wide analysis of non-poly adenylated antisense ncRNAs, a strong correlation with the abundance of antisense ncRNA, the relative amount of DNA methylation at gene promoters, and the suppression of sense/mRNA gene expression from these promoters in human cells was observed (Watanabe *et al.*, 2010). Taken together these new insights have begun to suggest that there is a vast array of ncRNAs in the human cell, some of which are active epigenetic modulators of gene transcription.

Antisense ncRNA expression has been observed previously in genome-wide based studies. In one study, the authors observed that the suppression of long antisense ncRNAs could result in two eventualities; either the sense counterpart was discordantly increased or concordantly decreased in expression (Katayama *et al.*, 2005). This study clearly suggests that there is far more complexity than has previously been appreciated when it comes to ncRNA mediated modes of gene regulation in human cells. While the concordant versus discordant antisense ncRNA regulatory pathways have been discussed little was known about their mechanism of action (Wahlestedt 2006). Recent work (Hawkins 2010) as well as previously published observations (Morris *et al.*, 2008; Yu *et al.*, 2008) suggest that the endogenous mechanism, whereby long antisense ncRNAs direct discordant regulation, is in essence the same pathway utilized by *de novo* derived small antisense RNAs or siRNAs to control gene transcription (Figs. 13.1 and 13.2).

While there are some subtle differences in the mechanism the general notion is that long antisense ncRNA-based discordant regulation is the pathway whereby smaller forms of ncRNA guide TGS in human cells, i.e. discordant regulation may in fact be the endogenous TGS pathway in human cells (Fig. 13.2).

Utilizing the endogenous ncRNA TGS pathway to stably modulate gene expression

The targeting of epigenetic complexes to the genome by the action of small or long antisense ncRNAs is presumed to be via the interaction between the antisense ncRNA and actively transcribed regions at it's homology containing target. The target could be a promoter-associated RNA (Han *et al.*, 2007) whereby the transcription of a gene is controlled (Fig. 13.1), or possibly interactions with complimentary non-coding RNAs that are acting as local chromatin scaffolds for genes (Kanhere *et al.*, 2010; Tsai, Manor *et al.* 2010) or intergenic regions (Khalil *et al.*, 2009; Tsai *et al.*, 2010). Others have proposed an RNA/DNA interaction might be occurring at the particular targeted promoters and occluding transcription (Petruk S 2006), although little data supporting this notion have been presented to date. It is clear that RNAPII has a ~1 nucleosome footprint at the TATAA of promoters (Boeger *et al.*, 2005) and two studies targeting TGS by targeting TATAA find suppression in the absence of changes in the local epigenetic marks (Janowski 2005; Napoli *et al.*, 2009), whereas those targeting small ncRNAs upstream of TATAA generally see a role for epigenetic complexes in the observed silencing (Morris *et al.*, 2004; Ting *et al.*, 2005; Weinberg 2005; Hoffman and Hu 2006; Kim *et al.*, 2006, 2008; Han, 2007; Pulukuri and Rao, 2007; Sugimoto *et al.*, 2007; Gonzalez *et al.*, 2008; Lim *et al.*, 2008; Suzuki *et al.*, 2008; Hawkins *et al.*, 2009; Turner *et al.*, 2009; Turunen *et al.*, 2009). Regardless of these differences, one commonality in all of these proposed examples are that the respective ncRNAs all appear to have a common theme that they have a direct impact on the accessibility and fidelity of the genome, i.e. access of the DNA to the cellular machinery and transcription.

The ability to specifically control epigenetic states has much potential with regards to future therapeutics. Most importantly, ncRNA mediated TGS has the ability to modulate local chromatin architecture in a manner that can be long-lasting (Suzuki 2005; Hawkins *et al.*, 2009; Turner *et al.*, 2009). The potential is immense with the notion that target specific permanent changes in gene expression can be instilled. But how can we harness the knowledge gained to date to specifically modulate TGS in human cells. One approach is to consider all the documented functional versus non-functional small ncRNA TGS experiments to date for patterns. When such an effort was carried out, using the target sequence, a unique pattern of purines was observed. This pattern was instilled in a computer code and several oncogene promoters assessed for small ncRNA directed TGS (Fig. 13.5A). Interestingly, a ~70% efficacy rate was observed when several different genes were targeted for TGS (Fig. 13.5B–E). When the top TGS inducing sites were contrasted an interesting common pattern in the target site emerged (Fig. 13.5F). This same algorithm could in theory also be used to target long antisense ncRNAs with siRNAs or small antisense ncRNAs to induce TGA. The notion being that those sites targeted by a putative long antisense ncRNA, specifically at a gene promoter would also utilize the same purine trend for functional targeting. Interestingly, a ~66% efficacy rate of TGA was observed when targeting putative antisense ncRNAs for several tumour-suppressor genes (Fig. 13.6A and B). The observed activation of TP53 appeared to correlate with a reduction in TP53 antisense ncRNA (Fig. 13.6C). Taken together these observations suggest that certain loci are susceptible to small ncRNA directed TGS and that these loci are under the endogenous regulation by long antisense ncRNAs, which when targeted can result in activation of gene transcription. It should be noted that those tumour-suppressor genes targeted with small ncRNAs (Fig. 13.6) based on the algorithm were targeted to undocumented antisense ncRNAs, i.e. simply presumed to exist based on the previous studies with Oct4, p21 and p15 (Morris *et al.*, 2008; Yu *et al.*, 2008; Hawkins and

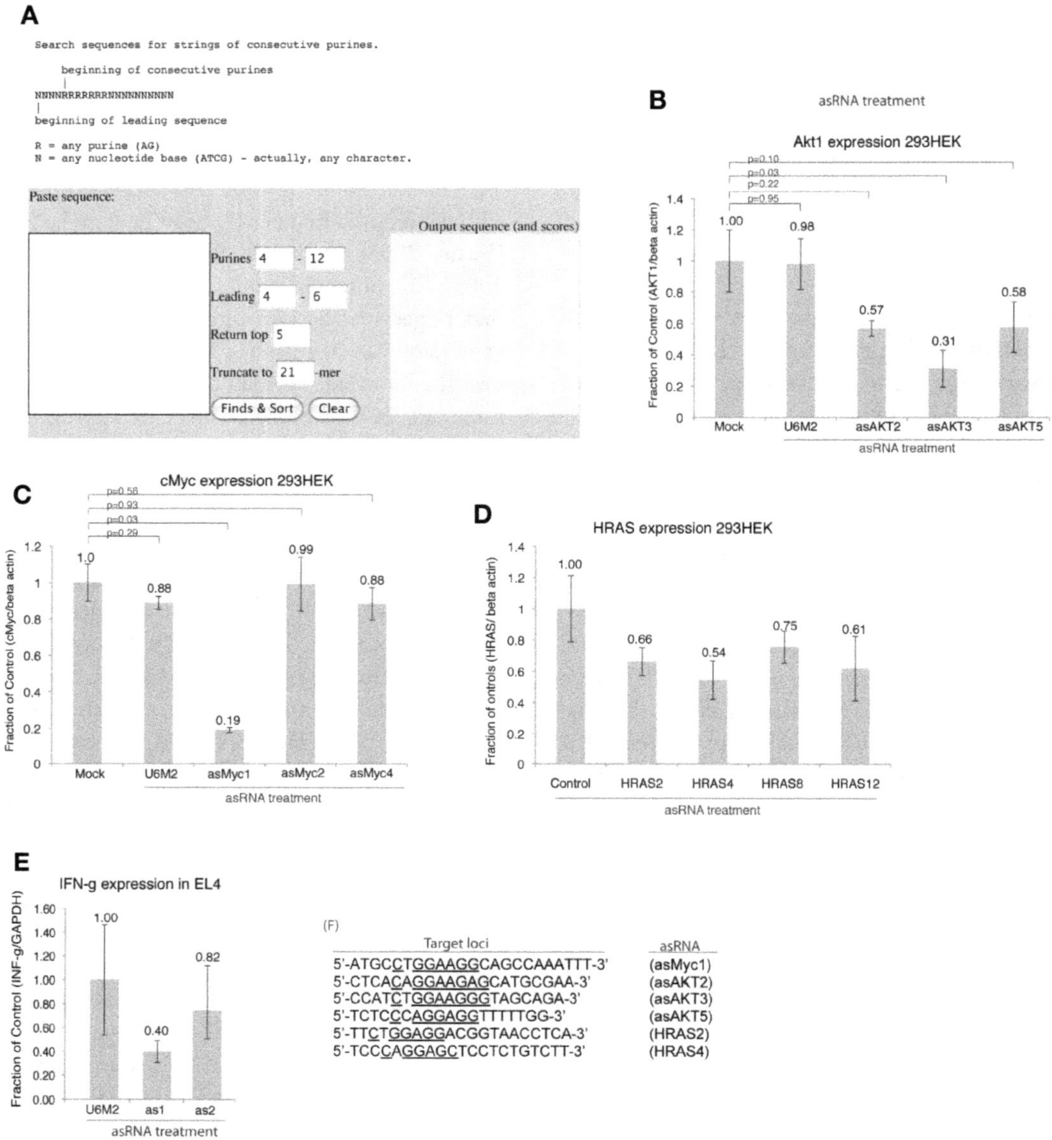

Figure 13.5 Algorithm designed TGS of oncogenes. (A) Computational based algorithm screening was carried out to determine TGS susceptible loci in gene promoters. The algorithm specifically identifies purine string and generates 21mers that contain the purine string ~4–6 bp from the 5′ end of the 21mer. Small antisense ncRNAs were generated based on hits from the algorithm to target (B) AKT-1, (C) cMyc, (D) HRAS, and (E) mouse INF-gamma. (F) A common code is recognized for the most functional algorithm designed small antisense ncRNAs. For (B–E) 293 HEK cells were transfected with small antisense ncRNA expressing U6M2 plasmids (~100 ng/10^5 cells) and for (E) mouse EL4 cells were transfected using the Neon electroporator system. (B–E) The averages of triplicate transfected cultures are shown with the standard error of the means and *P*-values from a paired *t*-test.

Morris, 2010). While very few antisense ncRNAs have been described too date these observations suggest that possibly every gene contains an antisense ncRNA counterpart. Taken together these observations suggest that it is possibly to a priori design small ncRNAs capable of specifically modulating either TGS or TGA. The power of this algorithm cannot be overstated as it suggest that

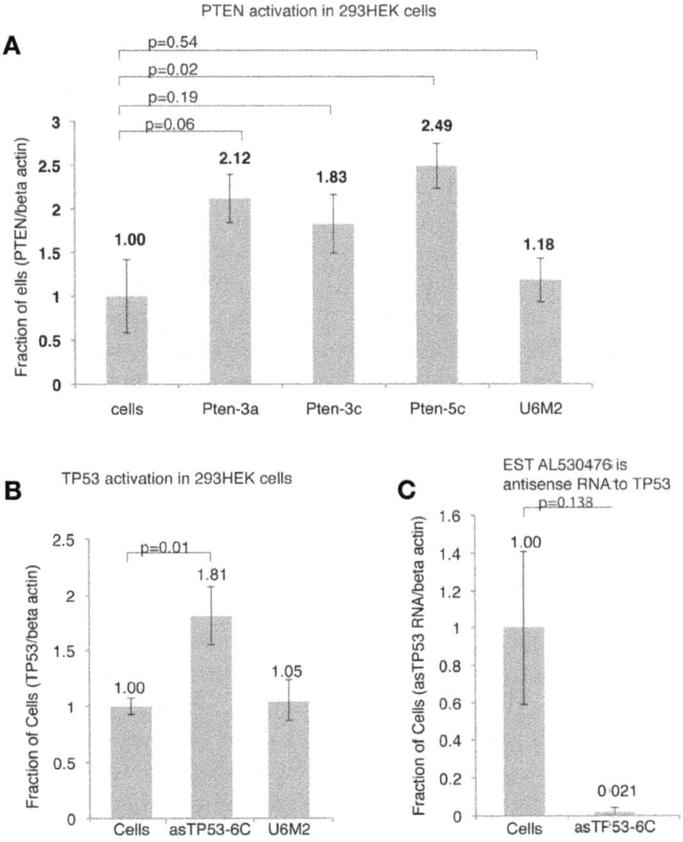

Figure 13.6 Algorithm designed TGA of oncogenes. Computational based algorithm screening was carried out to determine TGA susceptible loci in putative antisense ncRNAs for the tumour-suppressor genes (A) PTEN and (B) TP53 promoters. (C) A reduction in the antisense ncRNA EST AL530476 was observed in those cultures treated with the small ncRNA. For (A–C) 293 HEK cells were transfected with small antisense ncRNA expressing U6M2 plasmids (~100 ng/10^5 cells). The averages of triplicate transfected cultures are shown with the standard error of the means and *P*-values from a paired *t*-test.

we are one step closer to the development of an entirely new class of pharmacopoeia whereby we can now target epigenetic states to those genes deemed requiring transcriptional regulation. However, one should note that the main issue arising from such a notion is the aspect of delivery.

Conclusion

Over the last decade much hype has been generated around RNAi and the ability of double-stranded RNAs, siRNAs, shrines and miRNAs to regulate gene expression in a post-transcriptional manner (Tomari and Zamore, 2005). The observation that double-stranded RNAs can modulate mRNA expression post-transcriptionally is no doubt a paradigm shift. However, over the last ~7 years it has become increasingly apparent that RNAi is the proverbial tip of the iceberg with regards to RNA based regulation in human cells. Studies carried out to discern how siRNAs are able to modulate gene transcription in human cells have led to insights suggesting that antisense ncRNAs, be them short or long, can transcriptionally regulate gene expression via the specific targeting of epigenetic silencing complexes to loci in the genome.

One might however ask; why should these relatively low expressed natural antisense transcripts be of any interest? The answer to this question lies in the fact that antisense ncRNAs appear in some instances to regulate gene transcription via the targeting of epigenetic changes to particular loci in the genome. This is noteworthy as for over four decades epigenetic changes to the genome have been observed in various cell types and cancers with little understanding as to what is guiding these complexes to their respective target loci. Epigenetics (DNA and histone methylation) has been mechanistically implicated in cancer but the mechanism of action for directing epigenetic changes was always missing. The emerging observations with antisense ncRNAs and lincRNA is suggestive of mechanisms explaining how epigenetic silencing complexes are targeted to particular loci in the genome. These observations present an entirely new discipline of scientific inquiry. This new area of inquiry has immense potential with regards to targeted interventions aimed at controlling the epigenetic state of particular genes. With regards to cancer therapeutics such a potential is immense. It is regrettable that so little emphasis has been placed in harnessing this newly emerging paradigm for therapeutic gain.

The targeted recruitment of epigenetic changes can also have a significant impact on the selective pressures the cell and/or organisms face as mounting evidence suggest that epigenetic changes are heritable and able to be transmitted across generations (Heijmans *et al.*, 2008; Franklin *et al.*, 2010). These observations argue that in essence ncRNAs possess the ability to instil long term changes in gene expression that are capable of being passed on to daughter cells. Such a notion runs counter to a Darwinian existence and argues that epigenetic changes, which are outside of the DNA, can have an impact on generations of daughter cells or offspring. Curiously this presumed model, for how genes are regulated by antisense ncRNAs or lincRNAs, insinuates a Mendellian existence might also be operative inside the human cell. In essence changes outside the DNA are affecting the expression of the DNA and these epigenetic changes can be passed to daughter cells and potentially to offspring. Such a notion brings forth the question as to what extent ncRNAs are involved in ongoing selective pressures and/or the cell responding to selective pressures. Such a paradigm would be envisaged to operate from the standpoint that a selective pressure (diet, environment, etc.) results in increased expression of a particular subset of ncRNAs which in turn affects the epigenetic architecture of downstream targets ultimately resulting in enhanced cell survival due a particular gene expression profile. This particular expression profile is then conserved in the respective daughter cells, which exhibit an enhanced fitness relative to other cells. However, such a scenario might be a bit forthcoming as maybe the fundamental role of miRNAs is to in essence govern the relative ratios of antisense ncRNAs to sense mRNAs and prohibit such profound changes. Such a notion runs counter to the current dogma of how miRNAs functionally operate in the human cell but could also be a functional role for miRNAs that has not yet been appreciated.

Supporting a notion for miRNAs as governs of longer forms of ncRNA expression is the observation that to date miRNA mediated targeting of transcripts has been thought of as being a relatively potent suppressor of gene expression, but that this suppression is transient, lasting until new transcripts are generated or the mRNA complexed miRNAs are degraded. This is a feature that may lend miRNAs to function more as a stoichiometric regulators of general RNA abundance, allowing for the cell to augment gene expression accordingly based on particular selective pressures (Morris 2009; Herranz and Cohen 2010; Rigoutsos and Furnari 2010). Such a scenario might also allow for transient ncRNA-based responses to be initiated without permanent epigenetic changes being instilled on the cell. Thus, if favourable, those cells expressing higher proportions of particular ncRNAs can outcompete their neighbours and the expression persists long enough to modulate particular epigenetic changes that become instilled in the chromatin of the dominant cells. There is precedence for such regulation when considering the androgen receptor and specific patterns of antisense ncRNAs being expressed in response to androgen receptor activation (Louro *et al.*, 2007). Supportive of miRNA regulation of ncRNAs is the observation that double-stranded RNAs can actually target both sense or antisense strands (Wei *et al.*, 2009), and maybe the reason miRNAs are abundantly expressed and linked to several forms of cancer in human cells is due to their role in regulating bidirectionally transcribed discordantly regulated loci (discussed in detail in Morris, 2009). As the majority of those discordantly regulated bidirectionally transcribed genes detected and/or characterized to date appear to be involved in cell cycle, the loss of which can result in oncogenic transformation, it is not too far of a stretch to envisage a role for miRNAs in regulating antisense ncRNAs. In fact, one example has been recently described whereby miRNAs have been shown to be involved in the control of pseudogene RNAs for the cell cycle-related gene p16 (Rigoutsos and Furnari, 2010).

While much progress has been gained over the last few years with regards to the realization that a complex paradigm is active in Human cells whereby vast species of ncRNAs are active epigenetic regulators of gene expression, many questions remain, such as 'What governs the expression of antisense ncRNAs?', 'Are miRNAs, or other species of small RNAs such as tiRNAs (Taft *et al.*, 2009), actually involved in regulating both sense and antisense ncRNAs?', and 'To what extent are ncRNAs involved in the ability of the cell to either detect and/or respond to selective pressures?'. Furthermore, what dictates a gene being under ncRNA-based regulation and how pronounced are these forms of regulation in the human cells? These are but only a few of the questions that will no doubt one day be addressed and hopefully be implicit in reshaping our current models of ncRNA-based modes of gene regulation.

Acknowledgements

This project is funded by NIH R01HL083473 and NIH R01AI084406 to K.V.M., and by R01CA151574-01 and R01CA153124-01 to P.K.V.

References

Alberts, B., Bray, D., Lewis, J., Raff, M., Roberts, K., and Watson, J.D. (1994). Molecular Biology of the Cell, 3rd edition (Garland Science: New York).

Boeger, H., Bushnell, D.A., Davis, R., Griesenbeck, J., Lorch, Y., Strattan, J.S., Westover, K.D., and Kornberg, R.D. (2005). Structural basis of eukaryotic gene transcription. FEBS Lett. *579*, 899–903.

Brideau, C.M., Kauppinen, K.P., Holmes, R., and Soloway, P.D. (2010). A non-coding RNA within the Rasgrf1 locus in mouse is imprinted and regulated by its homologous chromosome in trans. PLoS ONE *5*, e13784.

Camblong, J., Iglesias, N., Fickentscher, C., Dieppois, G., and Stutz, F. (2007). Antisense RNA stabilization induces transcriptional gene silencing via histone deacetylation in *S. cerevisiae*. Cell *131*, 706–717.

Cho, D.H., Thienes, C.P., Mahoney, S.E., Analau, E., Filippova, G.N., and Tapscott, S.J. (2005). Antisense transcription and heterochromatin at the DM1 CTG repeats are constrained by CTCF. Mol. Cell *20*, 483–489.

Ebralidze, A.K., Guibal, F.C., Steidl, U., Zhang, P., Lee, S., Bartholdy, B., Jorda, M.A., Petkova, V., Rosenbauer, F., Huang, G., *et al.* (2008). PU.1 expression is modulated by the balance of functional sense and antisense RNAs regulated by a shared cis-regulatory element. Genes Dev. *22*, 2085–2092.

Franklin, T.B., Russig, H., Weiss, I.C., Graff, J., Linder, N., Michalon, A., Vizi, S., and Mansuy, I.M. (2010). Epigenetic transmission of the impact of early stress across generations. Biol. Psychiatry *68*, 408–415.

Fuks, F., Burgers, W.A., Brehm, A., Hughes-Davies, L., and Kouzarides, T. (2000). DNA methyltransferase Dnmt1 associates with histone deacetylase activity. Nat. Genet. *24*, 88–91.

Fuks, F., Burgers, W.A., Godin, N., Kasai, M., and Kouzarides, T. (2001). Dnmt3a binds deacetylases and is recruited by a sequence-specific repressor to silence transcription. Embo J. *20*, 2536–2544.

Fuks, F., Hurd, P.J., Deplus, R., and Kouzarides, T. (2003a). The DNA methyltransferases associate with HP1 and the SUV39H1 histone methyltransferase. Nucleic Acids Res. *31*, 2305–2312.

Fuks, F., Hurd, P.J., Wolf, D., Nan, X., Bird, A.P., and Kouzarides, T. (2003b). The methyl-CpG-binding protein MeCP2 links DNA methylation to histone methylation. J. Biol. Chem. *278*, 4035–4040.

Gonzalez, S., Pisano, D.G., and Serrano, M. (2008). Mechanistic principles of chromatin remodeling guided by siRNAs and miRNAs. Cell Cycle *7*, 2601–2608.

Greger, I.H., Aranda, A., and Proudfoot, N. (2000). Balancing transcriptional interference and initiation on the GAL7 promoter of Saccharomyces cerevisiae. Proc. Natl. Acad. Sci. U.S.A. *97*, 8415–8420.

Gupta, R.A., Shah, N., Wang, K.C., Kim, J., Horlings, H.M., Wong, D.J., Tsai, M.C., Hung, T., Argani, P., Rinn, J.L., *et al.* (2010). Long non-coding RNA HOTAIR reprograms chromatin state to promote cancer metastasis. Nature *464*, 1071–1076.

Guttman, M., Amit, I., Garber, M., French, C., Lin, M.F., Feldser, D., Huarte, M., Zuk, O., Carey, B.W., Cassady, J.P., *et al.* (2009). Chromatin signature reveals over a thousand highly conserved large non-coding RNAs in mammals. Nature *458*, 223–227.

Han, J., Kim, D., and Morris, K.V. (2007). Promoter-associated RNA is required for RNA-directed transcriptional gene silencing in human cells. PNAS *104*.

Han, J., Kim, D., and Morris, K.V. (2007). Promoter-associated RNA is required for RNA-directed transcriptional gene silencing in human cells. Proc. Natl. Acad. Sci. U.S.A. *104*, 12422–12427.

Hannon, G.J. (2002). RNA interference. Nature *418*, 244–251.

Hannon, G.J., and Rossi, J.J. (2004). Unlocking the potential of the human genome with RNA interference. Nature *431*, 371–378.

Hawkins, P.G., and Morris, K.V. (2010). Transcriptional regulation of Oct4 by a long non-coding RNA antisense to Oct4-pseudogene 5. Transcription *1*, 165–175.

Hawkins, P.G., Santoso, S., Adams, C., Anest, V., and Morris, K.V. (2009). Promoter targeted small RNAs induce long-term transcriptional gene silencing in human cells. Nucleic Acids Res. *37*, 2984–2995.

Heijmans, B.T., Tobi, E.W., Stein, A.D., Putter, H., Blauw, G.J., Susser, E.S., Slagboom, P.E., and Lumey, L.H. (2008). Persistent epigenetic differences associated with prenatal exposure to famine in humans. Proc. Natl. Acad. Sci. U.S.A. *105*, 17046–17049.

Herranz, H., and Cohen, S.M. (2010). MicroRNAs and gene regulatory networks: managing the impact of noise in biological systems. Genes Dev. *24*, 1339–1344.

Hoffman, A.R., and Hu, J.F. (2006). Directing DNA methylation to inhibit gene expression. Cell. Mol. Neurobiol. *26*, 425–438.

Huarte, M., Guttman, M., Feldser, D., Garber, M., Koziol, M.J., Kenzelmann-Broz, D., Khalil, A.M., Zuk, O., Amit, I., Rabani, M., *et al.* (2010). A large intergenic noncoding RNA induced by p53 mediates global gene repression in the p53 response. Cell *142*, 409–419.

Janowski, B.A., Huffman, K.E., Schwartz, J.C., Ram, R., Hardy, D., Shames, D.S., Minna, J.D., and Corey D.R. (2005). Inhibiting gene expression at transcription start sites in chromosomal DNA with antigene RNAs. Nat. Chem. Biol. *1*, 210–215.

Janowski, B.A., Huffman, K.E., Schwartz, J.C., Ram, R., Nordsell, R., Shames, D.S., Minna, J.D., and Corey, D.R. (2006). Involvement of AGO1 and AGO2 in mammalian transcriptional silencing. Nat. Struct. Mol. Biol. *13*, 787–792.

Jeffery, L., and Nakielny, S. (2004). Components of the DNA methylation system of chromatin control are RNA-binding proteins. J. Biol. Chem. *279*, 49479–49487.

Kanhere, A., Viiri, K., Araujo, C.C., Rasaiyaah, J., Bouwman, R.D., Whyte, W.A., Pereira, C.F., Brookes, E., Walker, K., Bell, G.W., *et al.* (2010). Short RNAs are transcribed from repressed polycomb target genes and interact with polycomb repressive complex-2. Mol. Cell *38*, 675–688.

Katayama, S., Tomaru, Y., Kasukawa, T., Waki, K., Nakanishi, M., Nakamura, M., Nishida, H., Yap, C.C., Suzuki, M., Kawai, J., *et al.* (2005). Antisense transcription in the mammalian transcriptome. Science *309*, 1564–1566.

Khalil, A.M., Guttman, M., Huarte, M., Garber, M., Raj, A., Rivea Morales, D., Thomas, K., Presser, A., Bernstein, B.E., van Oudenaarden, A., *et al.* (2009). Many human large intergenic noncoding RNAs associate with chromatin-modifying complexes and affect gene expression. Proc. Natl. Acad. Sci. U.S.A. *106*, 11667–11672.

Kim, D.H., Villeneuve, L.M., Morris, K.V., and Rossi, J.J. (2006). Argonaute-1 directs siRNA-mediated transcriptional gene silencing in human cells. Nat. Struct. Mol. Biol. *13*, 793–797.

Kim, D.H., Saetrom, P., Snove, O., Jr., and Rossi, J.J. (2008). MicroRNA-directed transcriptional gene silencing in mammalian cells. Proc. Natl. Acad. Sci. U.S.A. *105*, 16230–16235.

Klase, Z., Kale, P., Winograd, R., Gupta, M.V., Heydarian, M., Berro, R., McCaffrey, T., and Kashanchi, F. (2007). HIV-1 TAR element is processed by Dicer to yield a viral micro-RNA involved in chromatin remodeling of the viral LTR. BMC Mol. Biol. *8*, 63.

Knee, R., and Murphy, P.R. (1997). Regulation of gene expression by natural antisense RNA transcripts. Neurochem. Int. *31*, 379–392.

Latos, P.A., and Barlow, D.P. (2009). Regulation of imprinted expression by macro non-coding RNAs. RNA Biol. *6*.

Lee, J.T. (2009). Lessons from X-chromosome inactivation: long ncRNA as guides and tethers to the epigenome. Genes Dev. *23*, 1831–1842.

Li, K., Blum, Y., Verma, A., Liu, Z., Pramanik, K., Leigh, N.R., Chun, C.Z., Samant, G.V., Zhao, B., Garnaas, M.K., *et al.* (2009). A non-coding antisense RNA in tie-1 locus regulates tie-1 function in vivo. Blood *6*, 100–106.

Li, L.C., Okino, S.T., Zhao, H., Pookot, D., Place, R.F., Urakami, S., Enokida, H., and Dahiya, R. (2006). Small dsRNAs induce transcriptional activation in human cells. Proc. Natl. Acad. Sci. U.S.A. *103*, 17337–17342.

Lim, H.G., Suzuki, K., Cooper, D.A., and Kelleher, A.D. (2008). Promoter-targeted siRNAs induce gene silencing of simian immunodeficiency virus (SIV) infection in vitro. Mol. Ther. *16*, 565–570.

Loewer, S., Cabili, M.N., Guttman, M., Loh, Y.H., Thomas, K., Park, I.H., Garber, M., Curran, M., Onder, T., Agarwal, S., *et al.* (2010). Large intergenic non-coding RNA-RoR modulates reprogramming of human induced pluripotent stem cells. Nat. Genet. *42*, 1113–1117.

Louro, R., Nakaya, H.I., Amaral, P.P., Festa, F., Sogayar, M.C., da Silva, A.M., Verjovski-Almeida, S., and Reis, E.M. (2007). Androgen responsive intronic non-coding RNAs. BMC Biol. *5*, 4.

Mattick, J.S., and Makunin, I.V. (2006). Non-coding RNA. Hum. Mol. Genet. *15 Spec No 1*, R17–29.

Matzke, M.A., and Birchler, J.A. (2005). RNAi-mediated pathways in the nucleus. Nat. Rev. Genet. *6*, 24–35.

Matzke, M.A., Matzke, A.J., Pruss, G.J., and Vance, V.B. (2001). RNA-based silencing strategies in plants. Curr. Opin. Genet. Dev. *11*, 221–227.

Morris, K.V. (2005). siRNA-mediated transcriptional gene silencing: the potential mechanism and a possible role in the histone code. Cell Mol .Life Sci. *62*, 3057–3066.

Morris, K.V. (2009a). Long antisense non-coding RNAs function to direct epigenetic complexes that regulate transcription in human cells. Epigenetics *4*, 296–301.

Morris, K.V. (2009b). Non-coding RNAs, epigenetic memory, and the passage of information to progeny. RNA Biol *6*, 242–247.

Morris, K.V. (2009c). RNA-directed transcriptional gene silencing and activation in human cells. Oligonucleotides *19*, 299–306.

Morris, K.V., and Vogt, P.K. (2010). Long antisense non-coding RNAs and their role in transcription and oncogenesis. Cell Cycle *9*, 2544–2547.

Morris, K.V., Chan, S.W., Jacobsen, S.E., and Looney, D.J. (2004). Small interfering RNA-induced transcriptional gene silencing in human cells. Science *305*, 1289–1292.

Morris, K.V., Santoso, S., Turner, A.M., Pastori, C., and Hawkins, P.G. (2008). Bidirectional transcription directs both transcriptional gene activation and suppression in human cells. PLoS Genet. *4*, e1000258.

Napoli, S., Pastori, C., Magistri, M., Carbone, G.M., and Catapano, C.V. (2009). Promoter-specific transcriptional interference and c-myc gene silencing by siRNAs in human cells. EMBO J. *28*, 1708–1719.

Omoto, S., and Fujii, Y.R. (2005). Regulation of human immunodeficiency virus 1 transcription by nef microRNA. J. Gen. Virol. *86*, 751–755.

Omoto, S., Ito, M., Tsutsumi, Y., Ichikawa, Y., Okuyama, H., Brisibe, E.A., Saksena, N.K., and Fujii, Y.R. (2004). HIV-1 nef suppression by virally encoded microRNA. Retrovirology *1*, 44.

Pauli, A., Rinn, J.L., and Schier, A.F. (2011). Non-coding RNAs as regulators of embryogenesis. Nat. Rev. Genet. *12*, 136–149.

Petruk, S., Sedkov, Y., Riley, K.M., Hodgson, J., Schweisguth, F., Hirose, S., Jaynes, S.B., Brock, H.W., Mazo, A. (2006). Transcription of bxd noncoding RNAs promoted by trithorax represses

Ubx in cis by transcriptional interference. Cell *127*, 1209–1221.

Place, R.F., Li, L.C., Pookot, D., Noonan, E.J., and Dahiya, R. (2008). MicroRNA-373 induces expression of genes with complementary promoter sequences. Proc. Natl. Acad. Sci. U.S.A. *105*, 1608–1613.

Pulukuri, S.M., and Rao, J.S. (2007). Small interfering RNA directed reversal of urokinase plasminogen activator demethylation inhibits prostate tumor growth and metastasis. Cancer Res. *67*, 6637–6646.

Rigoutsos, I., and Furnari, F. (2010). Gene-expression forum: Decoy for microRNAs. Nature *465*, 1016–1017.

Sugimoto, K., Kage, H., Aki, N., Sano, A., Kitagawa, H., Nagase, T., Yatomi, Y., Ohishi, N., and Takai, D. (2007). The induction of H3K9 methylation by PIWIL4 at the p16Ink4a locus. Biochem. Biophys. Res. Commun. *359*, 497–502.

Suzuki, K., Shijuuku, T., Fukamachi, T., Zaunders, J., Guillemin, G., Cooper, D., and Kelleher, A. (2005). Prolonged transcriptional silencing and CpG methylation induced by siRNAs targeted to the HIV-1 promoter region. J. RNAi Gene Silenc *1*, 66–78.

Suzuki, K., Juelich, T., Lim, H., Ishida, T., Watanebe, T., Cooper, D.A., Rao, S., and Kelleher, A.D. (2008). Closed chromatin architecture is induced by an RNA duplex targeting the HIV-1 promoter region. J. Biol. Chem. *283*, 23353–23363.

Taft, R.J., Pheasant, M., and Mattick, J.S. (2007). The relationship between non-protein-coding DNA and eukaryotic complexity. Bioessays *29*, 288–299.

Taft, R.J., Glazov, E.A., Cloonan, N., Simons, C., Stephen, S., Faulkner, G.J., Lassmann, T., Forrest, A.R., Grimmond, S.M., Schroder, K., *et al.* (2009). Tiny RNAs associated with transcription start sites in animals. Nat. Genet. *41*, 572–578.

Tan, Y., Zhang, B., Wu, T., Skogerbo, G., Zhu, X., Guo, X., He, S., and Chen, R. (2009). Transcriptional inhibition of Hoxd4 expression by noncoding RNAs in human breast cancer cells. BMC Mol. Biol. *10*, 12.

Ting, A.H., Schuebel, K.E., Herman, J.G., and Baylin, S.B. (2005). Short double-stranded RNA induces transcriptional gene silencing in human cancer cells in the absence of DNA methylation. Nat. Genet. *37*, 906–910.

Tomari, Y., and Zamore, P.D. (2005). Perspective: machines for RNAi. Genes Dev. *19*, 517–529.

Tsai, M.C., Manor, O., Wan, Y, Mosammaparast, N., Wang, J.K., Lan, F., Shi, Y., Segal, E., and Chang, H.Y. (2010). Long noncoding RNA as modular scaffold of histone modification complexes. Science *329*, 689–693.

Tufarelli, C., Stanley, J.A., Garrick, D., Sharpe, J.A., Ayyub, H., Wood, W.G., and Higgs, D.R. (2003). Transcription of antisense RNA leading to gene silencing and methylation as a novel cause of human genetic disease. Nat. Genet. *34*, 157–165.

Turner, A.M., De La Cruz, J., and Morris, K.V. (2009). Mobilization-competent Lentiviral Vector-mediated Sustained Transcriptional Modulation of HIV-1 Expression. Mol. Ther. *17*, 360–368.

Turunen, M.P., Lehtola, T., Heinonen, S.E., Assefa, G.S., Korpisalo, P., Girnary, R., Glass, C.K., Vaisanen, S., and Yla-Herttuala, S. (2009). Efficient regulation of VEGF expression by promoter-targeted lentiviral shRNAs based on epigenetic mechanism: a novel example of epigenetherapy. Circ. Res. *105*, 604–609.

Vire, E., Brenner, C., Deplus, R., Blanchon, L., Fraga, M., Didelot, C., Morey, L., Van Eynde, A., Bernard, D., Vanderwinden, J.M., *et al.* (2005). The Polycomb group protein EZH2 directly controls DNA methylation. Nature *439*, 871–874.

Wahlestedt, C. (2006). Natural antisense and noncoding RNA transcripts as potential drug targets. Drug Discov. Today *11*, 503–508.

Wang, K.C., Yang, Y.W., Liu, B., Sanyal, A., Corces-Zimmerman, R., Chen, Y., Lajoie, B.R., Protacio, A., Flynn, R.A., Gupta, R.A., *et al.* (2011). A long noncoding RNA maintains active chromatin to coordinate homeotic gene expression. Nature *472*, 120–124.

Watanabe, Y., Numata, K., Murata, S., Osada, Y., Saito, R., Nakaoka, H., Yamamoto, N., Watanabe, K., Kato, H., Abe, K., *et al.* (2010). Genome-wide analysis of expression modes and DNA methylation status at sense-antisense transcript loci in mouse. Genomics *96*, 333–341.

Wei, J.X., Yang, J., Sun, J.F., Jia, L.T., Zhang, Y., Zhang, H.Z., Li, X., Meng, Y.L., Yao, L.B., and Yang, A.G. (2009). Both strands of siRNA have potential to guide posttranscriptional gene silencing in mammalian cells. PLoS One *4*, e5382.

Weinberg, M.S., Villeneuve, L.M., Ehsani, A., Amarzguioui, M., Aagaard, L., Chen, Z.X., Riggs, A.D., Rossi, J.J., and Morris, K.V. (2006). The antisense strand of small interfering RNAs directs histone methylation and transcriptional gene silencing in human cells. RNA *12*, 256–262.

Yu, W., Gius, D., Onyango, P., Muldoon-Jacobs, K., Karp, J., Feinberg, A.P., and Cui, H. (2008). Epigenetic silencing of tumour suppressor gene p15 by its antisense RNA. Nature *451*, 202–206.

Index